T0178400

Association for Women in Mathematics Series

Volume 18

Series Editor
Kristin Lauter
Microsoft Research
Redmond, Washington, USA

Association for Women in Mathematics Series

Focusing on the groundbreaking work of women in mathematics past, present, and future, Springer's Association for Women in Mathematics Series presents the latest research and proceedings of conferences worldwide organized by the Association for Women in Mathematics (AWM). All works are peer-reviewed to meet the highest standards of scientific literature, while presenting topics at the cutting edge of pure and applied mathematics, as well as in the areas of mathematical education and history. Since its inception in 1971, The Association for Women in Mathematics has been a non-profit organization designed to help encourage women and girls to study and pursue active careers in mathematics and the mathematical sciences and to promote equal opportunity and equal treatment of women and girls in the mathematical sciences. Currently, the organization represents more than 3000 members and 200 institutions constituting a broad spectrum of the mathematical community in the United States and around the world.

More information about this series at http://www.springer.com/series/13764

Susan D'Agostino • Sarah Bryant
Amy Buchmann • Michelle Craddock Guinn
Leona Harris
Editors

A Celebration of the EDGE Program's Impact on the Mathematics Community and Beyond

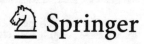

Editors
Susan D'Agostino
Science Writing Advanced
Academic Program
John Hopkins University
Baltimore, MD, USA

Sarah Bryant
Department of Mathematics
and Computer Science
Dickinson College
Carlisle, PA, USA

Amy Buchmann
Department of Mathematics
University of San Diego
San Diego, CA, USA

Michelle Craddock Guinn
Department of Mathematics
and Computer Science
Belmont University
Nashville, TN, USA

Leona Harris
Division of Sciences and Mathematics
University of the District of Columbia
Washington, DC, USA

ISSN 2364-5733   ISSN 2364-5741   (electronic)
Association for Women in Mathematics Series
ISBN 978-3-030-19488-8   ISBN 978-3-030-19486-4   (eBook)
https://doi.org/10.1007/978-3-030-19486-4

© The Author(s) and the Association for Women in Mathematics 2019
This work is subject to copyright. All rights are reserved by the Publisher, whether the whole or part of the material is concerned, specifically the rights of translation, reprinting, reuse of illustrations, recitation, broadcasting, reproduction on microfilms or in any other physical way, and transmission or information storage and retrieval, electronic adaptation, computer software, or by similar or dissimilar methodology now known or hereafter developed.
The use of general descriptive names, registered names, trademarks, service marks, etc. in this publication does not imply, even in the absence of a specific statement, that such names are exempt from the relevant protective laws and regulations and therefore free for general use.
The publisher, the authors, and the editors are safe to assume that the advice and information in this book are believed to be true and accurate at the date of publication. Neither the publisher nor the authors or the editors give a warranty, express or implied, with respect to the material contained herein or for any errors or omissions that may have been made. The publisher remains neutral with regard to jurisdictional claims in published maps and institutional affiliations.

This Springer imprint is published by the registered company Springer Nature Switzerland AG.
The registered company address is: Gewerbestrasse 11, 6330 Cham, Switzerland

This book is dedicated to Sylvia Bozeman, Rhonda Hughes, and all of the Women Math Warriors of the EDGE Program.

Organization

Program Chairs

Susan D'Agostino	John Hopkins University, United States
Sarah Bryant	Dickinson College, United States
Amy Buchmann	University of San Diego, United States
Michelle Guinn	Belmont University, United States
Leona Harris	University of the District of Columbia, United States
Dimana Tzvetkova	Springer, United States

Contents

Introduction 1
Susan D'Agostino

Part I   The Enhancing Diversity in Graduate Education (EDGE)
         Program

The EDGE Program: 20 Years and Counting 9
Sarah Bryant and Jessica Spott

EDGE Through the Years 19
Sarah Bryant, Amy Buchmann, Michelle Craddock Guinn,
Susan D'Agostino, and Leona Harris

Twenty Years of Enhancing Diversity in Graduate
Education from the Perspectives of Five Dynamic Women
Who Have Led the Program 31
Farrah J. Ward and Leona Harris

Second-Generation Programs: The Far-Reaching Impact of EDGE 47
Rachelle C. DeCoste

Difficult Dialogues in the Midwest: A Retrospective on the Impact
of EDGE at Purdue University 55
Alejandra Alvarado, Donatella Danielli, Rachel Davis, Zenephia Evans,
and Edray Herber Goins

The Long-Lasting Impact of EDGE: Testimonials from the EDGE
Community 67
Sarah Bryant, Amy Buchmann, Michelle Craddock Guinn,
Susan D'Agostino, and Leona Harris

Part II   Mathematics Inclusivity and Outreach Work

**Academic Preparation for Business, Industry,
and Government Positions** 79
Alejandra Alvarado and Candice R. Price

**Striking the Right Chord: Math Circles Promote (Joyous)
Professional Growth** 89
Lance Bryant, Sarah Bryant, and Diana White

**The Women in Mathematics Symposia: An Organic Extension
of the EDGE Program** 99
Amy Buchmann, Yen Duong, and Ami Radunskaya

**The Career Mentoring Workshop: A Second-Generation
EDGE Program** 119
Rachelle C. DeCoste

**Emotional Labor in Mathematics: Reflections on Mathematical
Communities, Mentoring Structures, and EDGE** 129
Gizem Karaali

Part III   Mathematics Teaching

Experiencing Mathematics Abroad 149
Michelle Craddock Guinn and Bradford Schleben

**Change Is Hard, But Not Impossible: Building Student Enthusiasm
for Inquiry-Based Learning** 167
Jill E. Jordan

**Linear Algebra, Secret Agencies, and Zombies: Applications
to Enhance Learning and Creativity** 175
Carolyn Otto

Part IV   Mathematics Research

A Multivariate Rank-Based Two-Sample Test Statistic 197
Jamye Curry, Xin Dang, and Hailin Sang

Sufficient Conditions for Composite Frames 213
Wojciech Czaja and Karamatou Yacoubou Djima

**Mathematical Modeling of Immune-Mediated Processes
in Coagulation and Anticoagulation Therapy** 237
Erica J. Graham and Ami Radunskaya

**Trigonometric-Type Functions Derived from Polygons Inscribed
in the Unit Circle** 277
Torina Lewis

An Extension of Wolfram's Rule 90 for One-Dimensional Cellular
Automata over Non-Abelian Group Alphabets 289
Erin Craig and Eirini Poimenidou

A Preliminary Exploration of the Professional Support Networks
the EDGE Program Creates 317
Candice R. Price and Nina H. Fefferman

A Model for Three-Phase Flow in Porous Media
with Rate-Dependent Capillary Pressure...................................... 327
Kimberly Spayd and Ellen R. Swanson

An Invitation to Noncommutative Algebra.................................... 339
Chelsea Walton

Part V Mathematical Lives

Taking a Leap Off the Ivory Tower: Normalizing Unconventional
Careers 369
Karoline P. Pershell

A Mathematician's Journey to Public Service 377
Carla D. Cotwright-Williams

Reflections on the Challenges of Mentoring.................................. 385
Carol Wood

Author Index... 389

About the Editors

Susan D'Agostino is a Taylor Blakeslee fellow of the Council for the Advancement of Science Writing at John Hopkins University who earned her PhD in Mathematics at Dartmouth College. A math and science writer, her essays have been published in *Scientific American, Financial Times, Undark Magazine, Nature, Ms., Chronicle of Higher Education, Times Higher Education, Math Horizons, Mathematics Teacher*, AAC&U's *Liberal Education*, and AAUP's *Academe*, among other publications. Her forthcoming book, *How to Free Your Inner Mathematician* (Oxford University Press, 2020), delivers engaging mathematical content while providing reassurance that mathematical success has more to do with curiosity and drive than innate aptitude. Her website is www.susandagostino.com.

Sarah Bryant is a mathematician, educator, and advocate for broadening participation in math. Since earning her PhD in Mathematics at Purdue University, she has held visiting teaching positions at Gettysburg College, Dickinson College, and Shippensburg University. As Project Manager of NSF ADVANCE STEM-UP PA, she supported faculty women in STEM through recruitment and advancement initiatives. Since her time as an EDGE participant in 2002, she has repeatedly returned to the EDGE Program as a mentor, speaker, and instructor. She is a co-founder and co-director (with her husband Lance) of the Shippensburg Area Math Circle. Her website is www.sarahbryantmath.com.

Amy Buchmann is an assistant professor in the Department of Mathematics at the University of San Diego. Her research is in the field of mathematical biology, and she is particularly interested in computational biofluids. She was an EDGE participant in 2010 and returned as a mentor in 2012–2014. After participating in EDGE, she earned her doctorate in Applied and Computational Mathematics and Statistics from the University of Notre Dame. Before joining the faculty at the University of San Diego, she was a postdoctoral fellow in the Department of Mathematics and Center for Computational Science at Tulane University.

Michelle Craddock Guinn is an associate professor in the Mathematics and Computer Science Department at Belmont University. She earned her PhD in Mathematics at the University of Mississippi. Her area of study was functional analysis. After graduation, she accepted a post-doctoral position at the US Military Academy in West Point, NY, and later was awarded the Davies Fellowship which allowed her the time to conduct research in image processing at the US Army Research Laboratory. She was a participant in EDGE in 2004 and a mentor in 2008.

Leona Harris is chair of the Division of Sciences and Mathematics and an associate professor of Mathematics at the University of the District of Columbia. She earned her PhD in Applied Mathematics from North Carolina State University and completed a postdoctoral fellowship at the Environmental Protection Agency. She is the executive director of the National Association of Mathematicians and a co-founder of the Infinite Possibilities Conference. Throughout her career, she has forged strong partnerships with faculty colleagues and professionals in education, industry, and government through academic support programs, interdisciplinary research, curriculum and program development, and external grant initiatives. Her website is www.leonaharrisphd.com.

Introduction

Susan D'Agostino

We call ourselves Women Math Warriors. When we first gathered in 1998, only 13% of US PhDs in the mathematical sciences were awarded to American women, of which only 2% were underrepresented women [1]. We are community members of the Enhancing Diversity in Graduate Education (EDGE) Program, an American Mathematical Society "program that makes a difference" in improving diversity in the US mathematical profession [2]. In *A Celebration of the EDGE Program's Impact on the Mathematics Community and Beyond*, we report on our own program, mathematics outreach and inclusivity work, mathematics teaching, mathematics research, and mathematical lives.

In Part I, we offer a broad overview of the EDGE Program. Sarah Bryant (see editor signature for EDGE involvement) and Jessica Spott (EDGE 2018 administrative support) deliver a comprehensive summary of data, impacts, and outcomes of the EDGE Program's multifaceted approach to mentoring a diverse group of women in pursuit of advanced degrees in the mathematical sciences. The editorial board offers a photographic journey of EDGE Program participants, mentors, local coordinators, instructors, and leaders from 1998 through 2018. Farrah Jackson (EDGE 1999 participant, 2003 and 2004 mentor) and Leona Harris (see editor signature for EDGE involvement) offer an inside account of the EDGE Program's history as told by the original founders and past and present program directors. Rachelle DeCoste (EDGE 1998 participant, 2002 mentor, 2015 instructor) presents the first curated list of local, state, and national leadership positions in which the EDGE community members have served. Alejandra Alvarado (EDGE 2002 participant, 2006 mentor, 2013 instructor, 2016 local coordinator), Donatella

S. D'Agostino (✉)
John Hopkins University, Baltimore, MD, USA
e-mail: sdagpst2@jhu.edu

© The Author(s) and the Association for Women in Mathematics 2019
S. D'Agostino et al. (eds.), *A Celebration of the EDGE Program's Impact on the Mathematics Community and Beyond*, Association for Women in Mathematics Series 18, https://doi.org/10.1007/978-3-030-19486-4_1

Danielli (EDGE Mentoring Cluster Leader, Indiana), Rachel Davis, Zenephia Evans (EDGE 2016 Difficult Dialogues Facilitator), and Edray Herber Goins (EDGE 2016 local coordinator) discuss the positive impact the EDGE Program had on Purdue University, the 2016 host institution. In addition, the editorial board presents testimonials describing the EDGE Program's long-lasting impact on participants' lives.

In Part II, we report on mathematics inclusivity and outreach work undertaken by EDGE community members. Alejandra Alvarado (EDGE 2002 participant, 2006 mentor, 2013 instructor; 2016 local coordinator) and Candice Price (EDGE 2012 mentor) provide context and advice for the nearly half of mathematicians who seek employment in business, industry, or government either for first post-doctoral positions or mid-career pivots. Lance Bryant, Sarah Bryant (see editor signature for EDGE involvement), and Diana White argue that math circles—known for benefiting the K–12 students they serve—foster meaningful and spirited professional growth among mathematicians who heed the math circle call. Amy Buchmann (see editor signature for EDGE involvement), Yen Duong (EDGE 2010 participant), and Ami Radunskaya (EDGE 1998–2002, 2009–2011, and 2018 instructor, 2008 local coordinator, 2012–present co-director) explore the history, motivations, data, reflections, and impacts of the Women in Mathematics Symposia, a successful collection of annual, regional women-in-mathematics conferences. Rachelle DeCoste (EDGE 1998 participant, 2002 mentor, 2015 instructor), founder of the Career Mentoring Workshop, examines her efforts, lessons learned, and successes in diversifying the mathematics community by way of a program targeting women completing mathematics doctorates. Gizem Karaali (EDGE 2008 instructor) reports on both positive and negative ways that mathematics faculty are asked to provide emotional labor, arguing that mentoring programs, such as EDGE, help diminish the negative and reinforce the positive.

In Part III, we provide a window into contemporary undergraduate mathematics teaching. Michelle Craddock Guinn (see editor signature for EDGE involvement) and Bradford Schleben discuss a summer academic program for US students studying in Europe and Australia that emphasizes cultural understanding and promotes mathematics as a universal language. Jill Jordan (EDGE 1999 participant) offers advice for undergraduate mathematics faculty seeking to transform student attitudes regarding inquiry-based curriculums from skepticism to enthusiasm and confidence. Carolyn Otto (EDGE 2006 participant) introduces engaging activities and assignments for a project-based linear algebra class in which the students are considered members of a "Zombie Containment Task Force."

In Part IV, we showcase new developments in mathematics research. Jamye Curry (EDGE 2009 participant), Xin Dang, and Hailin Sang offer a new multi-variate, rank-based test statistic for determining whether two samples hail from the same population. Karamatou Yacoubou Djima (EDGE 2008 participant; 2013 mentor) and Wojciech Czaja present a result on composite wavelet frames, a tool for representing data at increasingly precise resolution. Erica Graham (EDGE

2006 participant, 2010 mentor) and Ami Radunskaya (EDGE 1998–2002, 2009–2011, and 2018 instructor; 2008 local coordinator; 2012 – present Co-Director) introduce a mathematical model of deep vein thrombosis that identifies contributing factors to embolus formation, an important step for informing clinical treatment. Torina Lewis (EDGE 2008 participant) introduces a new class of periodic functions known as "geometric polygon functions" that may be useful in matroid theory. Erin Craig and Eirini Poimenidou (EDGE 2006 and 2013 local coordinator, 2002, 2014, 2016 instructor) extend Wolfram's Rule 90 for one-dimensional cellular automata over non-abelian group alphabets and applies the finding to automata over dihedral groups. Candice Price (EDGE 2012 mentor) and Nina Fefferman discuss preliminary results exploring EDGE Program network organization metrics, establishing a foundation for understanding features essential for participant success. Kimberly Spayd and Ellen Swanson (EDGE 2006 participant, 2013 instructor) extend Hayes and LeFloch's work by deriving a model for a three-phase flow in porous media with rate-dependent capillary pressure, research that helps engineers and environmentalists understand accidental pipeline leaks contaminating soil and water supplies. Chelsea Walton (EDGE 2012, 2013, 2014, 2015 instructor) delivers an engaging introduction to noncommutative algebra appropriate for advanced undergraduate and graduate students.

Finally, in Part V, we tell engaging mathematical stories about some EDGE community members' lives. Karoline Pershell's (EDGE 2003 participant, 2008 mentor) narrative of her post-PhD trajectory, including a leap from academe into public, private, and not-for-profit work, demonstrates the value of an ongoing self-reflection in defining personal success. Carla Cotwright-Williams (EDGE 2001 participant, 2005 mentor) examines how her childhood in a service-oriented family set her on a course for a public-service career as a mathematician at the US Social Security Administration and the NASA Ames Research Center. Carol Wood (EDGE 2004 instructor) draws on her experiences teaching EDGE students to discuss the challenges of finding a just-right balance between encouraging and preparing students to succeed in mathematics.

As publishing this volume was an it-takes-a-village endeavor, we have many people and organizations to thank, including Ami Radunskaya for suggesting that we edit this volume; Sylvia Bozeman, Rhonda Hughes, Ulrica Wilson, Ami Radunskaya, and Raegan Higgins for the ongoing inspiration and EDGE Program leadership; EDGE Program sponsors for generously providing us financial and moral support (see https://www.edgeforwomen.org/our-sponsors/); Dimana Tzvetkova at Springer for the early encouragement and ongoing assistance; Dahlia Fisch at Springer for the assistance in the final stretch; anonymous peer-reviewers for their time and expertise; allies at the Association for Women in Mathematics, the National Association of Mathematicians, and the Society for the Advancement of Chicanos/Hispanics and Native Americans in Science; and, of course, the Women Math Warriors of the EDGE Program whose mathematics research, teaching, leadership, testimonials, and rich professional and personal lives are the subjects of this book.

"I stood at the border, stood at the edge and claimed it as central, claimed it as central and let the rest of the world move over to where I was," [3] Toni Morrison once said in an interview about her books. Morrison's statement captures the spirit of what we have worked to accomplish in our book. That is, we claim the EDGE Program as central and invite readers to move over to where we are. Editing this uncommon volume of papers during the Me Too and Black Lives Matter era— a period that has overlapped with a need for Marches for Science—has provided deeply meaningful work. Be inspired by the intelligent, thoughtful, and bold writing of EDGE Program community members. Then, go and affect positive change in the world.

Respectfully,

Susan D'Agostino

Editor in Chief

On behalf of the editorial board:

Sarah Bryant, Editor (EDGE 2002 participant; 2005 and 2006 mentor; 2012, 2015, 2018 instructor; 2014–2016 EDGE foundation executive director)

Amy Buchmann, Editor (EDGE 2010 participant, 2012–2014 mentor)

Michelle Craddock Guinn, Editor (EDGE 2004 participant, 2008 mentor)

Susan D'Agostino, Editor in Chief (EDGE 1998 participant)

Leona Harris, Editor (EDGE 2008 and 2018 instructor)

From left: Susan D'Agostino, Michelle Guinn, Amy Buchmann, Leona Harris, Sarah Bryant

References

1. A. M. Society, "Annual Survey of the Mathematical Sciences PhDs Awarded," American Mathematical Society, 1998. [Online]. Available: http://www.ams.org/profession/data/annual-survey/phds-awarded. [Accessed 1 March 2019].
2. American Mathematical Society, "American Mathematical Society Mathematics Programs That Make a Difference 2007," 2007. [Online]. Available: http://www.ams.org/profession/prizes-awards/ams-supported/emp-citation2007 . [Accessed 28 February 2019].
3. T. Morrison, Interviewee, *Toni Morrison Uncensored, Princeton, NJ : Films for the Humanities & Sciences.* [Interview]. 2003.

Part I
The Enhancing Diversity in Graduate Education (EDGE) Program

The EDGE Program: 20 Years and Counting

Sarah Bryant and Jessica Spott

Abstract The EDGE (Enhancing Diversity in Graduate Education) Program is designed to strengthen the ability of women and minority students to successfully complete graduate programs in the mathematical sciences. The founders of EDGE sought to empower women entering graduate school in mathematics. In this chapter, we provide some statistics about EDGE participants and the impacts and outcomes of EDGE on these women and the field of mathematics.

1 EDGE: A Multi-Faceted Mentoring Program

The EDGE (Enhancing Diversity in Graduate Education) Program was started in 1998 by Dr. Sylvia Bozeman and Dr. Rhonda Hughes. EDGE began with a shared vision of supporting women entering mathematics PhD programs. However, as EDGE has grown and participants have finished graduate degrees and pursued careers, the scope of EDGE programs has also grown. The ultimate goal of EDGE is to enhance the diversity of leadership in the mathematics community. In support of this goal, EDGE runs a comprehensive mentoring and training program including a summer session, reunion conference, mentoring clusters, and other activities. These activities further the EDGE goal of empowering women through all developmental stages of their professional identities (Fig. 1).

The EDGE program offers comprehensive mentoring for women, approximately half of whom are underrepresented minorities, pursuing careers in the mathematical sciences. EDGE activities are designed to provide ongoing support toward their academic development and research productivity at several critical stages of their

S. Bryant (✉)
Dickinson College, Carlisle, PA, USA
e-mail: bryants@dickinson.edu

J. Spott
Texas Tech University, Lubbock, TX, USA
e-mail: jessicaspott@gmail.com

© The Author(s) and the Association for Women in Mathematics 2019
S. D'Agostino et al. (eds.), *A Celebration of the EDGE Program's Impact on the Mathematics Community and Beyond*, Association for Women in Mathematics Series 18, https://doi.org/10.1007/978-3-030-19486-4_2

9

Fig. 1 EDGE Postcard by Tai-Danae Bradley

careers-entering graduate students, advanced graduate students, postdocs, and early career mathematicians. Brief descriptions of these activities are given below:

Summer Session The EDGE Summer Session is a 4-week residential program for women entering graduate programs in the mathematical sciences that consist of two main workshops in analysis and algebra. The workshops are immersion experiences that simulate the fast pace of studying graduate-level mathematics. Advanced graduate students serve as mentors. The summer session also includes mini-courses in vital areas of mathematical research, guest lectures by short-term visitors from academia and industry, group problem-solving sessions, and activities that foster network building and professional development.

Annual Reunion Conference Each year during the EDGE Summer Session, this 2-day reunion conference includes activities designed to mentor graduate students and recent doctoral graduates. Both the current and previous year's summer session participants attend the conference along with EDGE participants who have recently earned their PhDs.

Research Mini-Sabbaticals As part of its objective to support the research activities of women in mathematics, EDGE funds extended visits (2–6 weeks) for advanced graduate students, recent PhDs, and junior faculty to work with a collaborator in their research area. These research visits advance the research and intellectual growth of the recipient and increase their visibility in the mathematics research community.

Regional Mentoring Clusters Small groups of women in mathematics (undergraduates, graduate students, junior faculty, and senior faculty) in close geographical proximity meet 2–3 times a semester to facilitate tiered mentoring. Currently, EDGE sponsors cluster activities in southern California, Georgia, Indiana, North Carolina, Iowa, and the mid-Atlantic region.

Special Sessions and Regional Research Symposia EDGE has hosted special sessions at multiple Joint Math Meetings and Association for Women in Mathematics Conferences. The EDGE regional research symposia grew from the EDGE clusters out of an expressed interest for more direct interaction between women mathematicians at all stages in their careers. The goal of the regional symposia is to strengthen the network of female mathematicians in a particular geographical region, and to encourage collaborations and mentoring relationships.

2 EDGE Summer Session: A Program That Makes a Difference

Every summer, the EDGE cohorts form a diverse group, with talented students selected from across the nation. From 1998 to 2018, there have been 255 participants from more than 150 undergraduate and 100 graduate institutions. The participants are recruited with an eye toward racial and ethnic diversity, and the data show 48.2% of EDGE participants are White while 51.8% are from Underrepresented Minority Groups. Over 40% of EDGE participants are from 4-year colleges and over 23% from historically black colleges and universities (HBCUs). With such a variety of unique educational backgrounds and experiences, the opportunity for growth through community building and the exchanging of ideas is endless.

Though EDGE began by alternating Summer Sessions between Spelman College and Bryn Mawr College, in 2003, the pattern changed and EDGE Summer Session was at Pomona College. Since then, colleges and universities across the country have become hosting partners with EDGE (see Fig. 2). Additional hosting sites now include North Carolina A&T University, New College of Florida, North Carolina State University, Florida A&M University, Harvey Mudd College, Howard University, Purdue University, Mills College, and Texas Tech University.

Fig. 2 Map of EDGE summer session locations; Undergraduate and Graduate Institutions of participants indicated (note: locations outside continental US not shown)

At each location, the EDGE directors and a local coordinator work with a team to ensure all students have access to housing, food, and individual study/classroom space while on campus for the month of June. Local hosts also aim to provide an enjoyable experience centered on holistic student success and professional identity development through formal and informal interactions between the summer session students and EDGE mentors, faculty, presenters, and staff.

Elizabeth Sharp, Interim Vice President of Diversity, Equity & Inclusion at Texas Tech University spoke of the impact of hosting the 2018 EDGE Summer Session. "It was an honor for Texas Tech University to have the opportunity to host EDGE and we are proud to have been part of a program with such a strong track-record of impact on women in the field of mathematics. The presence of EDGE on our campus created a strong energy and further signaled our commitment to equity in the field of mathematics."

During one weekend of the Summer Session, local coordinators host a reunion conference for EDGE participants from the previous years' cohort to reconnect, debrief, and pass along advice to the new EDGE cohort. Returning students share their experiences with the next EDGE cohort by telling of their highs and lows of their first year of graduate school. One student explained: "Getting through that first year was such a rush of emotions … I'm excited to hopefully pass on some encouragement."

EDGE Summer Session Fast Facts: 1998–2018

People Involved

255 participants across 20 sessions (median cohort size is 14)
48 mentors (32 former EDGE Summer Session participants)
42 instructors (11 former EDGE Summer Session participants or mentors)

Participant Race/Ethnicity

48.2% White; 34.9% African American/Black; 9.0% Hispanic/Latina, 5.1% Asian/Pacific Islander; 2.7% Multiracial

Undergraduate Institutions Among Participants

153 colleges and universities, from 38 states and the Virgin Islands

Undergraduate Institution Profiles

44.1% are Historically Black Colleges and Universities, Hispanic Serving Institutions, or Women's Colleges; 40.8% are Baccalaureate Colleges

(continued)

> **Graduate Institutions Among Participants**
>
> 96 colleges and universities, from 39 states and four additional countries (Canada, Germany, France, and England)
>
> **Top 5 Most-Frequent Undergraduate Institutions**
>
> Spelman College, Bryn Mawr College, Pomona College, The College of New Jersey, and Xavier University
>
> **Top 5 Most-Frequent Graduate Institutions**
>
> North Carolina State University, University of Nebraska–Lincoln, Iowa State University, University of Maryland–College Park, and University of Kentucky

3 EDGE Women Become Mathematicians and Leaders

With so many programs and activities, it is impossible to capture the impact of EDGE in a single statistic. The initial goal of the founding directors —to support women entering and persisting in graduate studies towards a PhD in mathematics—has certainly produced measurable success. Since 1998, 255 women have participated in the EDGE summer session. To date, more than 90 EDGE participants have earned PhDs, and the directors expect to celebrate the 100th EDGE PhD in 2019. In fact, the PhD completion rates among EDGE alumnae have been above 60% (and as high as 80%) for 6 of the 10 cohorts between 2000 and 2010 (Note: There was no EDGE Summer Session in 2007).

As mentioned in the previous section, the EDGE participants are a diverse group. The first African-American EDGE participant earned her PhD in 2005. From 2005 to 2015 (the most recent AMS report on PhD completion), EDGE alumnae account for 20% of all African-American women graduating with doctorates in mathematics. In 2015, EDGE alumnae were 45% of that demographic. Tracking one decade of EDGE (2000–2010) with a total of 130 participants (123 whose outcomes are known and shown below), we have the following statistics regarding PhD and Master's completion (Fig. 3).

In this 10-year window, more than 90% of participants earned a graduate degree or are persisting in their studies. The overall rate of PhD completion for EDGE participants was 55% in this same 10-year period. The last published national data on PhD completion rates put the overall national completion rate at 51%, and much lower for women and underrepresented minorities [2]. This supports the claim that EDGE has created a successful model for preparing women for graduate school in mathematics.

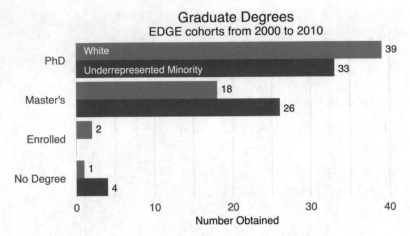

Fig. 3 Graduate degrees for EDGE cohorts 2000–2010

EDGE focuses not only on the future academic successes of its participants; it also strives to develop the whole person to be successful. EDGE participants gain the knowledge and skills needed for them to survive graduate school. But equally significant is the emotional support and confidence EDGE offer students encouraging them to thrive, both in graduate school and in their future careers.

> I left EDGE with a feeling of empowerment. I was confident that I belonged where I was. I deserved to be in the program that I got into … I think that without that, I would have had more self-doubt. But it was such a strong feeling of empowerment when I left the summer session and that continued on. I think that helped me know I was where I was supposed to be.—2017 EDGE participant describes the impact of EDGE on her success.

During the EDGE Summer Session, participants experience the blending of friendly and professional training, which helps them transition from undergraduate programs to graduate school. One 2017 participant explains this adaptation of professional mentors, "A lot of people have supportive people in their lives but I hadn't had supportive *math* people in my life …" Through decades of experience, the directors of EDGE have identified this support system as one of the most essential components for learning and helping students develop their professional identity.

The Summer Session addresses the cognitive, skills-based, and attitudinal needs and growth for EDGE participants through the workshops, mini-courses, and networking activities. Addressing each of these needs helps students anticipate what to expect in their doctoral programs and increases their learning and likeliness to succeed. These three needs have been addressed repeatedly through aspects of the EDGE Summer Session and continue to be further enforced in various EDGE activities. Because of the life-long relationships that are developed through the EDGE network, EDGE participants are continually supported throughout distinct phases in their careers. EDGE 2006 participant Ellen Swanson reiterates, "For me, more than 10 years after being an EDGE participant, the EDGE connection continues to run strong."

EDGE women are not only graduating with Masters and PhDs, but have also become influential leaders in mathematics and related fields. EDGE alumnae are:

- Teaching at institutions across the United States. Approximately 110 EDGE alumnae are now teaching the next generation of mathematicians;
- Advocating for sound STEM policy, including advising state government on K-12 education, developing math curriculum, attending AWM on the Hill events, and serving as AAAS Fellows;
- Involved in a wide range of outreach initiatives including leading Math Circles, organizing Sonia Kovalevsky Days, and organizing and participating in conferences and workshops aimed at broadening participation in mathematical sciences.

3.1 Diverse Careers of EDGE Alumnae

EDGE alumnae thrive in a wide variety of roles, inside and outside of academia. We include here a few profiles, to highlight the breadth of the EDGE network.

Sarah Bleiler-Baxter (EDGE 2006) earned an MA in Mathematics and PhD in Mathematics Education from the University of South Florida. She is an Associate Professor at Middle Tennessee State University, where her scholarly work focuses on undergraduates' learning of mathematical proof.

Yen Duong (EDGE 2010) earned a PhD from University of Illinois at Chicago. She was a 2018 AAAS Mass Media Fellow, writing for the *Raleigh News & Observer*. She currently covers healthcare for *North Carolina Health News* and does freelance science writing.

Chandra Erdman (EDGE 2002 & 2003) earned an MA in Statistics from Columbia University and PhD in Statistics from Yale University. She was a Principal Researcher at the US Census Bureau and is now a Statistician at Google.

Heather Harrington (EDGE 2006) earned a PhD from Imperial College London. She is now a Royal Society University Research Fellow and Associate Professor at University of Oxford, leading the Algebraic Systems Biology Group and co-directing the Centre for Topological Data Analysis.

Talithia Williams (EDGE 2000) earned an MA in Mathematics from Howard University and PhD in Statistics from Rice University. She is a Professor at Harvey Mudd College. She was the host of the six-part PBS series "NOVA Wonders" and is the author of *Power in Numbers: The Rebel Women of Mathematics*.

4 Recognition of Excellence

In 2018, EDGE received the Presidential Award for Excellence in Science, Mathematics and Engineering Mentoring (PAESMEM), awarded by the White House Office of Science and Technology Policy (OSTP) and the National Science Foundation (NSF) (Fig. 4). This honor recognizes the critical roles mentors play outside the traditional classroom in the academic and professional development of the future STEM workforce. To be eligible for nomination, individuals and organizations must exhibit exemplary mentoring sustained over a minimum of 5 years.

> This award helps make people aware that there are concrete things we can do to increase diversity in the mathematical sciences—things that work. Not just the number of women in mathematics but women in mathematics from diverse backgrounds.—Ami Radunskaya (EDGE co-director), quote from [1].

EDGE first received national recognition when it was honored in 2007 by the American Mathematical Society as a "Program that Makes a Difference." This award was established in 2005, to highlight programs that aim to bring more persons from underrepresented backgrounds into the pipeline to advanced degrees in mathematics and professional success. These programs must be replicable models that have achieved documentable success.

Since EDGE has acquired the national recognition as an exemplary program for mentoring success through graduate studies, the founders and directors have also received multiple awards and honors. Among many accolades, it is notable that Ami Radunskaya, Rhonda Hughes, and Sylvia Bozeman have each earned the AAAS Mentor Award.

Fig. 4 Receiving the Presidential Award for Excellence in Science, Mathematics, and Engineering Mentoring. June 26, 2018. L to R: Raegan Higgins, Ulrica Wilson, Rhonda Hughes, Sylvia Bozeman, Ami Radunskaya

5 Planning for the Future of EDGE

From its founding in 1998 until 2017 (with the exception of 2007), the EDGE Program was made possible by funding from National Science Foundation, the National Security Agency, the Andrew W. Mellon Foundation, and Microsoft Research. The EDGE directors worked in close coordination with program officers committed to the goals of EDGE to ensure funding for the program.

In 2013, in response to an overwhelming push from former EDGE participants, the directors established a 501(c)3 nonprofit organization, the Sylvia Bozeman and Rhonda Hughes EDGE Foundation to raise money to support EDGE activities and better ensure the longevity of the program. The EDGE Foundation advisory board includes current and former directors, along with mathematicians and stakeholders with a shared commitment to the continuation of the mission of EDGE.

Since its establishment, the EDGE Foundation has partnered with several colleges and universities and other entities. Sponsors include Texas Tech University, Spelman College, American Mathematical Society, National Security Agency, Bryn Mawr College, Pomona College, Cal Poly Pomona, University of Washington Applied Mathematics, Cornell University, Worcester Polytechnic Institute, University of Nebraska–Lincoln, University of Washington, Henry Luce Foundation, and Springer (publisher of this work). Some academic sponsors have benefitted from their partnership with EDGE by visiting the Summer Session with plans to adopt successful strategies for retaining women in PhD programs; by recommending incoming PhD students to the EDGE Summer Session; and by receiving applications from EDGE participants whose initial choice of a graduate program was not the correct fit.

EDGE co-director Ami Radunskaya is currently co-PI of NSF INCLUDES: WATCH US grant (NSF 1649365). This project seeks to increase and diversify the number of professional mathematicians in the United States by identifying and proliferating best practices and known mechanisms for increasing the success of women in mathematics graduate programs, particularly women from underrepresented groups.

Raegan Higgins, an EDGE 2002 participant and current Associate Professor of Mathematics at Texas Tech University, became co-director of EDGE in 2017. She says, "As we look to the future, we aim to have EDGE everywhere. We want people, institutions, and businesses to learn from and to implement the ideas and practices that EDGE has used to increase and to maintain the number of women and minorities in mathematics. EDGErs are thriving in this community and we want the world to know how it can contribute to this success."

Informed Consent All participants in this manuscript gave informed consent for the use of their quotes.

Acknowledgments The authors would like to thank Lance Bryant for his contributions to the data analysis contained in this chapter. Thanks to Tai-Danae Bradley for the use of the EDGE artwork on page 1. The authors also would like to thank Ami Radunskaya, Raegan Higgins, and the volume editors for their suggestions which improved the manuscript.

References

1. Abraham, Sneha (2018). Math Mentoring on the EDGE: Program Co-Directed by Prof. Ami Radunskaya Receives Presidential Award. *Pomona News*. Retrieved from www.pomona.edu/news/2018/07/19-math-mentoring-edge-program-co-directed-prof-ami-radunskaya-receives-presidential-award
2. Council of Graduate Schools. (2004). Ph.D. Completion and Attrition: Numbers, Leadership, and Next Steps, Washington, DC: Council of Graduate Schools summarizes the research literature on doctoral attrition and completion that has informed this project. Retrieved from www.phdcompletion.org

EDGE Through the Years

Sarah Bryant, Amy Buchmann, Michelle Craddock Guinn, Susan D'Agostino, and Leona Harris

Abstract This chapter offers a collection of group photos from each of the EDGE summer sessions held between 1998 and 2018.

S. Bryant
Dickinson College, Carlisle, PA, USA
e-mail: bryants@dickinson.edu

A. Buchmann (✉)
University of San Diego, San Diego, CA, USA
e-mail: abuchmann@sandiego.edu

M. C. Guinn
Belmont University, Nashville, TN, USA
e-mail: Michelle.Guinn@belmont.edu

S. D'Agostino
John Hopkins University, Baltimore, MD, USA
e-mail: sdagpst2@jhu.edu

L. Harris
University of the District of Columbia, Washington, DC, USA
e-mail: leona.harris@udc.edu

© The Author(s) and the Association for Women in Mathematics 2019
S. D'Agostino et al. (eds.), *A Celebration of the EDGE Program's Impact on the Mathematics Community and Beyond*, Association for Women in Mathematics Series 18, https://doi.org/10.1007/978-3-030-19486-4_3

EDGE 1998 held at Bryn Mawr College

EDGE 1999 held at Spelman College

EDGE 2000 held at Bryn Mawr College

EDGE 2001 held at Spelman College

EDGE 2002 held at Bryn Mawr College

EDGE 2003 held at Pomona College

EDGE 2004 held at Spelman College

EDGE 2005 held at North Carolina A&T State University

EDGE 2006 held at New College of Florida

EDGE 2008 held at Pomona College

EDGE 2009 held at Spelman College

EDGE 2010 held at North Carolina State University

EDGE 2011 held at Florida A&M University

EDGE 2012 held at Pomona College

EDGE 2013 held at New College of Florida

EDGE 2014 held at Harvey Mudd College

EDGE 2015 held at Howard University

EDGE 2016 held at Purdue University

EDGE 2017 held at Mills College

EDGE 2018 held at Texas Tech University

Twenty Years of Enhancing Diversity in Graduate Education from the Perspectives of Five Dynamic Women Who Have Led the Program

Farrah J. Ward and Leona Harris

Abstract In preparation for the completion of this book, we had the pleasure of interviewing the five dynamic "women math warriors" who have served as co-directors of the EDGE Program throughout its 20 years of existence, in an effort to hear firsthand about the vision and impact of the EDGE Program. The account below provides an overview of the history, contributions, and hopes for the future of the EDGE Program as seen through the eyes of its beloved founders and directors.

1 In the Early Years a Partnership is Born: Sylvia Bozeman and Rhonda Hughes

It is hard to imagine a more perfect pair of founders for the Enhancing Diversity in Graduate Education (EDGE) Program than Dr. Sylvia Bozeman, who was a Professor of Mathematics at the prestigious Spelman College, a historically black women's college in Georgia, and Dr. Rhonda Hughes, who was a Professor of Mathematics from the prestigious Bryn Mawr College, a liberal arts women's college in Pennsylvania.

Sylvia and Rhonda first met in the late 1980s at a Joint Mathematics Meeting in Atlanta, Georgia. The late Dr. Lee Lorch, who felt that they should know each other because they were both worried about the same issues, introduced them to each other and the rest is history! Rhonda remembers that it snowed in Atlanta that year and the city "shut down." They both recall meeting for dinner during the conference and continuing to talk over the phone about their shared interests and concerns, and potential ways in which they could collaborate with each other.

F. J. Ward
Elizabeth City State University, Elizabeth City, NC, USA
e-mail: fjward@ecsu.edu

L. Harris (✉)
University of the District of Columbia, Washington, DC, USA
e-mail: leona.harris@udc.edu

© The Author(s) and the Association for Women in Mathematics 2019
S. D'Agostino et al. (eds.), *A Celebration of the EDGE Program's Impact on the Mathematics Community and Beyond*, Association for Women in Mathematics Series 18, https://doi.org/10.1007/978-3-030-19486-4_4

From the beginning it was clear that these two women would cultivate a true, long-term partnership. Sylvia and Rhonda, both department chairs of mathematics departments at their respective women's colleges, quickly realized that they shared a common passion for finding ways to increase the numbers of mathematics majors in their departments and, in particular, get more women involved in mathematics.

The EDGE Program was not their first collaboration to support women mathematicians. In the early 1990s, there was a national influx of Research Experiences for Undergraduates (REUs) around the country that focused on getting rising juniors and seniors involved in undergraduate research. Sylvia and Rhonda decided that they wanted to create a program that focused on getting freshman and sophomore students, who had demonstrated promise in calculus courses, interested in mathematics and prepared for REUs and careers in the mathematical sciences.

After many hours of brainstorming, planning, and grant-writing, Sylvia and Rhonda secured funding for their first collaborative project, the Bryn Mawr–Spelman Summer Mathematics Program, which targeted promising, freshman and sophomore women at Spelman, Bryn Mawr and area colleges. In this program, Rhonda and Sylvia built in experiences for students that would develop their mathematical confidence, attract them to the mathematics major, introduce them to research, and prepare them for later participation in a summer REU.

From the beginning of their partnership, Sylvia and Rhonda were always focused on fairness and equity. As such, Sylvia recalls that they made a conscious effort to refer to the program as "the Bryn Mawr–Spelman Summer Mathematics Program for two years" and then they "called it the Spelman–Bryn Mawr Summer Mathematics Program," for the next 2 years so as not to show favoritism to any one of their institutions. This successful initiative, which lasted for 4 years (1992–1996), created a space for Sylvia and Rhonda to grapple with what they referred to as "some of the real issues in the math community" and to develop a strong professional relationship as well as a close friendship. Their bond, built on mutual respect, shared goals, and a common vision for the field of mathematics led to lots of discussions about next steps and future collaborations.

Towards the end of the grant cycle for the Spelman–Bryn Mawr Summer Mathematics Program, Rhonda and Sylvia reviewed a report from a conference sponsored by the National Science Foundation (NSF) that focused on improving graduate education in the mathematical sciences. The outcomes from the NSF conference solidified their desire to work on a different effort that focused squarely on what was happening in graduate school. Having witnessed firsthand several mathematically talented women from both Spelman and Bryn Mawr, who were strong and well-prepared, leave their graduate programs after the first year of study, Sylvia and Rhonda knew that finding ways to build and increase support for women as they transitioned from undergraduate to graduate school was the key ingredient to helping them persist through their graduate programs.

2   The Development of the EDGE Program: Key Ingredients for Success

The initial goal for the new program was clear, "to help students make it through their first year of graduate school and go back for the second year," says Sylvia. We wanted to create, recalls Rhonda, an "academic bridge program" for women that realistically simulated what students would experience during their first year of a mathematics graduate program with an effort to build in features that would focus on increasing retention and persistence.

To ensure that the summer program they were developing would accurately reflect the graduate experience, Rhonda and Sylvia drew from their own experiences but also solicited advice from others, including current and recent graduate students. One such advisor was Diana (Dismus) Campbell, who was a mathematics graduate student at Rutgers University at the time and also a graduate of Spelman College. Diana had been one of Sylvia's students at Spelman and was then a mentor for the Spelman–Bryn Mawr Summer Mathematics Program. Both Sylvia and Rhonda recall that Diana provided valuable insight from a student's perspective for the EDGE Program and that she was very instrumental in the design of the program.

Rhonda and Sylvia were deliberate and purposeful when designing each component of the EDGE Program. They spent a significant amount of time creating the EDGE experience (including graduate-level course content, near-peer mentoring, bonding experiences, group work, and networking opportunities), and working on a grant proposal to secure funding for the program. They had the proposal written when they realized that they still needed to come up with a name for the program that would stand out. Rhonda and Sylvia both recall spending numerous hours on the phone, "stretched out across the bed" toiling over the perfect name for the program that would have a catchy acronym. They went back and forth, tossing around different possibilities for names that would lead to a good acronym: "one or two letters at a time" until the perfect name arose, Enhancing Diversity in Graduate Education (EDGE).

Although Sylvia and Rhonda developed what they believed was an outstanding proposal for a new innovative program to support women, during a critical time in the mathematics community, the EDGE Program was not initially funded by the National Science Foundation (NSF). It might have been that their vision for the program was too farfetched for the review panel at the time. Although they did not secure grant funding initially, Sylvia and Rhonda were very concerned that even some of their "best students were struggling," and they knew that something had to be done. They felt very strongly that their program could help fix this national problem, so they persisted in the grant acquisition process because they knew they were on to something. Luckily there was one NSF program officer, Lloyd Douglas, who believed in the vision for the EDGE Program and shopped the proposal around to various program officers at NSF searching for ways to fund EDGE. In the end, Rhonda and Sylvia received grant funds from both NSF and the National Security Agency (NSA) to support the EDGE Program. They both recall that the funding amount was reduced which resulted in the need for them to restructure the program from 6 to 4 weeks in order to cut the budget. In subsequent years, the Andrew W. Mellon Foundation provided generous and crucial support to the program, thanks to its program director Dr. Danielle Carr.

Rhonda and Sylvia were able to draw from their own personal experiences in graduate school when developing the EDGE Program. Rhonda can remember struggling in her graduate program and was even asked to leave after the first year. "I had to regroup and try again. The leap was challenging but I figured out how to approach and understand mathematics, and my performance improved." Rhonda recalls questioning whether she "belonged" even though she pursued her graduate degree at the same school where she received her undergraduate degree. After careful reflection, Rhonda emphatically stated "I don't think I could have created the program had I not struggled myself." Sylvia experienced firsthand how "exposure" made the graduate experiences a bit "uneven." Within her first weeks of graduate school, Sylvia realized that many of her classmates had been exposed to more mathematics while they were undergraduates than she had and several of them had even utilized their graduate textbooks during their undergraduate studies. Sylvia's experience was not unique, and they both understood that it was essential to simulate this environment during the EDGE Program. In order to replicate the mathematical diversity of many of the graduate schools in the nation, they intentionally selected women from a variety of institutions ranging from small liberal arts colleges to historically black colleges and universities to research-intensive universities. Understanding that different people with different backgrounds would experience the rigors of graduate work in different ways, they sought out to ensure that the program was built in a way that would equip students with the tools needed to successfully adjust to the graduate school culture.

Graduate-Level Courses and Near-Peer Mentorship. When designing the course format for the program, courses were intentionally designed so that "the material started off slowly with topics many students had already seen as an undergraduate and gradually advanced so that by the time the four weeks were up, the participants were experiencing graduate – level content," says Sylvia. In addition, since participants were exposed to the faster pace of graduate school in classes with peers who were mathematically and racially diverse, Sylvia and Rhonda knew that it would be important to focus on group work and group dynamics. In order to address this issue, they deliberately created mandatory group study sessions with the goal of creating a space for students to learn to successfully work with their peers, a skill they knew was essential to success in graduate school.

Having graduate students serve as mentors for EDGE participants (known as EDGErs) was a key ingredient of the EDGE Program from the very beginning. The mentors were chosen to serve as role models for student participants and to serve as liaisons between the students and the instructors. The ability for EDGErs to interact with current graduate students who could readily relate to what they were feeling was crucial. Mentors were at the center of the study sessions and not only assisted

EDGErs with their coursework but worked with students to interpret questions and develop problem-solving strategies.

"We did not want the students to compete with each other when they came in the summer even though we knew there were students at different levels," Sylvia explains, so it was extremely important that homework assignments were not graded. Instead, instructors would give considerable feedback on the student's choices of proof strategy, technique, and structure. Sylvia wanted to emphasize to participants that "everybody had gaps somewhere," and the goal was not to "compete with the people across the table," but instead for participants to "focus on filling in their gaps and making themselves stronger and ready for those graduate classes." They wanted an individual student to understand that she was really only competing with herself and they wanted them to be able to clearly answer the questions: What knowledge and skills did you have entering the program? What do you know now, at the end of 4 weeks, as you leave the program?

Bonding Experiences. Understanding the significance of the mentor/mentee bond, Rhonda and Sylvia built in Thursday night dinners out, where mentors and mentees bonded over dinner without any directors or faculty involved. While helping students adapt to the rigor of graduate-level mathematics was important, they knew that the inclusion of nonacademic components in EDGE was equally important. At the end of the second week of the program, EDGE participants created and performed in a talent show. On the surface, the talent show may have seemed like a way to entertain the previous cohort during their reunion weekend, but its inclusion was quite deliberate. Students who had a stronger mathematical foundation were able to stand out in class, but the talent shows provided a good opportunity for students who may have been struggling initially in their courses to "demonstrate their strengths in other areas." Rhonda proclaimed, "the talent show provided an affirmation of their strengths, reminding people of areas where they had special talents." Rhonda stressed the importance of this component and recalled that it gave students a "sense of belonging." Sylvia pointed out that the talent shows helped to keep students encouraged and created respect for abilities beyond mathematics.

3   The Early Years of the EDGE Program

While Rhonda and Sylvia integrated many high-impact practices into the EDGE Program, there were unforeseen challenges. They realized it was not enough to assemble a diverse group of students; there was a need to intentionally address communication among students from different backgrounds. So, in the second year of the EDGE Program, Rhonda ad Sylvia enlisted a sociologist, Barbara Carter, from Spelman College, to design and facilitate a seminar entitled "Difficult Dialogues," where social, racial, economic, gender, and cultural differences were addressed.

Another challenge arose as participants coped with the heavy course load and their struggles to complete their homework assignments in a short period of time.

Rhonda recalls calling Sylvia one day to discuss how some students were in tears and to determine whether they needed to make modifications to the coursework. In the end, no major adjustments to the amount and level of coursework were made. Rhonda and Sylvia knew that the feeling of being overwhelmed was a natural experience for many first-year graduate students and they saw the value of students experiencing and grappling with those emotions in a supportive, nurturing environment.

As the Co-Founders and first Co-Directors of the EDGE Program, Sylvia and Rhonda enlisted lots of people to help them with this effort along the way. Diana Campbell's role in the early years of the EDGE Program cannot be overstated. She worked with Rhonda and Sylvia from the conception of EDGE and she served as their Administrative Director for many years. Sylvia proclaims, "we really owe her," as she reflects on the past and remembers all that Diana contributed to the program. Both Sylvia and Rhonda credit Diana for having lots of good ideas that were used to help develop the components of the program, including coming up with the idea of calling EDGE Program participants, "EDGErs." They recalled the fact that they had "little money" for the program in the early years and they often joked that "our national headquarters were in Diana's bedroom."

Sylvia and Rhonda also note that Ann Dixon, a Bryn Mawr alumna, cared deeply about women's issues and was always willing to help with some of the administrative duties. For many years, and to this day, Ann designed and maintained the EDGE website, posting news, facilitating the exchange of information within the EDGE community, and giving access to a new group of students each summer. Nona Smith, the grants administrator at Bryn Mawr and a steadfast supporter of the program, helped the program run smoothly.

4   The "Baltimore 10" Help the Co-Founders Make Key Decisions on the Future of EDGE

Similar to the Bryn Mawr–Spelman Summer Mathematics Program, the EDGE Program alternated between Bryn Mawr and Spelman for the first 4 years. After operating the EDGE Program for 4 years, Rhonda and Sylvia invited a group of key supporters, dubbed the "Baltimore 10," to a meeting in Baltimore to discuss the future directions of the program. Rhonda notes that it was during this meeting that they "re-upped" and decided to "go national" and seek out institutions nationwide that would be interested in hosting the EDGE Program in future years. The ability to host EDGE across the country would increase visibility among students and professionals, so that the EDGE network would have an opportunity to grow and ultimately be seen as an outstanding STEM mentoring program.

5 Who are the EDGE Co-Founders?

When asked to reflect on their own personal and professional personalities and what it was like to work with each other as colleagues, here is what Sylvia and Rhonda had to say:

Sylvia considers herself to be a "servant leader" who patterns herself after her mother who served others in the community and the church. "I had a great mentor in Etta Falconer. Dr. Falconer was so good at bringing people together and getting them to work together. I learned everything I know about managing programs and people from Etta Falconer." When asked about her personality, Sylvia says, "I am kind of low-key."

"Rhonda is much more experimental! She is willing to step out there and try things. She is bolder; much bolder than I am," Sylvia proclaims. "I learned a lot from Rhonda as we worked together on the EDGE Program. She would set a goal that I would be hesitant to set . . . . I finally learned to step out there with Rhonda. We did so much together. We became very close through our experiences. It was really a gift to me to be able to work with her like that. It helped me to work with other people."

Some would consider Rhonda to be an extrovert, but Rhonda explains, "I am really an introvert but I have another side that I can call upon when needed. Personally I identify with and aspire to be like people who stand up for truth and justice, people who fight against the status quo. Lee Lorch and Mary Gray were role models for me, as they always called out things that they deemed to be unjust. I would describe myself as passionate."

Rhonda points out that she "learned diplomacy from Sylvia. Sylvia is very diplomatic. She has deeply held beliefs and she delivers things with style and grace." Rhonda jokingly describes their relationship as "good cop/bad cop" and says, "I was the more feisty one, the more outspoken of the two of us." Rhonda notes, "I could say this is wrong" about some injustice and Sylvia would have the "more diplomatic approach. We really worked well together."

6 The Glue: Ulrica and Ami, Intentional Choices and Long-Term Commitment

When deciding to continue the EDGE Program, Sylvia and Rhonda knew that they would need to solicit the help of others who would have a long-term commitment to the program.

Dr. Ami Radunskya, who was an Assistant Professor of Mathematics at Pomona College, was invited by Rhonda and Sylvia to work with the EDGE Program as a real analysis instructor in 1998, the first year of the summer program. Ami was a junior faculty member and in her third year as an Assistant Professor. At the time, she had never met Rhonda or Sylvia, but was recommended to them by Dr. Victor

Donnay, one of Rhonda's Bryn Mawr colleagues, as someone who would be good for the program because she had been actively involved in other diversity initiatives. Ami distinctly remembers getting tenure the year after she started working with EDGE. She recalls spending a lot of time working with various summer programs with missions that she cared about and she remembers being concerned about how that would affect her chances for tenure, because she "did not have a lot of publications" at the time.

Having been involved in a number of summer programs for women in mathematics, Ami recalls that "EDGE immediately stood out as one of the only ones that really cared about racial diversity." The community aspect of the EDGE Program really kept Ami coming back each year. "I very much wanted to work for women in math because as a grad student, postdoc, and junior faculty, I needed that kind of community for myself. EDGE created the kind of community that I wanted to belong to."

Ami's mathematical knowledge, enthusiasm, zest for life, dedication to diversity, and belief in the ability of all students made her an ideal EDGE instructor; and Sylvia and Rhonda kept inviting her to come back. "Faculty had to spend a lot of time and a lot of emotional energy keeping students encouraged but yet making them work and not letting up; they had to demand something from them and yet keep them encouraged and Ami was good at that," recalls Sylvia. "We kept inviting her back because she did such a good job and she kept coming back... She was just really good with the students."

Ami knew she had "concrete strategies for navigating the complex web of one's graduate school path" and she was eager to share them with EDGErs. She recalls that when she was a graduate student at Stanford, there were no women faculty and that the "graduate student lounge was overtly an unfriendly place for women." She recalls organizing a group of graduate students to "clean up the space" so that women would feel comfortable there and it turned out to be a "better place for everybody." The point that she wanted to stress here was that you shouldn't "feel that you have to accept things the way they are, even though you are just a first year grad student with no power," and that there are "allies that are there to help you."

Ami was the mother of a small child while in graduate school and she remembers having trouble attending study sessions on campus and struggling to support her family on her graduate stipend. She recalls inviting her classmates to her house to study and "luring" them there by cooking them dinner. She also recalls going to the graduate director and explaining why she needed more money to be able to stay in school. He found more funding for her, but also gave her "more work to do." From these experiences, she shares the following advice with EDGErs and other students that she mentors: "Seek out your allies; build your own community; ask for what you need; and don't beat yourself up if your circumstances are different from others."

In addition to transferring her knowledge to the EDGE participants, Ami used the EDGE Program to fill a void in her own life. Ami recalls that prior to EDGE, she didn't have any mentors and she used her time at EDGE to seek mentorship from Sylvia and Rhonda. Ami recalls having many "long talks with both Sylvia and Rhonda about teaching, life, relationships, and parenting" and attributes their mentorship to her overall personal and professional growth. "I was impacted by their stories." She recalls Sylvia inviting her to church when EDGE was held at Spelman and remembers what a welcoming experience that was. Ami has continued to go to Sylvia for advice on interpersonal relationships and she notes that "her advice is always spot on!" Ami credits Rhonda for giving her advice on how to deal with curricular issues and administrative problems when she was department chair at Pomona, and she notes that Rhonda is good at giving "concrete, no – nonsense advice about how to move through your career."

Ami recalls being the first person to be selected to "run the program at an institution that wasn't Spelman or Bryn Mawr" as a Local Coordinator in 2003. "It was a big deal for me. I felt very honored that they trusted me to run it at Pomona." Ultimately, Ami would go on to run the program again at Pomona in 2008 and 2012.

Ulrica Wilson was a third-year graduate student at Emory University when she joined EDGE as a graduate mentor in 1999, the second year of the program. Ulrica had graduated from Spelman College with a BS in Mathematics. Sylvia mentored Ulrica while she was a student at Spelman and invited her to work with the Spelman–Bryn Mawr Summer Mathematics Program. Having persisted through her own struggles in graduate school, Sylvia and Rhonda felt that Ulrica was a natural choice for a graduate mentor in the early years of the EDGE Program. After completing a Master's degree at the University of Massachusetts at Amherst, Ulrica went back to Spelman to teach while Sylvia was still the department chair. Sylvia recalls, "she taught for a year or two but Dr. Falconer wasn't having that . . . she had to go back to graduate school." So Ulrica went to Emory to pursue her PhD.

While at Emory, Ulrica returned as a mentor every summer and she contributed greatly to the EDGE Program with her experience, mathematical knowledge, laid-back personality, and positive spirit. Ulrica recalls that the mentorship that she received from Spelman professors impacted how she mentored EDGE participants, noting that she tried to "mimic the experience."

Ulrica credits the EDGE Program with providing her with the opportunity to gain mentorship as a young graduate student. She recalls being mentored, not only by Sylvia and Rhonda, but also by the numerous instructors and lecturers with whom she interacted with throughout the program. According to Ulrica, EDGE provided her with one of the first professional experiences where she engaged with her research outside of the confines of her department, a key component to her mathematical development. Ulrica notes, "I was very clear about how I had benefitted from being in that space. I chose to be there each year." She credits the program for increasing her "capacity to engage at a high level at Emory."

Ulrica also recalls a time when she needed funding to travel to France to give a research presentation as a graduate student and Rhonda found funds to supplement what her department could pay. She emphasized that this experience was one of the reasons that the EDGE Program began providing travel funds to its participants to attend conferences in future years.

Once Ulrica graduated from Emory in 2004 with a PhD in Mathematics, she started a postdoc at Claremont McKenna College, and in 2005, Dr. Ulrica Wilson began serving as an abstract algebra instructor in the EDGE Program for the first time. Ulrica distinctly remembers being excited because "it felt like a promotion." She also recalls that when she was applying for academic positions, Sylvia and Rhonda helped her "navigate through the different offers" and she accepted a position at Morehouse College as Assistant Professor of Mathematics in 2007.

Ulrica and Ami were invited to be instructors each year, for many years, during the second 2 weeks of the program: Ami teaching Analysis and Ulrica teaching Abstract Algebra. Rhonda notes that Ami and Ulrica "were consistent. They were always going above and beyond. They brought an energy that we didn't always have. Both were the two anchors teaching for many years." Ulrica recalls that Rhonda and Sylvia gave her and Ami the ability to manage certain parts of the program and that they were able to introduce new traditions to the program. One such tradition was the concept of calling EDGErs "Women Math Warriors (WMW)" that Rhonda made clear "was all Ami" and joked that "she and Sylvia did not have the personalities to come up with that." Sylvia noted that throughout the years "Ulrica and Ami became good friends; good colleagues... they knew how to work together and they both took a real interest" in the EDGE Program.

7 Transition in Power: Passing the Torch to Ami and Ulrica

When it was time for Rhonda and Sylvia to retire from their colleges, they realized that they needed to consider what was going to happen with the EDGE Program. Sylvia says, "we were just moving along and all of a sudden we realized 'ok' we are about to retire and we want the EDGE Program to be in some good hands; and here are two people (Ami and Ulrica) who have proven themselves to be conscientious about it and capable and dedicated to the program; and they will be a good team! We have seen them work together." After deciding that Ulrica and Ami were ideal people to turn the program over to, Sylvia and Rhonda decided to set up a four-way conference call to tell them that they wanted them to take over the EDGE Program.

Ami remembers the phone call distinctly and recalls being "surprised that they would trust me with their baby. It's like someone calls you and you won a big a prize." Although Ami did not view herself as an administrator, she agreed to serve as one of the next Co-Directors, along with Ulrica, because she really believed in EDGE. "I had been working on the program every summer. I didn't think about the burden of finding funding. I felt like I could do all the other parts. It brought me a lot of joy. I also loved working with Ulrica. We seemed like a good combination. I knew that Rhonda and Sylvia would still be around and lots of other people would be around. I knew that there was already a community that would be there for us. I was just shifting roles."

Ulrica was also "caught off guard" when Sylvia and Rhonda called to discuss the offer but she agreed to serve as one of the next Co-Directors because it "felt like the right fit... like the natural thing to do." She also felt that there were things that she could contribute that would help to grow the outcomes of the program.

Ulrica distinctly remembers there being an "official hand-off" in 2011 when Rhonda and Sylvia handed off a silver platter to Ami and Ulrica to mark the official transition in power. When looking back, Rhonda jokes that Ulrica was replacing her as the "feisty one" and that Ami was replacing Sylvia as the "peace-maker." Sylvia pointed out that under this new leadership, increased attention was given to supporting the research efforts of advanced EDGErs. They also led the creation of the Sylvia Bozeman and Rhonda Hughes EDGE Foundation for long-term support of the Program.

8 The Impact of EDGE on Its Participants After the Summer Program

When Ami and Ulrica became Co-Directors, Ulrica says that one of their first projects was to work on marketing materials. They wanted to make sure that potential funders "understood that EDGE was much more than the four weeks." So they worked hard to capture the informal aspects of EDGE in an effort to show that the EDGE Program represented a true community and that the mentoring aspects of the program lasted well beyond the summer program.

In their marketing materials, they also wanted to capture the variety of ways in which EDGErs receive help from the EDGE community to overcome obstacles. For example, when EDGErs were unsuccessful with their preliminary exams and doubted themselves, the EDGE staff was always right there encouraging them along the way; or when EDGErs encountered unsupportive programs and contemplated leaving graduate school, the EDGE staff would help them to find new graduate programs.

The support system that EDGE provides EDGErs during graduate school is a key component of its success. If EDGErs feel isolated during their early years of graduate school, EDGE provides students with a diverse network of women at all levels that are prepared to assist them during crucial transition periods.

9 A New Era Begins When an EDGEr Becomes a Co-Director

By 2017, Ami and Ulrica had been serving as Co-Directors for 5 years and Ulrica was interested in focusing more heavily on her role as Vice President of the Sylvia Bozeman and Rhonda Hughes EDGE Foundation, a position that allowed her to continue to work with Ami, the President of the Foundation. As such, it was time for Ulrica to find someone to replace her as Co-Director. After a short discussion, Ami and Ulrica felt that Dr. Raegan Higgins would be a great choice. Raegan was an Associate Professor of Mathematics at Texas Tech University at the time; she was a 2002 EDGE participant, she had taught in the program as one of the real analysis instructors from 2014 to 2017, and she had been expressing a strong interest in hosting EDGE at Texas Tech. Ami recalls that when Raegan kept coming back to teach in the summer, "she had very strong ideas about how things should be run; she had a lot of energy; she seemed to be in the right place in her career; she really wanted to host EDGE at her institution, she had been talking about it for a while; and she just came to mind right away as someone who would be great for the role." Ulrica agreed and approached Raegan to ask her to serve EDGE in this new capacity.

Raegan recalls getting a vague email from Ulrica asking her to set up a time to talk about her role in EDGE. When they spoke over the phone, Ulrica explained to Raegan that she was preparing to end her term and Raegan remembers being caught "off guard" when Ulrica said, "we have met and we think it would be good if you were a Co-Director." Raegan's initial reaction was "Yes!" She remembers being excited that they trusted her to assume this role, but she also remembers asking a lot of questions about expectations, obligations, funding concerns, and time commitments, given that she had two small children and that she was actively doing research and writing in preparation for applying for Full Professor. Raegan also wanted to know how long Ami would be serving. Ulrica assured her that Ami would remain Co-Director for several years and that she would still be around to help, just in a different capacity, as she was planning to stay on to help write grant proposals and seek new funding sources.

Raegan recalls that as an undergraduate, she often looked to her peer mentor, Kiandra Johnson, for advice when deciding on the next steps in her mathematical path. After completing EDGE in 2001, Kiandra shared her experiences during EDGE with Raegan and strongly encouraged her to apply for the program the next year; and in 2002, Raegan went on to participate in the program. The impact of the EDGE Program on Raegan's mathematical development was clear. When talking about what helped her get through graduate school, Raegan credits "the Lord, her parents, her husband, and then EDGE"; and she jokes about interchanging the order of her "husband and EDGE," in terms of their impact on her career since she was introduced to the EDGE Program first. Raegan vividly recalls that when she took real analysis in graduate school, she often contacted Ami, who had been her real analysis instructor in the EDGE Program, for help with some of her assignments. She distinctly remembers, "copying and faxing Ami proofs" that she had written in order to get assistance. "Ami would critique the proofs, provide feedback and fax them back" to her and she credits Ami for helping her get through the course and for helping her to become a better mathematician.

Thrilled about the opportunity to work with Ami in this capacity, Raegan agreed to serve as Co-Director, and in 2017, she served as a Co-Director and as a real analysis instructor. Raegan felt strongly that this would be a great way to give back to a community that gave so much to her. Raegan proclaims, "when I think about just how impactful EDGE has been for me getting to the PhD," it was easy for her to say "Yes!"

Raegan appreciated the opportunity to serve as Local Coordinator in 2018. She felt that as a Co-Director it was important to host EDGE at her home institution at least once. In addition, as a faculty member, bringing EDGE to Texas Tech was important to her at that time in her career, because it would allow her colleagues to see who she really is, why she has made certain choices in her professional life, and what impacts and motivates her.

10   EDGE's Contributions to the Mathematics Community

During our interviews with all of the EDGE Co-Directors, past and present, we asked them to share what they believed were the most important contributions of the EDGE Program to the mathematics community.

Rhonda believes that "we showed the mathematics community that it can be diverse in a completely natural, organic way" and she expressed her excitement about the "diverse group of women moving into leadership positions" who have come out of the EDGE Program. She takes great pride in their efforts to select a diverse group of participants and believes that those efforts "created an ecosystem that is very different from the math community we all knew as students. We have shown that this is what success can look like."

Sylvia points to the "supportive community that EDGE women have created" and is very proud of the "large percentage of African-American women who earned PhDs in mathematics who went through the EDGE Program." Sylvia is excited about what EDGErs are doing now and she notes that the impact of the EDGE Program is best demonstrated through what Rhonda has coined the "EDGE 2nd Effects," programs created by EDGErs that have the same spirit and philosophy of EDGE. These programs carry on the legacy of EDGE by continuing to promote diversity and inclusion within the mathematics community.

Ami credits the program for helping to "produce women who have become movers and shakers in the math community."

Ulrica is happy to see that there are "grad programs operating differently because of the influence of EDGE."

Raegan expressed her deep appreciation for the Co-Founders, Sylvia and Rhonda, who "recognized the need for EDGE" in the mathematics community and then "did something about it."

11   EDGE's Impact on Its Co-Founders and Co-Directors

We also asked all of the Co-Directors, past and present, to share with us what they've learned from EDGE and how it had impacted their lives. This is what they shared:

Sylvia explains that "it is difficult to explain how amazing it is to see all of these young women and their accomplishments. When we get together for reunions at the Joint Mathematics Meetings and I see them at conferences, I realize that all of these women have PhDs that they may not have earned if it had not been for the EDGE

Program. They may not have finished!" She emphasizes what they were thinking when they started the program: "When we conceived of the EDGE Program Rhonda and I decided that we not only wanted students to earn their degrees, we also did not want them to be 'beaten up' in the process, that is, to be so disillusioned with the mathematics community that they didn't want to ever see research again and they didn't ever want to do anything in the mathematics community ... It is wonderful to see them excited and willing to take on professional roles. They are willing to do research together; they are willing to talk at conferences; and they are willing to accept leadership roles in the mathematics community. Just to see all of them out there doing what they want to do; that is very rewarding for me. I am proud not only of the students in the EDGE community, but also proud of the mentors and the teachers, the people who were young and came into EDGE and were impacted by the program. They developed new colleagues. It just warms my heart to see it." Sylvia also notes that the EDGE Program has "extended my professional circle tremendously and it is extended among a diverse group of people."

Rhonda says emphatically, "we cannot predict with certainty 'who's got the right stuff' to be successful in mathematics. So many EDGErs have risen from what they would describe as 'struggling.' To me EDGE proves that with hard work and persistence, students can catch up and ultimately prevail!" Rhonda notes that EDGE has afforded her great friendships and she exclaims, "it is my life's work. It is the most important thing that that I've done in my professional life."

Ulrica shares, "I have learned what's possible, in particular I learned how to administer ideas. It really trained me to go from an idea, to communicating the idea to other people, to tweaking it, implementing it, revising it, to it running on its own. This is something I keep doing in other areas of my professional life. Starting with that blank piece of paper and being able to dissect what this is, sell it to other people, and get funding for it. EDGE is where I started to do that."

Ulrica stressed the impact of EDGE on her career and explains, "when I got into this, I don't think that I anticipated taking on so many administrative roles in my professional life. When I went to get my PhD and started a tenure track position, I really expected to be a more traditional faculty member and focus solely on my research and teaching. I had not anticipated doing administrative roles at all, but it started here and other opportunities have followed." She credits EDGE for introducing her to the national mathematics community. She believes that it is the root of how she knows so many people across the discipline.

When reflecting on what she's learned, Ami explains, "I have learned that together, even a small group working together, can have a big impact. The fact that at least a quarter of all African American women with PhDs in math, in many of the past years, have come through EDGE. That's huge. It is just a small program but it really has a big impact in changing the demographics of the math community, which is an important thing to change. I learned that a small group of people, if they have the passion and work together, can affect real change. Don't think you can't just because there are a few of you. That has really empowered me." Ami also notes that she has learned that "there is some commonality in the challenges that women face but there are also so many differences. I see how different it is for African

American women than it is for White women or Latino women . . . there are all sorts of layers of challenges, so there are times when you can't just lump all women together when addressing those challenges." Finally, she points out that "EDGE has provided me with a lot of my math friends and collaborators. My math life has been hugely impacted. Now I have people I can work with, organize conferences with . . . it's fantastic . . . A pretty good chunk of my professional community comes from the EDGE community and in my private life, I am fortunate to count some of my very good friends among people I met through EDGE."

Raegan explains that the EDGE Program has taught her "to be tenacious." She is impacted by the "scope and the range of all our experiences and the range of our successes and failures" and she is excited by how EDGErs continue to "rally together" to support one another. She points out that through the program, EDGErs learn that their graduate studies and professional life will be hard, but that "there is someone in this community that is going through (or has gone through) the same level of difficulty" who will be able to help you get through your circumstances.

There is no doubt that the vision of these five dynamic leaders lives on in the EDGE community, as captured by 2018 EDGE participant Gabby Angeloro when she wrote, "EDGE has revitalized my passion for math. EDGE has re-energized my stamina to do hard work. I feel like I am a part of something larger than myself. It fuels me to achieve my dream of earning a PhD in mathematics."

Second-Generation Programs:
The Far-Reaching Impact of EDGE

Rachelle C. DeCoste

Abstract EDGE community members undertake many diverse initiatives. These diverse programs and activities are one way that the EDGE program is having an impact on the mathematics community beyond the successful completion of PhDs among participants. This article includes a wide sample of particular efforts by the community and serves as evidence of the broad reach and as a reference for others who wish to do outreach in the mathematics community.

Many in the EDGE community are actively working to diversify mathematics through a variety of activities, programs, websites, etc. A wide variety of activities, as seen in the many examples below, encourage women, students, and faculty of color, first-generation students, and LGBTQ mathematicians, among others. Other initiatives are aimed at exposing the larger public to mathematics and mathematicians to help challenge the idea of who can be a mathematician. These programs are vast in their audiences: middle-school girls, high-schoolers, undergraduates, graduate students, and even professional mathematicians. The examples provided below were collected through self-reporting by EDGE community members. They are examples of the good work that the EDGE founders, directors, and community members are doing beyond EDGE and also should serve as a valuable resource. There are likely many initiatives and individuals whom we have missed and this list will continue to grow in the years and decades ahead. In these examples, however, we see how EDGE participants create, direct, and lead initiatives aimed at diversifying mathematics beyond EDGE. They are second-generation EDGE activities.

The EDGE community, as discussed in this article, includes all student participants, graduate student mentors, and professional mathematicians who have taught in the program. As a way to distinguish the roles, we offer the following key:

R. C. DeCoste (✉)
Wheaton College, Norton, MA, USA
e-mail: decoste_rachelle@wheatoncollege.edu

© The Author(s) and the Association for Women in Mathematics 2019
S. D'Agostino et al. (eds.), *A Celebration of the EDGE Program's Impact on the Mathematics Community and Beyond*, Association for Women in Mathematics Series 18, https://doi.org/10.1007/978-3-030-19486-4_5

47

E## denotes EDGE Participant with year, M## graduate student mentor, I## instructor.

Many programs serve multiple purposes, but we attempt to organize them by target audience or key activity for the initiative.

1 Activities for Girls (K-12)

Many programs exist to encourage K-12 girls' interest in mathematics. Sonia Kovalevsky (SK) Days occur all over the country and are typically one-day conferences to engage high-school girls in mathematics. They were not begun by the EDGE community; however, several members of the community have founded or played significant roles in such programs at their home institutions.

Sonia Kovalevsky Days, Omayra Ortega E01, I12 (founder). University of Iowa, Arizona State, Pomona College. Focus of the day is on high-school students and the transition to college; middle-school students are also welcome. The day includes an undergraduate panel, a keynote by professional woman in mathematics or related field, and hands-on activities. The goal is to share the fun of mathematics while allowing girls to build community to learn they are not alone.

SK High School and Middle School Mathematics Day, Carolyn Otto E06 (co-organizer), University of Wisconsin-Eau Claire.

FEMMES Capstone, Ziva Myer E11, M15 (presenter), Duke University. FEMMES (Females Excelling More in Math, Engineering and Science) provides outreach programs in STEM for middle-school girls.

Shippensburg (PA) Math Circle, Sarah Bryant E02, I15,18 (co-founder, co-director). Program that shares the joy and creativity with fourth and fifth graders in a rural area with 60% free/reduced lunch rate.

2 Seminars/Classroom Activities/Research Opportunities/Conferences for Undergraduates

Math Mentoring, Christine Berkesh E04 (co-founder), Duke University. Program for undergraduate women in mathematics with emphasis on peer mentoring and fostering connections between undergraduates, graduate students, postdocs, and faculty.

"*Mathematician Mondays*", Carolyn Otto E06, University of Wisconsin-Eau Claire. Short introductions to living mathematicians at the start of class once a week, with a focus on women and underrepresented groups.

"*Celebrating Women in Mathematics*" *Colloquium*, Carolyn Otto E06 (organizer), University of Wisconsin-Eau Claire. A STEM colloquium organized for undergraduates in conjunction with the local SK Day. A diverse group of four to six women in mathematics from a variety of careers talk to the students about their mathematical journey, including their challenges and their successes. 2018 Speakers included Dr. Syvillia Averett (College of Coastal Georgia) and Dr. Evelyn Lamb (Roots of Unity Blog, *Scientific American*).

Math Modeling Competition, Kamila Larripa E01, Humboldt State University. Over 20 students annually are mentored through the process of preparing for and participating in the Consortium for Mathematics and its Application's international multi-day competition, the Mathematical Contest in Modeling; women students are a majority of the participants. Activities continue throughout the year. Student mentors have been funded by a grant to run problem-solving sessions and a coding boot camp. An additional outside speaker is brought in to mentor female undergraduates. An increase in students attending graduate school has been observed since these efforts have started.

Hidden No More Lecture Series, Alison Marr E02, Southwestern University. A speaker series focusing on women from underrepresented minority/ethnic groups talking about their personal journeys in mathematics and the mathematics they do. The series will continue and expand to include speakers from all STEM fields.

Aqua Squad, Candice Price M12, University of San Diego. A research group that features problems in social justice, viewed through the lens of mathematics that includes undergraduate students of color.

EQUIP: Embracing Quantitative Understanding and the Inquiry Process, Alison Marr E02 (co-creator, co-director), Southwestern University. A program to strengthen math skills and make connections between math and other STEM fields for first-year students from backgrounds underrepresented in STEM (defined as racial/ethnic minorities, first-generation students, and those from challenged socioeconomic backgrounds).

Peer Supplemental Instruction (PSI), Jamye Curry E09 (co-founder), Georgia Gwinnett College. Support students in a successful transition from high school to college-level STEM courses. The program provides collaborative learning opportunities for students enrolled in gateway courses. PSI Student Leaders prepare lesson plans using STEM-centered active learning strategies with the aim of students' learning skills and understanding of the material in their classes. Assessment of the program has revealed gains in student grades and leader knowledge of course concepts. In addition, both leaders and student participants gain new skills and competencies that should contribute to their success in STEM education and ensuing careers.

Washington Directed Reading Program, Samantha Fairchild E15 (co-organizer), University of Washington. Each undergraduate is paired with a graduate student for a math reading experience. The goal is to help those in underrepresented groups build a relationship with a mathematician and learn what it means to do mathematics.

Young Women in Mathematics (YWM), Raegan Higgins E02, I14–17, EDGE Co-director, Texas Tech University (TTU). A unique opportunity for TTU women to empower, motivate, and support one another in a field where they may face obstacles due to their gender.

Association for Women in Mathematics Student Chapter, Emille Davie Lawrence E01 (founder), University of San Francisco.

Pacific Coast Undergraduate Math Conference, Alissa Crans M03, I08,12 (co-organized 2005–2015). Undergraduate Conference that earned the American Mathematical Society Programs That Make a Difference Award.

Association for Women in Mathematics at the Technion (*WoMathTech*), Arielle Leitner E09, Israel Institute of Technology. First organization for Israeli women in mathematics; currently, in Israel, women comprise approximately 15% of undergraduate students and 5% of the faculty in mathematics.

Wheaton College Summit for Women in STEM. Rachelle DeCoste E98, M02, I15 (co-founder, co-organizer), Wheaton College (MA). One-day summit for over 200 regional undergraduates, faculty, and other STEM professionals. Day includes panel, keynote, undergraduate research talks, informal networking, and community-building.

3 Conferences and Programs for Graduate Students and Professional Mathematicians

Women in Mathematics Regional Symposia (*WiM-*), Conferences that highlight female speakers, with the aim of creating regional professional and personal networks of women mathematicians, encouraging students to present in welcoming environments, and discussing challenges faced by women mathematicians.

- Alissa Crans M03, I08,12, Amy Radunskaya EDGE co-director. Southern California.
- Amy Buchmann E10, Yen Duong E10. Midwest, Texas and the Carolinas.

WiSCon (*Women in Symplectic and Contact Geometry/Topology*) *at ICERM*, Ziva Myer E11, M15 (co-organizer). Research Collaboration Conference for Women to build a network of women in the field.

Women in Noncommutative Algebra and Representation Theory (*WINART*), Chelsea Walton I12–15.

- (Contact organizer) Banff International Research Station, Banff, Canada, April 2016.
- (Contact organizer and research group co-leader) WINART2, University of Leeds, Leeds, UK, May 2019.

Underrepresented Students in Topology and Algebra (*USTARS*), Candice Price M12 (co-founder, co-organizer). Conference to showcase the research of underrepresented graduate students. The aim of the conference is to cultivate research and mentoring networks among such students.

Designing for Equity by Thinking in and about Mathematics, Juliana Belding E03 (member). NSF-funded professional development program that addresses racially based inequities in secondary mathematics education in the United States.

Career Mentoring Workshop for Women in Mathematics (CaMeW). Rachelle DeCoste E98, M02, I15 (founder, director). A 3-day workshop for women entering their final year in graduate school that aims to support women as they search for their first postdoctoral position.

4 Organization of Research Sessions at National Meetings

EDGE Paper Sessions at the JMM (co-organizers)

- Shanise Walker E12; Laurel Ohm E13.

  - MAA Contributed Paper Session "The EDGE Program: Pure and Applied Talks by Women Math Warriors", 2019.
  - MAA Contributed Paper Session "20th Anniversary-The EDGE Program: Pure and Applied Talks by Women", 2018.

- Candice Price M12; Amy Buchmann E10.

  - AMS Special Session on Pure and Applied Talks by Women Math Warriors Presented by EDGE, 2017.
  - MAA Contributed Session on Pure and Applied Talks by Women Math Warriors Presented by EDGE, 2016.
  - Pure and Applied Talks by Women Math Warriors Presented by EDGE, 2015.
  - Pure and Applied Talks by Women Math Warriors Presented by EDGE, 2014.

Sessions at AWM Research Symposium

- Alejandra Alvarado E02; Candice Price M12 (co-organizers) "EDGE-y Mathematics: A Tribute to Dr. Sylvia Bozeman and Dr. Rhonda Hughes", 2017.
- Kathleen Ryan E08 (co-organizer); Research from the "Cutting EDGE", and (co-editor) Advances in the Mathematical Sciences: Research from the 2015 Association for Women in Mathematics Symposium.

AMS-AWM Special Session on Women in Symplectic and Contact Geometry and Topology, JMM 2018, Ziva Myer E11, M15 (co-organizer).

MAA Town Hall Meeting on Goals for Minority Participation in Mathematics, MathFest 2013, Alissa Crans M03, I08,12; Talithia Williams E00 (co-organizers). Report: https://www.maa.org/news/maa-mathfest-2013-town-meeting-on-goals-for-minority-participation-in-mathematics.

MAA Town Hall Mathematical Mamas—Being Both Beautifully, MathFest 2018, Emille Davie Lawrence E01; Erin Militzer E04 (co-organizers).

5 Establishing Networks

Math Mamas Facebook Group, Emille Davie Lawrence E01 (creator and administrator). Facebook group to provide support and advice for self-identifying mothers who are in mathematics. Currently, the group has over 600 members.

Women and Non-binary Researchers in Noncommutative Algebra and Representation Theory, Chelsea Walton I12–15 (creator and manager). Site and email listserv that is a resource for (cis and trans) women and non-binary researchers in Noncommutative Algebra and Representation Theory. http://women-in-ncalg-repthy.org/.

EDGE Ambassadors Initiative, Keisha Cook E14, M17 (co-founder and organizer). Provide role models to women at Historically Black Colleges and Universities (HBCUs) by sending EDGE participants to visit mathematics departments at HBCUs to give talks, participate in panel discussions and discuss EDGE, and pursuing graduate degrees in mathematics.

6 Public Awareness of Mathematics or Increasing Diversity in Mathematics

Mathematically Gifted and Black, Erica Graham E06; Raegan Higgins E02, I14–17, EDGE Co-director; Candice Price M12; Shelby Wilson E06 (co-creators). A website devoted to recognizing and highlighting the work of Black mathematicians. A new profile is posted each day during Black History Month. http://mathematicallygiftedandblack.com/. In 2018, the AMS printed a poster highlighting the historical contributions of Black mathematicians and one featuring a selection of living mathematicians featured on the website.

NOVA Wonders, Talithia Williams E00 (host). Six-part 2018 PBS series that considers some of science's biggest questions and the scientists behind the research to answer them.

Power In Numbers: *The Rebel Women of Mathematics*, Talithia Williams E00 (author). 2018 book containing biographies of women in mathematics, both historic and current.

Journal of Humanistic Mathematics Special Issue on Mathematics and Motherhood, Emille Davie Lawrence E01 (co-editor). Volume 8, Issue 2 (July 2018).

"Own your Body's Data" TED Talk, Talithia Williams E00. TED talk on collecting and using data on the human body that has been viewed over 1.5 million times.

PBS Infinite Series, Tai-Danae Bradley E14 (host). PBS web series on mathematics and science.

New Hampshire STEM Education Task Force, Susan D'Agostino E98 (appointed member). Served the NH Governor who requested recommendations and help implementing programs designed to modernize STEM education in the state.

Math3ma, Tai-Danae Bradley E14 (creator). Website to help students transition from undergraduate to graduate mathematics through the sharing of the author's own experiences. https://www.math3ma.com/.

7 Other

Dr. Gertrude Geraets Endowed Fund Stacy Hoehn E04, Franklin College. Provides $1000 award to undergraduate, with preference going to a female majoring in mathematics or computing, to encourage more women to persist in these majors. The award recognizes Dr. Geraets who received her Ed.D. during a time when few women were earning doctorates.

Association for Women in Mathematics, Executive Director, Karoline Pershell E03.

Golden Anniversary Campaign for the National Association for Mathematicians (*NAM*), Emille Davie Lawrence E01 (committee member). In honor of NAM's 50th anniversary in 2019, the committee aims to raise money to endow all of NAM's programs.

Undergraduate Mathematics Major, Susan D'Agostino E98 (founder), Southern New Hampshire University. Founded a mathematics major at a university with a large first-generation population.

Associate Department Head for Equity and Diversity, Miriam Freedman E00, Penn State. An advocate in the Chemistry Department for developing a diverse faculty and graduate student body.

Difficult Dialogues in the Midwest: A Retrospective on the Impact of EDGE at Purdue University

Alejandra Alvarado, Donatella Danielli, Rachel Davis, Zenephia Evans, and Edray Herber Goins

Abstract EDGE 2016 was held at Purdue University from June 6 through July 2, 2016. This meeting brought together some 20 participants from the EDGE program along with alumni from the Midwest EDGE Cluster at Purdue, graduate students from the Purdue Chapter of AWM, and undergraduate students from the local REU titled PRiME (Purdue Research in Mathematics Experience). In this article, we discuss the impact of EDGE 2016 on Purdue University, as told by the Midwest EDGE Cluster and AWM advisor Donatella Danielli; Mini-Course leader Rachel Davis; Difficult Dialogue leader Zenephia Evans; and local organizers Alejandra Alvarado and Edray Goins.

A. Alvarado
Eastern Illinois University, Department of Mathematics and Computer Science, Charleston, IL, USA
e-mail: aalvarado2@eiu.edu

D. Danielli
Purdue University, Department of Mathematics, West Lafayette, IN, USA
e-mail: danielli@math.purdue.edu

R. Davis
University of Wisconsin at Madison, Department of Mathematics, Madison, WI, USA
e-mail: rachel.davis@wisc.edu

Z. Evans
Purdue University, Office of the Dean of Students, West Lafayette, IN, USA
e-mail: zevans@purdue.edu

E. H. Goins (✉)
Pomona College, Department of Mathematics, Claremont, CA, USA
e-mail: edray.goins@pomona.edu; ehgoins@mac.com

© The Author(s) and the Association for Women in Mathematics 2019
S. D'Agostino et al. (eds.), *A Celebration of the EDGE Program's Impact on the Mathematics Community and Beyond*, Association for Women in Mathematics Series 18, https://doi.org/10.1007/978-3-030-19486-4_6

1   Introduction

It was a sunny July afternoon in West Lafayette, Indiana. Edray Goins was escorting Ulrica Wilson around Purdue University, showing her the campus before she addressed the Department of Mathematics with the talk entitled *Division: In an Algebra, In a Career, and In Research Mathematics*. She had just finished a lunchtime meeting with a handful of tenure-track faculty of color who were from the College of Science.

Little did Ulrica know that Edray had ulterior motives for inviting her to visit Purdue. He wanted to convince her that the EDGE Program should be held at the West Lafayette campus one day. She had just finished teaching at EDGE 2011 as the Algebra Instructor, and she would take over from Sylvia Bozeman and Rhonda Hughes as one of the two EDGE co-directors beginning with EDGE 2012 at Pomona College. Since the inception of the EDGE in 1998, the program had been hosted at a variety of locations, but never at a tier one university in the Midwest. Purdue University had a lot to offer EDGE, but Edray wanted to bring the program to West Lafayette for what EDGE would offer to Purdue. The mission of EDGE is in its name: Enhancing Diversity in Graduate Education. The EDGE Program has always assembled together a diverse group of women for 4 weeks in the summer, and Edray's goal was to combine this activity with other communities at Purdue striving to address issues involving inclusion and access.

It would take five more years before EDGE would come to Indiana. After extensive work and preparation, EDGE 2016 was held at Purdue University from June 6 through July 2, 2016. This article contains reflections from five individuals who ran EDGE 2016 (titles during summer 2016 in parentheses):

- Edray Goins, local organizer (Associate Professor of Mathematics at Purdue)
- Alejandra Alvarado, local organizer (Assistant Professor of Mathematics at Eastern Illinois University)
- Donatella Danielli, leader of the Indiana EDGE Mentoring cluster (Professor of Mathematics at Purdue)
- Rachel Davis, leader of a mini-course at EDGE on using the Sage cloud (now CoCalc) for computations. (Golomb Visiting Assistant Professor at Purdue)
- Zenephia Evans, leader of "Difficult Dialogues," two 2-h workshops to help prepare the participants for their first year of graduate school (Director of the Science Diversity Office at Purdue).

2   Reflections from Edray Goins and Alejandra Alvarado, Local Organizers

Structure of the Program As local organizers, we were responsible for coordinating four faculty (Maia Averett, Raegan Higgins, Eirini Poimenidou, and Shelby Wilson), one instructor (Rachel Davis, who lead a course entitled "SAGE: System

Fig. 1 EDGE 2016 group photo from June 8, 2016

for Algebra and Geometry Experimentation"), one "Difficult Dialogues" facilitator (Zenephia Evans), three mentors (Chassidy Bozeman, Angelica Gonzalez, and Stefanie Wang), and 14 students (Alicia Arrua, Sarah Chehade, Zaynab Diallo, Genesis Islas, Meghan Malachi, Zonia Menendez, Kirsten Morris, Erica Musgrave, Nida Kazi Obatake, McCleary Philbin, Amanda Reeder, Stephanie Reyes, Morgan Strzegowski, and Sarah Yoseph).

The students would take classes for the first 2 weeks (Linear Algebra and Analysis), have a short break for the "Reunion Weekend," then resume with classes for the final 2 weeks (Abstract Algebra and Measure Theory). The mentors and students worked, lived, and ate together on campus. There were also eight invited speakers (Christine Berkesch, Sarah Bryant, Deidra Coleman, Donatella Danielli, Piper Harron, April Harry, Erin Militzer, and Carmen Wright). We made a conscious effort to have the 20 or so EDGE 2016 participants interact with the Purdue Chapter of AWM as well as the REU PRiME (Purdue Research in Mathematics Experience). Students from these three groups had lunch together and attended the seminars together, with the hope that they would connect through the shared experience of being in an immersive mathematics environment. We are thankful to the supportive faculty and staff who worked with us to ensure that the participants had a challenging and fulfilling experience (Fig. 1).

The Impact on Purdue We were well aware that EDGE was unlike anything that the Department of Mathematics had seen before. Purdue has only graduated one African American woman in mathematics (Kathy Lewis in 1999), and has never graduated a US Latina in mathematics. The department has some 60 tenure-track and tenured faculty, yet only five are women. The department hosts some 30 colloquium speakers every year, yet we are lucky if three are women. Professor Rodrigo Bañuelos remarked to us after Piper Harron gave her talk during the program: "That was a remarkable speech. She spoke about the perils of being a graduate student, doing mathematics, and being pregnant at the same time. I imagine this is the first time anyone has ever spoken about being pregnant in this colloquium room." We suspect Rodrigo is completely correct.

We are not aware of whether the number of applications from domestic students, women, or underrepresented minorities has increased at Purdue since 2016–2017. However, we believe having EDGE at Purdue for those 4 weeks showed the mathematics faculty what a diverse department could look like. Mathematics graduate student Joan Ponce (one of Edray's mentees) put it this way: "I think the EDGE program is an amazing opportunity. I just wish it was longer."

3 Reflections from Donatella Danielli, Mentoring Leader

EDGE's Mentoring Clusters Mentoring is a crucial component of the EDGE experience. EDGE students are mentored by the EDGE summer faculty, the advanced graduate student assisting with the program, and the directors. As an expansion of one-on-one mentoring, in 2005 EDGE created regional Mentoring Clusters for women in the mathematical sciences, with financial support originally provided by an NSF ADVANCE grant. The goal of this structure is to advance women in academia at the three fundamental levels of graduate school, junior faculty and senior faculty, by creating mentoring networks among small groups of women in relatively close geographical proximity. The guiding principle is that, through periodic gatherings and frequent communication, the Clusters would facilitate the mentoring of junior women by senior women and the mentoring of graduate students by those in the other two groups. Hence, such a network would assist the younger two groups in advancing their professional goals while relying on the expertise of senior faculty. Although many of the students and junior faculty in the Clusters have been past participants in EDGE, other women mathematicians, at different stages of their career, have also joined in.

Currently there are seven regional EDGE Clusters: California, Georgia, Indiana, Iowa, Mid-Atlantic, Minnesota, and North Carolina. In addition, there is a non-regional Cluster focusing on Mathematics Education. Since its inception, I have had the pleasure of being the Leader of the Indiana Cluster, whose members have been affiliated with Indiana University, Notre Dame University, Purdue University, and the University of Illinois at Chicago. At any given time, the Cluster has had seven to ten active members, and we have strived to meet at least once a year. For some of the members this is a considerable effort, given that the various institutions are not in very close proximity to each other. However, I find remarkable how everybody is always enthusiastic to carve the time out of their own busy schedule to convene with their fellow EDGErs, to keep in touch and exchange ideas.

The Cluster gatherings and relationships have provided the graduate students with a forum to discuss issues on many academic and non-academic topics that impact their progress. Through the Cluster, junior faculty have raised issues related to finding early professional opportunities and negotiating the responsibilities of

Fig. 2 Donatella Danielli and Alejandra Alvarado on June 7, 2016

the early stages of their careers. Even if the Cluster members have very different backgrounds, some topics discussed in our conversations seem to be of recurring interest, such as:

- General assessment of how female graduate students and junior faculty feel treated in their respective departments;
- Balancing family and career;
- Comparison of teaching loads and course requirements in different departments;
- Entering graduate school a few years after college graduation;
- Transferring from one graduate program to another;
- How to choose an advisor.

Moreover, all the graduate students have expressed repeatedly their appreciation for the EDGE program, which helped to make their transition to graduate school much easier (Fig. 2).

Purdue's Mentoring Cluster In recent years, some of the Cluster meetings have been held in conjunction with the Women in Math Day at Purdue University, which is an annual initiative of the Department of Mathematics at Purdue. The spirit of this event is to provide an opportunity for women members of the department (faculty and students alike) to interact with each other and with a prominent female mathematician, who delivers the scientific highlight of the day, the Jean E. Rubin Memorial Lecture. Other activities of the day typically include a luncheon for all women members of the department, as well as female faculty members and graduate students from other disciplines at Purdue. The luncheon is followed by an informal meeting during which students and faculty can interact with each other and with the distinguished guest. The synergy between the two events has been beneficial for the Cluster members in several respects. In fact, it has allowed them to discuss their concerns with distinguished members of the mathematical community in a very relaxed and informal setting; it has expanded their professional network; it has inspired them to create similar events in their own institution. The Indiana Mentoring Cluster is an important support system and networking opportunity for its members. It is our hope and goal that it would constitute a crucial component in improving the status of women mathematicians in academia.

4   Reflections from Rachel Davis, Short Course Instructor

Sage and EDGE I knew Alejandra because we both had Edray Goins as a mentor during our time as postdocs at Purdue. During the Midwest Women in Mathematics Symposia, modeled after successful WIMS held in Southern California, Alejandra shared her idea that the EDGE 2016 mini-course topic could be mathematical computations in Sage. I was eager to join the team and introduce the students to Sage.

Sage stands for "System for Algebra and Geometry Experimentation" [5]. It is a computer algebra system founded by William Stein, a mathematician affiliated with the University of Washington. Computation itself has a history at Purdue–the first Department of Computer Sciences in the USA was established at Purdue University in 1962 (Fig. 3).

Sage covers mathematical computations from subjects ranging from calculus and statistics to algebra, combinatorics, graph theory, numerical analysis, and number theory. Sage is open-source, i.e., the computer code underlying the computational functions is shared openly, so that users have the ability to view and to improve

Fig. 3 Rachel Davis on June 8, 2016

the design of the software. There are also Sage Days conferences devoted to software development by interested mathematicians. (See [8] for more information and to get involved.) During the EDGE mini-course, participants signed into the SageMathCloud, which now goes by CoCalc (Collaborative Calculation in the Cloud). The hope was that Sage and EDGE would pair nicely with each other. In particular, Sage computations can help EDGE participants by providing intuition for mathematical research questions.

A Lasting Connection It is impressive to me that so many mathematicians involved in EDGE have returned to EDGE. As of 2016, Alejandra Alvarado has held *multiple* roles in EDGE as participant, mentor, speaker, instructor, and local organizer. In 2016, most of the mentors, instructors, and even guest speakers were former participants or closely related to the program. This involvement of mathematicians at different stages in EDGE is one way that EDGE has been able to excel at mentoring. Previous EDGE participants and advanced graduate students have gained practical experience facing the challenges of graduate school. In this way, EDGE has built an inspiring community of mathematicians and shines a spotlight on their accomplishments.

5   Reflections from Zenephia Evans, "Difficult Dialogues" Leader

Why Is EDGE Necessary? One of the featured news articles describing the panel session, *The Gender Gap In Mathematical and Natural Sciences from a Historical Perspective*, hosted during the International Congress of Mathematicians (ICM) meeting in August of 2018, discussed the lack of women in the mathematical realm [1]. The panelists noted that only one woman has earned the Fields Medal since it was established in 1936, and only 15% of the featured 200 speakers were female over the 9 day program. The lack of women is a frequent conversation topic at conferences, in mathematics departments, and many other places. EDGE remains necessary because the program aims to remove the barriers that may prohibit substantial increases of women in the mathematical arena.

Empowered and Informed by Data A version of the "Difficult Dialogues" workshop has been held at every EDGE summer session since 1999. During the EDGE session at Purdue University, I worked with the organizers to design 2-h sessions that would address the needs of women as they make the transition from undergraduate to graduate programs. We were greatly influenced by the book *Successful STEM Mentoring Initiatives for Underrepresented Students* [2]. We aimed to highlight the numbers of women who pursue mathematics; to explore the self-awareness of the participants; to stress the importance of creating a strong network of allies, advocates, and mentors; to share tips to deal with Imposter Syndrome; and to encourage participants to develop a plan that would aid in their success as graduate students.

A. Alvarado et al.

The attendees learned of the historical numbers of newly minted PhDs, as recorded by various surveys conducted by the American Mathematical Society [6], and were encouraged to monitor the numbers during their career. We stressed this as important because the data can allow the women to have a supporting narrative to bring awareness to this issue on the national and local levels. We did not present the data of women enrolled in each of the graduate programs among participants, but we did showcase the Data Digest at Purdue University [7], and challenged the participants to seek and know the data that is available for their perspective graduate programs. The College of Science at Purdue University has enrolled approximately 31.4% of women in the graduate programs from 2008 to 2018. The total graduate enrollment in 2018 was 1297 students, where 411 (32%) were female—which is the highest number of women and the third highest percentage since 2008 (34%) during this time frame. See Fig. 4.

The numbers are important because the lack of critical mass in certain environments may hinder the positive development of self-awareness of females that are entering the field of mathematics and other science majors. One of the concepts that can lead to success is to be aware of the traits that one possesses and those which may pose a challenge to one's chosen academic path. Self-awareness will allow one to stand strong in the face of doubt and to provide a method to combat the negativity which may be encountered during graduate school. Researchers have

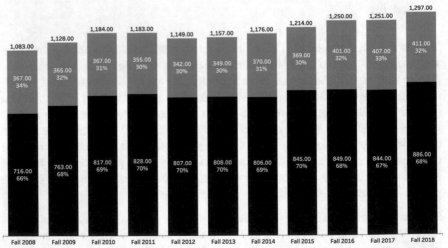

Fig. 4 Enrolled graduate students by gender in the College of Science at Purdue University (https://www.purdue.edu/datadigest/)

sited that persistence in STEM and other areas can be hindered if the individual does not have a sense of belonging in the spaces that they occupy. In order to have the sense of belonging, a level of self-awareness is necessary [4]. This awareness is heightened by knowing the requirements and being able to meet the requirements in the department, by earning the grades, passing qualifiers/comprehensive exams, and so forth.

Allies, Advocates, and Mentors As facilitators working with new graduate students, we wanted to stress the importance of being connected to a strong network of allies, advocates, and mentors. This network can be formed by a variety of people, including classmates, staff, faculty, and administrators. We wanted the participants to be mindful of the importance of networks that are necessary to assist in the navigation of new and unfamiliar academic terrain at their perspective institutions. Allies often are discovered in both on formal and informal settings at the college/university, conferences, and through normal day activities in the community. The EDGE participants have sole ownership in observing and monitoring their self-awareness in order to build an ally group that can motivate, encourage, and challenge them as they satisfactorily meet their program requirements (Fig. 5).

We discussed the importance of identifying advocates and selecting mentors that will support your success as a graduate student. This will require the participants to have conversations with former and current graduate students and others that interact with the potential mentors. Being able to engage in open and direct conversations with mentors to ensure success as a graduate student is essential but not trivial or

Fig. 5 Sylvia Bozeman, Zenephia Evans, and Edray Goins at the reunion weekend on June 18, 2016

easy. In sessions, we modeled this skill for the EDGE participants. We provided questions to ask about a potential mentor and practiced difficult conversations with potential mentors about needed time off, direction of the research, and funding for graduate school. Intentional allies, positive advocacy, and great mentorship can serve as connections to increase and build the self-awareness of the participants, whereas negative allies, advocates, and mentors can decrease the view of self and lead to development or enhancement of Imposter Syndrome (see below).

Imposter Syndrome Imposter Syndrome was first described in 1978 by clinical psychologists Clance and Imes [2] as a pervasive feeling of self-doubt, insecurity, or fraudulence and can occur once a person has been admitted to a prestigious university, received an award or promotion. Imposter Syndrome can strike anyone at any level of life. We wanted to ensure that the EDGE participants could define and be aware of the syndrome, learn some tips to deal with it, and develop a personal plan that will aid in their success when and if Imposter Syndrome strikes during the course of their graduate studies.

We read and discussed a *Scientific American* article by Hendriksen [3]. In particular, we reviewed the tips that have been developed to address Imposter Syndrome: know that feeling like a fraud is normal; remind yourself of what you have accomplished; tell a fan; seek out a mentor; teach; know that sometimes it is okay not to know what you are doing; praising efforts of kids; build in an expectation of initial failure. By presenting this content, we wanted to showcase the possibilities and the means needed to combat the experience of Imposter Syndrome.

Who Gets To Tell Your Narrative? During the 4 h of dialogue, we discussed difficult topics which may interfere with the success of the EDGE participants as they progress in the graduate programs. In the session "Who Gets To Tell Your Narrative?" we initiated the discussion by having the participants list the traits they possess that are needed to be mathematical trailblazers. We then asked them to recall and describe a situation when someone said or did something to get them to question the traits listed, and we walked through the negative reinforcements of the given narrative and discussed the possible outcomes which could result in knocking away at their self-confidence. We ended this part of the session by sharing ways which they could regain the positive reinforcements to combat the negativity of the narrative. In order for successful persistence in a field that is male-dominated, it is necessary to consistently monitor self-awareness, work to knowingly build a strong network of allies, advocates, and mentors, and understand Imposter Syndrome.

6   Concluding Remarks from Edray and Alejandra

Organizing and coordinating all details of the EDGE 2016 summer session at Purdue required planning that began in earnest during the Fall of 2015. We thank the many supportive people at Purdue and in the EDGE program that made this possible, including: Gregery Buzzard, Department Head of Mathematics; Hao Zhang,

Department Chair of Statistics; Rodrigo Bañuelos and Johnny Brown, Professors of Mathematics; Mark Ward, Associate Professor of Statistics; David Goldberg, Professor of Mathematics and Executive Director of the National Alliance for Doctoral Studies in the Mathematical Sciences (The Math Alliance); Ethan Kingery, of Purdue Conference Services; and the EDGE directors Ami Radunskaya and Ulrica Wilson.

During the 2011–2013 academic years, we worked together to organize summer speaker series where we brought a diverse group of early career female faculty to talk to students and faculty. We are so grateful that Ulrica visited one of those bright summers and (with thanks to this team) made the vision of hosting EDGE at Purdue a reality.

References

1. June Barrow-Green, Silvina Ponce Dawson, and Marie-Françoise Roy. The Gender Gap in Mathematical and Natural Sciences from a Historical Perspective. In: Proceedings of the International Congress of Mathematicians - 2018 (Sirakov, Boyan; Ney de Souza, Paulo and Viana, Marcelo eds.), World Scientific, pages 1073–1092. (2019).
2. Pauline Rose Clance and Suzanne Ament Imes. The imposter phenomenon in high achieving women: Dynamics and therapeutic intervention. Psychotherapy Theory Research & Practice, Volume15 Issue 3 (January 1978): pages 241–247.
3. Ellen Hendriksen. What Is Impostor Syndrome? (May 27, 2015). https://www.scientificamerican.com/article/what-is-impostor-syndrome/
4. Claude M. Steele. Whistling Vivaldi: How stereotypes affects us and what we can do. W. W. Norton & Company, 2011.
5. William Stein Sage: creating a viable free open source alternative to Magma, Maple, Mathematica, and MATLAB. *Foundations of computational mathematics, Budapest, 2011*, volume 403 of *London Mathematical Society Lecture Note Ser.*, pages 230–238. Cambridge Univ. Press, Cambridge, 2013.
6. Annual Survey of the Mathematical Sciences (AMS-ASA-IMS-MAA-SIAM). http://www.ams.org/profession/data/annual-survey/annual-survey
7. Purdue University Data Digest. https://www.purdue.edu/datadigest/
8. Upcoming SageDays workshops. https://wiki.sagemath.org/Workshops

The Long-Lasting Impact of EDGE: Testimonials from the EDGE Community

Sarah Bryant, Amy Buchmann, Michelle Craddock Guinn, Susan D'Agostino, and Leona Harris

Abstract The EDGE program has had a profound effect on hundreds of people in the mathematics community. Women who have been involved with the EDGE program were asked to briefly describe the impact the EDGE program has had on them. (Informed consent to publish was obtained for all quotes included in this chapter.) We received an outpouring of feedback. Common themes include inspiration, community, support, and the power of mentoring. We invite the reader to enjoy these insights into the meaningful connections formed through the EDGE program.

S. Bryant
Dickinson College, Carlisle, PA, USA
e-mail: bryants@dickinson.edu

A. Buchmann
University of San Diego, San Diego, CA, USA
e-mail: abuchmann@sandiego.edu

M. C. Guinn (✉)
Belmont University, Nashville, TN, USA
e-mail: Michelle.Guinn@belmont.edu

S. D'Agostino
John Hopkins University, Baltimore, MD, USA
e-mail: sdagpst2@jhu.edu

L. Harris
University of the District of Columbia, Washington, DC, USA
e-mail: leona.harris@udc.edu

© The Author(s) and the Association for Women in Mathematics 2019
S. D'Agostino et al. (eds.), *A Celebration of the EDGE Program's Impact on the Mathematics Community and Beyond*, Association for Women in Mathematics Series 18, https://doi.org/10.1007/978-3-030-19486-4_7

EDGE has revitalized my passion for math. EDGE has re-energized my stamina to do hard work. I feel like I am a part of something larger than myself. It fuels me to achieve my dream of earning a PhD in mathematics.—Gabby Angeloro (EDGE 2018 participant)

. . .

EDGE provided me with my first experiences learning collegiate level mathematics from someone who looked like me, thought like me, and liked some of the same things I like that weren't common among the peers I had experience with. And I got to see so many other black women concerned with social justice and representation on the same journey as me. It was incredibly affirming, rigorous, and continues to be an amazing resource.—Demara Austin (EDGE 2015 participant)

. . .

Some of my fondest memories studying mathematics were during the EDGE program in the summer of 2006. We worked together, made sense of problems, and prepared ourselves mentally for the realities of graduate school. I left there armed with confidence as a mathematician and with a support network of colleagues who were embarking on similar journeys. Without that support structure and the recognition that I was not 'alone' in my pursuit of a graduate degree in mathematics, I wonder if I would have persisted. When faced with difficult times during graduate school, I always thought back to my EDGE family. Just knowing that others were likely facing similar challenges gave me an extra layer of confidence to persist.—Sarah Bleiler-Baxter (EDGE 2006 participant)

. . .

EDGE has and continues to change my life. Everyday I remember that I have a community of women supporting me, rooting for me and watching and wanting me to succeed. Looking back at that community, I see women who represent my life goals and women who represent every stage along my journey. Knowing that other women have been exactly where I am and have overcome every struggle I currently experience is overwhelming and incredibly empowering. I see the woman I want to be in every EDGEr, and because of them I know I will accomplish my dreams.—Kelly Buch (EDGE 2017 participant)

. . .

Organizing a conference was a profound experience in my professional development that I can trace back to EDGE. Through the many challenges that come with graduate studies and a career in mathematics, I always find my experiences with EDGE to be incredibly energizing. I seek out any opportunity to be involved with the EDGE program because it consistently rejuvenates my interest in mathematics.—Amy Buchmann (EDGE 2010 participant; 2012–2014 mentor)

. . .

There are not words enough to capture the power of the EDGE program. I am so grateful for all of the amazing and supportive women in EDGE. In particular, the EDGE directors believed in me and provided instrumental support at every key transition point in my mathematical journey. I have seen how their investment of time, energy, and love has built a community of women who not only become mathematicians but Women Math Warriors.—Sarah Bryant (EDGE 2002 participant; 2005 and 2006 mentor; 2012, 2015, 2018 instructor)

. . .

In EDGE I have found mentors who care about every aspect of myself. I have had the privilege to be among people that are committed to my academic growth and personal success. I am not afraid to ask them for help, share my anxieties and celebrate with them moments of mathematical happiness.—Alejandra Castillo (EDGE 2017 participant)

. . .

EDGE prepared me for graduate math courses, but also welcomed me into an amazing network of women mathematicians! The cohort provided a support system that remains today. Learning from successful women, especially women of color, was truly inspirational. Giving back to EDGE has allowed me to inspire others and provide support for future women mathematicians.—Keisha Cook (EDGE 2014 participant; 2017 mentor)

. . .

Participation in EDGE changed my life. The opportunity to live and learn among likeminded peers all working towards a common goal was a totally different experience than I had in undergrad. I was a part of a community. I belonged. Interaction with the EDGE instructors, mentors, guest speakers, and fellow participants empowered me in ways that are immeasurable and priceless. I developed connection and bonds with the women of EDGE, which are strong and have withstood the test of time and distance. I've found support when I decided to pursue a non-traditional academic career. My EDGE support encouraged me because they recognized my aptitude and talent for this new environment. I wouldn't dare try to single out any one EDGEr for improving my life. There are too many, each of who I have special and unique connection!—Carla Cotwright-Williams (EDGE 2001 participant; 2005 mentor)

. . .

EDGE has inspired me to mentor undergraduates at my current institution, which mostly consists of under-represented, non-traditional students. I often share with my mentees my experience in EDGE and offer encouragement for them to continue on to graduate education upon receiving their undergraduate degrees.—Jamye Curry (EDGE 2009 participant)

. . .

When I participated in EDGE on the campus of Bryn Mawr College more than 20 years ago, I had no idea that I was joining a community of friends, colleagues, and collaborators in a sort of national mathematics department that would last throughout my career. I am grateful for Sylvia Bozeman's and Rhonda Hughes' foresight in creating the EDGE Program and for the inspiring community of women mathematicians that grew up around it.—Susan D'Agostino (EDGE 1998 participant)

. . .

This, I would say, is the EDGE model: receive and always, always, find a way to give back... Although I gained significant mathematical knowledge in the EDGE program, it is mainly for the network of women that I was able to connect with that I feel grateful... Between laughs and more serious heart-to-hearts, these incredible Women Math Warriors always remind me to never settle, whether for research, teaching or outreach and always having in mind to find ways in which I should pave the way of success for those who come after me.—Karamatou Yacoubou Djima (EDGE 2008 participant; 2013 mentor)

. . .

Before EDGE I had never thought about women in mathematics despite being one. Fun fact: EDGE was the first time I had been taught math by a woman in my life besides a substitute teacher in fifth grade. Now I find myself organizing women in math conferences, writing and talking about being a woman and mother in math, and mentoring any women and girl students I run into, all because of EDGE.—Yen Duong (EDGE 2010 participant)

. . .

EDGE gave me a solid foundation coming into grad school that helped me thrive in my first year. It also gave me an amazing support group and community of women that I know I will be connected with for the rest of my life. Lastly it gave me a lot of female role models at all stages in their mathematical careers that I can look up to. I am very grateful for everything that EDGE has provided me with.—Catherine Godfrey (EDGE 2017 participant)

. . .

The EDGE program hasn't stopped impacting me since I submitted my application. Even as a participant I didn't see how EDGE is a long-lasting community of support. It was hard to fathom how a quick summer program could change my life, but now it's impossible to fathom my life without it.—Shannon Golden (EDGE 2018 participant)

. . .

Through EDGE, I belong to an amazing and diverse community of women mathematicians that has invigorated me throughout the various stages of my career. EDGE has instilled within me a confidence in what I can accomplish; shown me the importance of doing good work and being active in my discipline; provided financial support and various professional opportunities; given me role models like none other; made me deeply grateful for the exceptional mentoring I have received along the way; and, inspired me to extend that mentoring to those who come after me. Basically, EDGE is the gift that keeps on giving.— Erica Graham (EDGE 2006 participant; 2010 mentor; 2019 instructor)

. . .

During the EDGE program, I didn't realize how influential this program would be to my mathematical journey. It didn't take long for me to recognize that EDGE would not only be an integral part of my graduate education but it has been throughout my career long after school, and not only professionally but also personally. EDGE is more than a summer program, but it is a family. There are several people I met through this program who have helped me study for qualifying exams, create research presentations, navigate the job market and how to achieve tenure and promotion successfully. The support I received from this program is outstanding, and I hope I can pay it forward to future mathematicians.—Michelle Craddock Guinn (EDGE 2004 participant; 2008 mentor)

. . .

EDGE provided me with confidence going into a PhD program, especially being the first in my family to get a PhD. It also provided a network–more like a family–of female mathematicians. We have supported each other through graduate school, postdoc transitions, research visit funding, tenure-track angst and work/life/motherhood balance. Even this last Joint Math Meeting, another EDGEr and I met up early morning (before 7am) to have a coffee catch up before all the activities began at the conference. I'm hugely indebted to the EDGE program and the women that have devoted their time and energy to make it so successful.—Heather Harrington (EDGE 2006 participant)

. . .

EDGE has provided me with tools to survive my first year of Grad school. It provided me with confidence that carried me when I was at my lowest points. I am eternally grateful to the support that EDGE has given me.—Micah Henson (EDGE 2017 participant)

. . .

The summer portion of EDGE was dress rehearsal for graduate school. That level of intense preparation greatly contributed to my success as a graduate student. The EDGE community is a HUGE part of my life and continues to contribute to my success.—Raegan Higgins (EDGE 2002 participant; 2014–2017 instructor; 2018 local coordinator; current co-director)

. . .

EDGE is a community of peers and mentors whom I would never have otherwise found! They bring me hope and inspiration when I am frustrated and don't see the point of completing my PhD. I truly do not believe I would finish grad school without EDGE!! Thank you to all, especially Raegan and Ami!!!—Austen James (EDGE 2017 participant)

. . .

The EDGE experience was overwhelming at times, but looking back, the chance to meet and work with so many different people for that intense period of time was such a valuable opportunity for me. Amid the whirlwind, I learned an appreciation for the larger math community, and as I continued on to graduate school and now a job in academia, I have continued to learn how important community is for mathematicians. There is something so special about making connections through a common interest in math. Following the example set by Sylvia and Rhonda, members of the EDGE community celebrate good times and help each other through difficult ones.—Jill Jordan (EDGE 1999 participant)

. . .

I am sincerely thankful for the EDGE program's profound and sustaining impact on my career, as well as to Rhonda, Ami, Sylvia, and Ulrica for their continuing mentorship, advice, and support. I received my Ph.D. in Applied Mathematics from UC Davis in 2006, was a Fulbright scholar in Mathematics in Switzerland, and currently teach at Humboldt State University on the beautiful and remote Northern California coast.—Kamila Larripa (EDGE 2001 participant)

. . .

The EDGE program replicates possible challenges that a person but more importantly a woman may face in a doctoral program. However, one distinct difference that the EDGE program incorporates are methods to overcome the challenges. . . . The dynamics of EDGE forced us to bind as a group and taught us that multiple levels of support may be necessary in graduate school as challenges occur. Every aspect of the program provided valuable lessons that are still relevant to me as a professor.—Torina Lewis (EDGE 2008 participant)

. . .

EDGE prepared me for graduate school not only academically, but also emotionally. The summer session allowed me to see my mistakes and failures as an opportunity for growth instead of a hindrance. During my first year of grad school, there have been many instances where I've had to pick myself up from a failure and keep going, which would have felt impossible without the encouragement of my fellow EDGErs. I formed some of my closest friendships in math because of EDGE. Thanks to these friends who have helped me navigate the difficult parts of graduate school, I am much more confident and secure in myself as a mathematician now. I still have a long way to go, but I'm confident that I'll be able to finish strong with the support of EDGE.—Jessie Loucks (EDGE 2018 participant)

. . .

I feel blessed that I was able to participate in EDGE 2017. It was intense but beneficial. It was an opportunity to improve our mathematical skills and knowledge. It is more than a 'one-month program.' It is a free lifetime membership to amazing resources and opportunities. It is my support system whenever I feel like giving up. Whenever I felt like I did not belong to my program, or I do not understand some assignments, I would quickly reach out to my fellow Edgers for support. It is a mentorship as well as a sisterhood for the Edgers. Our mentors are always there whenever we need help or feel discouraged. We need more women in this field and it starts with EDGE because they train Women Math Warriors.—Carmel Laetitia Mobio (EDGE 2017 participant)

. . .

I feel truly grateful and proud to be part of the EDGE community. It was valuable for me to have a cohort of women that was going through the challenges of graduate school with me and it is inspiring to see those who came before me and the amazing things they have accomplished. It is rare in academics to find a group that cheers for you in your successes, lifts you up in hard times, and provides support and mentorship at each step in your career. EDGE has been that group for me and I feel like I am a part of something truly special.— Molly Moran, (EDGE 2009 participant)

. . .

I am so grateful to be a part of the EDGE family. They have cheered me on throughout my graduate career and beyond. I enjoy passing that enthusiasm on to my EDGE mentees and watching them flourish! It's great to be a part of such a supportive community of women math warriors!—Ziva Myer (EDGE 2011 participant; 2015 mentor)

. . .

EDGE has been a constant positive force in my life since I first participated in 2001 at Spelman College. I had never in my life learned mathematics in an all-female environment and I never have since then. Learning in such a nurturing and comfortable environment gave me the skills necessary to not only survive but to flourish during my first years in grad school at the University of Iowa. I left the program with lifelong friends and collaborators who have grown to be my chosen family. I cherish my time with my fellow EDGErs.— Omayra Ortega (EDGE 2001 participant; 2012 instructor)

. . .

From my first year as a graduate student to my eighth year as a professor, I have always had my framed picture of my EDGE cohort within view of my desk. It is not something I notice everyday, but when I am struggling with my work and I look up and see it, it gives me reassurance that I will overcome my struggle. . . I came into EDGE as a female mathematician. One that was sometimes confident, sometimes unsure of herself and often worried about the future and the challenges ahead. I left EDGE as a fierce Woman Math Warrior.—Carolyn Otto (EDGE 2006 participant)

. . .

In the summer of 2003, I learned how underprepared I was for graduate school, I learned how to work with others who were smarter than me, and I saw that this work was hard for everyone. There was a transparency that was more than just a 'window into graduate school,' but was a transparency of ourselves to the other participants.—Karoline Pershell (EDGE 2003 participant; 2008 mentor)

. . .

EDGE has given me such joy, strength, inspiration and camaraderie through the years that I consider my involvement with the program, one of the highlights of my professional life. I enjoyed mentoring and being mentored and forming life long friendships.—Eirini Poimenidou (EDGE 2006 and 2013 local coordinator; 2002, 2014, 2016 instructor)

. . .

Even though I was not an EDGE participant, while I was in graduate school I was constantly introduced to and mentored by EDGErs! From Michelle Craddock Guinn, Emille Davie Lawrence, Rachelle DeCoste, Alejandra Alvarado, Carmen Wright, Omayra Ortega and Carla Cotwright-Williams I always felt supported by EDGE, even when I didn't know it.—Candice Price (EDGE 2012 mentor)

. . .

EDGE has been a consistent piece of my life as a professional mathematician, and the Southern California regional cluster helped me 'EDGify' my life during the academic year. . . For me, this is the joy of EDGE: watching our network not just grow, but thrive, with each node maintaining its own personality, while strengthening and growing new edges.—Ami Radunskaya (EDGE 1998–2002, 2009–2011, and 2018 instructor; 2008 local coordinator; current co-director)

. . .

EDGE has been such an amazing experience for me. I have gained so much from this program and math has only been a small part of it. The EDGE community is unparalleled and it has really inspired me. I am so grateful for the opportunity to be able to participate and I truly believe that I will be able to succeed because of this program. The immense support and encouragement from everybody involved is something I will cherish during my years of graduate school. I have a deep love for this program and I think me being here is such a blessing.—Lynnette Robinson (EDGE 2018 participant)

. . .

EDGE introduced me to an amazing group of women and gave me the support to not only survive but also thrive in my first year of grad school.—Rebecca Santorella (EDGE 2017 participant)

. . .

Since that time, I have remained in touch with many of the women I met through the EDGE network and they have kept me informed of numerous academic and professional opportunities. The EDGE Program has been very effective in diversifying STEM careers and EDGErs have been and still are very significant to the science and mathematics community. These women are inspiring trailblazers, leading in their chosen fields, serving as role models for girls and other women, and creating their own paths of success.—Martene L. Stanberry (EDGE 2005 participant)

· · ·

I received my Bachelor of Science degree from a small, liberal arts institution. I knew I wanted to learn more but I was already suffering from imposter syndrome due to my limited background. I was honored to be selected as a participant in EDGE 2006 at New College of Florida in Sarasota, FL. This experience didn't remedy my imposter syndrome but it did give me a network of amazing women whose help and support continue to mold my experiences within the math community. I attended North Carolina State University where I continued to experience the doubt of being able to fulfill the requirements of graduate school. But having a group of women who I thoroughly admired experiencing similar struggles and doubts as I was made me realize that this was just a piece of the challenge. Together we navigated through graduate school offering support and advice whenever possible. . . Ultimately, I chose to continue my career at a small liberal arts institution where I can provide support and encouragement to other young mathematicians doubting their abilities to continue their journey within the field of mathematics.—Ellen Swanson (EDGE 2006 participant; 2013 instructor)

· · ·

The EDGE program has impacted me in so many ways. As a student leaving undergraduate school, I was able to find connection in the mathematical community through meeting other students entering graduate school as well as awesome mentors at the EDGE program. While a graduate student, I reached out to mentors from the EDGE program for advice and was given great advice. It was through the EDGE program that I made connections with other women mathematicians that have proven to be long lasting.—Shanise Walker (EDGE 2012 participant)

· · ·

It was fantastic to work with other women, including many women of color, who were about to embark on their graduate school journeys. I was also honored to have the opportunity to work with other faculty who 'walk the walk' in efforts to increase diversity, inclusion, and equity of researchers and educators in the mathematical sciences. . . Being able to see myself in others–in students coming after me, in faculty clearing the path for me, and in peers with me along with the way–is a crucial part of my finding happiness and a sense of belonging in this job. . . . It is my humble wish to help clear the path so that EDGE program participants and other marginalized folks can see themselves, not through the muddied lens of others' biases or prejudices, but with the proper view of using one's talents (mathematics) to find happiness, community, and fulfillment with this work.—Chelsea Walton (EDGE 2012–2015 instructor)

· · ·

Thanks to the EDGE program, I discovered the value of learning mathematics in an all female setting. Since finishing my masters degree in mathematics, I have been teaching math and computer science at an all-girls school. In my classroom, I am inspired daily by the next generation of female thinkers and leaders. I have also recently returned to graduate school to complete my PhD in STEM education. My dissertation and future research will be centered around girls in mathematics and computer science. I often refer back to my EDGE days, when I realized the importance of being a part of a network of strong women to support and encourage one another.—Jennifer Rowe Webster (EDGE 2002 participant)

. . .

It is through my EDGE network that I have learned how to be a mathematician, a wife, a mother, a teacher and how to exist at the intersections of all those things. Consistently over the past 12 years, the EDGE network has provided both a net for me to fall down on and a ladder for me to climb up. In short, EDGE is deeply woven into the fabric of my life and will forever be one of the keys to my success.—Shelby Wilson (EDGE 2006 participant; 2009 mentor; 2014 and 2016 instructor)

. . .

EDGE is a community of women that have experienced or are about to experience the hardship required to get a PhD in math, especially as a woman. I have a plethora of amazing women who are inspiring and willing to be a mentor to me. It's great to know that I'm not alone.—Lyndsey Wong (EDGE 2017 participant)

. . .

I have been a part of the EDGE program since 2006 when I was a participant. Overall, I had a positive experience that summer. Of course, the difficulty was the actual work, and the inevitable feelings of inadequacy one faces in those situations. But it helped knowing I wasn't alone. I remember being impacted by seeing black women in pure mathematics. In the EDGE community, we are allowed to have our own path and story to tell. Sometimes EDGE provides emotional support to help us not feel alone, and sometimes it provides connections and opportunities that help us move our careers forward. I truly cherish the faith and belief that the EDGE community has shown in me.—Carmen Wright (EDGE 2006 participant; 2011 mentor)

. . .

EDGE mentally prepared me for the fact that I would struggle–a lot–in grad school, and gave me the confidence and support system I needed to know I'll overcome all of the struggles that come my way.—Jenna Zomback (EDGE 2017 participant)

. . .

Part II
Mathematics Inclusivity and Outreach Work

Academic Preparation for Business, Industry, and Government Positions

Alejandra Alvarado and Candice R. Price

Abstract According to the 2015 Annual Survey of the Mathematical Sciences, 1901 PhDs were awarded in the USA. The report shows that 52% of those recipients are working in academia. This is a decrease from the 2014 survey which stated that 56% of new doctoral recipients went on to academia. Thus, what support do we, as academics, provide to this growing population of business/industry, or government job seekers? The goal of this paper is to provide insight into programs tackling this question along with relevant information and advice for new PhDs interested in jobs outside of academia as well as those interested in making successful mid-career moves.

1 Introduction

Recently, a trend in academic conferences has been to include a panel session on business/industry or government (BIG) jobs for academics and recent graduates. Often these sessions include someone that started in a tenure track or tenured position before being seduced by the allure of industry.

At the Infinite Possibilities Conference in 2018 [9], there was such a panel session entitled, *Mid-Career Moves and New Opportunities*. The session included four female mathematicians with varied experiences working in government, industry, and academia. According to Dr. Carla Cotwright-Williams, who currently works for the Department of Defense, many first generation graduate students don't know what they are signing up for when they enter graduate school. Cotwright-Williams goes on to say that a career in academia can be very fulfilling, assuming

A. Alvarado (✉)
Naval Surface Warfare Center, Panama City Beach, FL, USA
e-mail: aalvarado2@eiu.edu

C. R. Price
University of San Diego, San Diego, CA, USA
e-mail: cprice@sandiego.edu

© The Author(s) and the Association for Women in Mathematics 2019
S. D'Agostino et al. (eds.), *A Celebration of the EDGE Program's Impact on the Mathematics Community and Beyond*, Association for Women in Mathematics Series 18, https://doi.org/10.1007/978-3-030-19486-4_8

the individual knows what it entails. If a student does not have professional role models growing up, they are at a disadvantage—not knowing their options or opportunities—*even* before entering college. This was the case with both authors of this paper. The assumption was that a PhD in mathematics only meant a job teaching at the university level. Many students are not exposed to the multitude of possible careers in STEM, much less the career opportunities within the mathematical sciences.

From the authors own experiences and through conversations with colleagues, they learned that many faculty themselves don't realize the vast opportunities within academia as well as in BIG; switching from one career to another is a possibility but requires preparation.

Another panelist Dr. Karoline Pershell, currently with the Association for Women in Mathematics, and Service Robotics and Technologies, noted that she had to be comfortable taking risks and not be afraid to fail in her career moves. After her postdoctoral position, panelist Dr. Maria Garcia took several years off due to a family situation that required her full attention. Her goal was always to return to work but was unsure when and where. After hiring a career coach, and networking, she received several job offers for many different types of positions. Although she had originally planned to stay in academia, she happily accepted a position at the US Bureau of the Census, where she has been for the last 20 years.

Another important piece of this story is the conversation around those underrepresented in mathematics. Included in the panelist presentations was a discussion on the lack of representation in the mathematical sciences and its impact on the question of career choice. How does a society encourage women to pursue PhDs when it appears that jobs are scarce? Data shows that women hold about a third of PhDs in the mathematical sciences, but only about 23% of women with STEM degrees actually work in STEM fields [18].

The IPC panel session inspired the theme of this paper. The potential for women to make significant contributions in the STEM workforce, specifically in higher paying careers and leadership positions, is vital. We need to insure women, especially women of color, have a seat at the table, and that their voices are heard. Perhaps exposure to more career opportunities for those with STEM degrees, mathematics in particular, will help shift the needle to a more balanced representation.

In this work, we explore several support networks for those with degrees in the mathematical sciences that are interested in BIG career opportunities. We conclude by providing advice collected from various resources.

2   Programs Supporting Interest in Business, Industry, or Government Positions

According to the most recent AMS Annual Survey, Report on the 2015–2016 New Doctoral Recipients, the number of new PhDs taking positions in BIG has increased to 495 this year compared to 409 two years prior. US academic hiring has decreased while US nonacademic hiring has increased, since 2012.

It should also be noted that while the overall unemployment rate of those who receive a PhD in the mathematical sciences is 5.9%, new doctorates from the Small Public Institutions reported the highest unemployment rate at 13.7% while new doctorates from the Biostatistics group have consistently reported the lowest unemployment. A 2016 Pilot Study, conducted by Dr. Amy Cohen, that looked at the transition from a research postdoctoral position into immediate employment, found that about a third went on to a tenure-track position, while 8% went into a BIG career [7]. As more mathematicians are shifting into positions in BIG, quite a few programs and groups have been formed around the primary mission of supporting mathematicians interested in business, industry, or government positions.

The *Business, Industry, and Government (BIG) Math Network* [3] brings together the mathematical sciences community to address several issues surrounding the connections between academia and positions outside of academia. The BIG Math Network is a collaborative effort between mathematical sciences societies, institutes, labs, businesses, government agencies, and academic partners. The goals for the network include bringing together the mathematical sciences community to build job opportunities for mathematicians; communicating the value of mathematical science in the workplace; cultivating connections between students, faculty, recruiters, and managers; increasing knowledge about internships and how to prepare for them; providing viable models for internship logistics (including timing, intellectual property, and training), and creating regional networks. The network realizes these goals by accomplishing three primary objectives:

1. The network's website includes information for students and departments, opportunities for job seekers, and blog posts from people with careers outside of academia.
2. The network has cosponsored career panels at conferences.
3. The network has created a tool to support departments to assess their current initiatives to connect with BIG and make strategic plans to do more. The network has also initiated the program *Math to Industry Bootcamp* at the Institute for Mathematics and its Applications in Minneapolis.

The Math to Industry Bootcamp is a 6-week summer program that provides about 30 graduate students the training and experience that is valuable in industry positions [3].

The Mathematics Association of America (MAA) program *Preparation for Industrial Careers in Mathematical Sciences* (PIC Math) prepares mathematical sciences undergraduate students for industrial careers by engaging them in research problems that come directly from industry by supporting faculty [16]. The PIC Math program has three specific aims:

1. Increase awareness among mathematical sciences faculty and undergraduates about nonacademic career options.
2. Provide research experience working on real problems from BIG.
3. Prepare students for industrial careers.

PIC Math provides a program that supports faculty by equipping them with content for a spring semester research and credit-bearing course focused on solving industrial problems. Each faculty participant is asked to assemble a team of three to five students and work with them to develop their problem solving, teamwork, and communication skills. Each team will choose from one of five problems that are real-world problems yet suitable for undergraduate students to work on. The resources for students and faculty participating in PIC Math include a series of training videos on techniques for generating solutions and decision aids useful for coping with "messy" real-world problems [16].

This program includes a 3-day summer training workshop for faculty at US institutions. This workshop provides participants with information on BIG careers to share with their students; guidance on developing BIG connections; exposure to problems that arise in industry; and often overlooked, training on how to help students develop skills that are valued by employers.

The BIG Math Network officially kicked off in the early 2017, while PIC Math received its first round of funding in 2013. But as far as longevity, the Society for Industrial and Applied Mathematics (SIAM) [17] has a long-standing representation of encouraging opportunities in industry. Its website includes a web page dedicated to organizations hiring mathematicians, profiles of various mathematicians who hold positions in BIG, and a download-able careers brochure.

While not explicitly a program for positions outside of academia, the Enhancing Diversity in Graduate Education (EDGE) has been a large source of support for its participants interested in positions outside of academia [8]. The EDGE program is a summer math program with the goal "of strengthening the ability of women students to successfully complete PhD programs in the mathematical sciences and place more women in visible leadership roles in the mathematics community" [8]. Being a part of the EDGE network provides participants with a network that includes women in positions outside of academia who can provide mentor-ship for BIG careers. In fact, Cotwright-Williams and Pershell are both members of the EDGE network.

Some companies have programs focused on recruiting those with PhD, thus giving applicants the opportunity to intern in industry during the summer or sabbaticals. One such program is run by the Institute for Defense Analyses (IDA) [10]. Since the 1950s, the IDA Center for Communications and Computing "has performed fundamental research in support of the National Security Agency's mission in cryptology," which includes both foreign signals intelligence and protecting the communications of the US Government [10]. The Center is a nonprofit entity consisting of the Centers for Communications Research with offices in Princeton, New Jersey (CCR-P), and La Jolla, California (CCR-L), and the Center for Computing Sciences in Bowie, Maryland (CCS). While the three offices have distinct areas of focus, they work closely with each other and share many overlapping research teams. For this paper, the most important collaboration occurs during the summer workshops, called SCAMPs. These workshops bring in academics and others to use a "team-style" approach to tackling several difficult problems each summer. The participants for these workshops are diverse in many ways: some come from the academic community while others from research organizations; there are many levels of experience ranging from seasoned researchers and distinguished faculty to advanced graduate students or exceptional undergraduate students; and disciplinary backgrounds can vary to include mathematics, computer science, statistics, physics, and electrical engineering. In a typical summer, the workshop has more than a hundred visitors across the three centers. The intense and collegial atmosphere is well known.

There also exist several programs that offer internship-like opportunities that post-PhD mathematicians can take advantage of. The American Association for the Advancement of Science (AAAS) offers visiting scholar positions and fellowship opportunities to "science and engineering professionals to participate in and develop leadership skills for government, policy-making, and mass media roles" [1]. The National Security Agency (NSA) offers sabbaticals ranging from 9 months to 2 years [13]. These visiting mathematicians have the opportunity to work on a variety of problems in different areas of mathematics. The Office of Naval Research is another government entity that has an internship-like appointment, the Summer Faculty Research Program [15]. Science and engineering faculty members can work at US Navy laboratories, on a recurring basis. The National Science Foundation (NSF) offers temporary/rotator programs where math PhDs can be temporary program directors and recommend which proposals to fund and have an influence on scientific direction, while still being affiliated with their current institution [12]. Usually after a year or two, participants in both of these visiting positions return to their institution with "new insights and experiences." Returning faculty have the opportunity to share their experiences and provide new knowledge about the diversity of career options for mathematicians with their students and peers.

3   Preparing Students for Careers in Business, Industry, or Government

Graduate students in the mathematical sciences work to advance the understanding of a relatively narrow field of study. Preparation for future careers is typically in an academic setting, for academic purposes, directed by academics. Thus many graduates aspire to receive faculty positions, specifically, tenure-track positions. But, the supply of newly trained PhDs outnumber faculty replacement needs. Hence, BIG employment offers alternative opportunities for these "surplus" graduates. According to a 2015 NSF-IPAM Mathematical Sciences Internship Workshop report [11], the number of PhDs in the USA has approximately doubled in the past 10 years, while the number of tenure-track positions is decreasing.

This has led to what is academia is calling, the "career diversity" movement [5]. These same issues are being seen not just in mathematics but across many disciplines. It has become increasingly important to begin training students, undergraduate and graduate, for diverse careers, rather than only training them for academia.

In [4], the authors state, "Many graduate students will continue to follow a traditional academic career path, but having the option to choose careers in industry and governmental organizations will benefit all of them." So the question becomes: how does one prepare students for careers outside of those offered in academia, especially when all you know is academia? Some programs have made this a priority at their institutions. They find that it is important to have more than one area of training. One institution and program that has been recognized for its efforts in the area is the University of Illinois Urbana-Champaign (UIUC) Department of Mathematics. UIUC has an NSF funded program that has been successfully helped place students in BIG positions through internships [11]. UIUC also offers a summer computational boot camp to their graduate students, with the goal "to teach practical computational mathematics techniques using Python programming in 2 weeks to someone with little or no programming experience." The results have been rewarding.

"At UIUC, several students who might not think of their thesis focus as applied or industrial mathematics topics have participated in internships, sometimes through on-campus collaborations with lab groups in other departments. For example, a combinatorist modeled infectious disease in sheep, in a veterinary medicine lab. A number theorist modeled ant colonies, in the entomology department. A functional analyst worked with an e-commerce analytics firm. A graph theorist worked with a financial trading firm and a student interested in differential equations worked on agricultural data analytics. One helpful mechanism for placing students in the internships is interviewing them about their interests outside of mathematics."

4 Advice for Mathematicians Interested in Careers in Business, Industry, or Government

Transitioning from academia to positions outside of academia is increasingly more common. According to the American Mathematical Society's 2015 Annual Survey of the Mathematical Sciences in the USA, 1901 PhDs were awarded [2]. The report shows that 52% of those recipients are working in academia. This is a decrease from the 2014 survey which stated that 56% of doctoral recipients went on to academia. This transition is not obvious nor is it smooth. For a job in academia, an applicant would highlight their individual achievements to stand out among a large number of candidates. Yet, according to *How to sail smoothly from academia to industry* "To beat the stiff competition, highlight your skills in collaboration, teamwork and meeting deadlines." Refocusing on collaboration, as opposed to individual achievements is more beneficial if one wants to enter the corporate world [14].

One of the authors of this work, Dr. Candice R. Price, spent 3 years in the mathematics department at the US Military Academy (USMA) in West Point, NY. This is an institution whose goal is to train future army officers and leaders in the USA. The mathematics curriculum at USMA, which includes mathematical modeling, calculus, and probability and statistics, has been structured intentionally to broaden the mathematical training of all West Point cadets. Because USMA is traditionally an engineering school, all cadets earn a Bachelors of Science with the goal of beginning a military career directly after graduation. Few graduates go on to graduate school and medical school. As an instructor at this institute, Price found that because the goal of the curriculum was to train students for specific jobs in the government, all of the math courses included real-world applications, public speaking, and professional writing. Price found that these intentional inclusion of these skills in the mathematics setting allowed the cadets to learn how to talk about mathematics in all setting, an important skill for any mathematician. When looking about at her own mathematical training, Price realized that it wasn't until her master's program were these skills introduced, and that was in the math education setting. On reflection, the inclusion of these techniques in the undergraduate mathematical curriculum at every stage allows students to hone the skills needed in any arena. This is one area that is being addressed by PIC Math programs. A benefit of this style of curriculum is that it allows for the opportunity to expose undergraduates to the many career options that mathematical training provides.

Several mathematicians who transitioned from academia echoed previous thoughts on the skills needed in their transitions. Dr. David Tello, formerly an assistant professor, transitioned to being an analyst at a financial institution, partly to spend more time with his ailing mother. He found that while his soft skills were lacking, his technical skills were excellent, and in the end was one of the reasons he was hired. Tello states, "Graduate programs need to concentrate on teaching these soft skills. Basic lectures on emotional intelligence, business writing, protocol in company meetings, and business etiquette are vital to survive the corporate world." Dr. Brie Finegold also transitioned from academia to industry as a research mathematician, also due to a family situation. Her research at the time was mostly theoretical and she had minimal programming experience. "However", Finegold states, "I realized that I could learn many of the things I needed on the job, and I demonstrated on interview that I was capable of thinking on my feet and that I was genuinely curious." Her problem-solving and writing skills acquired in academia were valuable in her new position.

In [6], Cohen noted "The health of the mathematical community requires that graduate students and early-career mathematicians see a broad range of paths to respected and satisfying careers, whether inside or outside academia" [6]. We hope that more mathematical science departments nationwide are encouraged to prepare students for all employment opportunities. To close out this work, we include some pointed advice, gathered from the mathematicians and references mentioned throughout this paper, on how to make a transition to BIG smoother:

• Attend panels or presentations by those in positions outside of academia. These are becoming more common at mathematical sciences conferences. Include professional development opportunities that provide information, training, or support for transitioning to positions in BIG.

- If currently in academia, teach courses that will make the transition easier and incorporate mathematical software. Computer programming is important but there is no need to be an expert.
- Make yourself visible on professional social media, such as LinkedIn, and include your resume. Network and seek out others in careers you find interesting.
- Seek the assistance from someone in BIG who can help turn your CV into a resume specifically for nonacademic positions.
- Current trends are conferences and workshops in data science, and applications of mathematics to political and social science. Explore a non-mathematical domain area and how mathematics is applied, through conferences.

Acknowledgements The authors would like to acknowledge the EDGE program, Rachel Levy, Suzanne Weekes, the Infinite Possibilities Conference, Leona Harris, Carla Cotwright-Williams, Karoline Pershell, Maria Garcia, the editors of this volume, and the anonymous reviewers.

References

1. American Association for the Advancement of Science, "https://www.aaas.org/", retrieved 22 August 2018.
2. American Mathematical Society, "http://www.ams.org/profession/data/annual-survey/annual-survey", retrieved 22 August 2018.
3. Fadil Santosa, *Big Math Network: Connecting Mathematical Scientists in Business, Industry, Government, and Academia*, " https://bigmathnetwork.org/", retrieved 22 August 2018.
4. Yuliy Baryshnikov, Lee DeVille, and Richard Laugesen, *Math PhD Careers: New Opportunities Emerging Amidst Crisis*, Notices of the AMS, March 2017, Vol. 64, no. 3, pp. 260–264.
5. Leonard Cassuto, *Can You Train Your Ph.D.s for Diverse Careers When You Don't Have One?*, 22 August 2018, "https://chroniclevitae.com/news/2092-can-you-train-your-ph-d-s-for-diverse-careers-when-you-don-t-have-one?cid=VTEVPMSED1"
6. Amy Cohen, *Disruptions of the Academic Math Employment Market*, Notices of the AMS, October 2016, Vol. 63, no. 9, pp. 1057–1060.
7. Amy Cohen, *Results from a 2016 Pilot Survey on Math Post-docs*, "https://arxiv.org/pdf/1703.06772.pdf", retrieved 27 January 2019.
8. The EDGE Program, "https://www.edgeforwomen.org/", retrieved 27 January 2019.
9. Infinite Possibilities Conference 2018, "http://www.msri.org/web/msri/activities/infinite", retrieved 27 January 2019.
10. IDA Center for Communications and Computing, "https://www.ida.org", retrieved 27 January 2019.
11. Rachel Levy, *2015 NSF-IPAM Mathematical Sciences Internship Workshop Report*, "http://www.ipam.ucla.edu/reports/2015-nsf-ipam-mathematical-sciences-internship-workshop-report/"
12. National Science Foundation, "https://www.nsf.gov/careers/rotator/", retrieved 22 August 2018.
13. National Security Agency, "https://www.nsa.gov/what-we-do/research/math-sciences-program/sabbaticals/", retrieved 22 August 2018.
14. Kendall Powell, *How to sail smoothly from academia to industry*, "https://www.nature.com/articles/d41586-018-03306-1", retrieved 22 August 2018.
15. Office of Naval Research, "https://www.onr.navy.mil/en/Education-Outreach/faculty/summer-faculty-research-program", retrieved 27 January 2019.

16. PIC Math - Preparation for Industrial Careers in Mathematical Sciences, "https://www.maa.org/programs-and-communities/professional-development/pic-math", retrieved 22 August 2018.
17. Society for Industrial and Applied Mathematics, "https://www.siam.org/", retrieved 22 August 2018.
18. Ryan Noonan, *Women in STEM: 2017 Update* US Department of Commerce, Economics and Statistics Administration, Office of the Chief Economist, November 13, 2017

Striking the Right Chord: Math Circles Promote (Joyous) Professional Growth

Lance Bryant, Sarah Bryant, and Diana White

Abstract Math Circles are extracurricular programs organized by mathematicians, aimed at introducing K-12 students or teachers to novel and interesting mathematics in a collaborative environment. While other authors have discussed the impact of Math Circles on participants, this article explores ways that Math Circles have substantial impact on the professional growth of faculty involved. We hope our experiences and commentary will inspire more faculty to become involved with Math Circles both locally and nationally.

1 Introduction

1.1 What Is a Math Circle?

Math Circles are extracurricular programs organized around mathematicians or mathematics enthusiasts collaborating with K-12 students and teachers. The informal atmosphere aims to encourage mathematical exploration and sustained problem solving. Math Circles should be fun, positive, and full of interesting problems. The content typically centers around low-threshold, high-ceiling problems connected to advanced mathematics. Math Circles for students emerged in Eastern Europe in the early twentieth century, and migrated to the USA along with professors that had these experiences in their youth [9]. The Boston, Berkeley, and San Jose Math

L. Bryant
Shippensburg University, Shippensburg, PA, USA
e-mail: lebryant@ship.edu

S. Bryant (✉)
Dickinson College, Carlisle, PA, USA
e-mail: bryants@dickinson.edu

D. White
University of Colorado Denver, Denver, CO, USA
e-mail: diana.white@ucdenver.edu

© The Author(s) and the Association for Women in Mathematics 2019
S. D'Agostino et al. (eds.), *A Celebration of the EDGE Program's Impact on the Mathematics Community and Beyond*, Association for Women in Mathematics Series 18, https://doi.org/10.1007/978-3-030-19486-4_9

Circles were among the first to appear in the USA during the 1990s. Today, about 200 Math Circle programs exist nationwide [9]. In this paper, we focus on Math Circles led by university or college faculty, so that we may outline some of the impacts on the professional growth of faculty who lead Math Circles.

Math Circles serve a variety of purposes. Goals include creating community, nurturing mathematical curiosity, developing an exploration/research mind-set, offering professional development (for teacher circles), and increasing STEM participation among underrepresented groups. Venues can be on-campus, giving the community greater access to a university or college, or in a school or community building to perhaps lessen the burden of participation. Schedules typically range from weekly to monthly, with Math Teachers' Circles tending to meet approximately monthly, whereas Math Circles for students often meet weekly. In short, Math Circles can be tailored to meet the local needs of those involved, and given such needs, there is likely an existing Math Circle that can serve as a model for creating a new one.

Though many Math Circles started via grassroots efforts, there has been enough sustained interest to lead to the development of national organizational structures. Two major mathematics research institutes, the Mathematical Sciences Research Institute (MSRI) and the American Institute of Mathematics (AIM), developed the National Association of Math Circles (NAMC) and the Math Teachers' Circle Network, respectively. These provide resources and support to the many Math Circles across the country. There is also a special interest group of the Mathematical Association of America dedicated to Math Circles (SIGMAA-MCST). Math Circle sessions and workshops for faculty have become standard fare at conferences such as MathFest and the Joint Mathematics Meetings. Thus, while the title of this subsection is "What is a Math Circle?" it is highly likely that many mathematicians are aware of them. We aim to address some of the ways these Math Circles are impactful, focusing on impact on the leaders rather than participants.

1.2 The Authors: Our Backgrounds and Motivation

Sarah and Lance Bryant are the founding directors of the Shippensburg Area Math Circle for 4th and 5th graders, with about fifteen participants per session and ten Saturday-morning meetings per year. The program started in 2014 as an after-school activity based on a desire to do some math outreach at their daughter's school. The Math Circle has expanded to a district-wide program housed in the Mathematics Department at Shippensburg University. Sarah and Lance were in the first cohort of the NAMC Math Circles Mentorship and Partnership (MC-MAP) program in 2015. They continued to be involved in the MC-MAP program in 2016 as returning Math Circle leaders and again in 2017 as Math Circle mentors.

Diana White runs the Rocky Mountain Math Circle Program, which consists of a regular Math Teachers' Circle, one or more Math Circles for students, a Math Circle Math Camp, and Julia Robinson Mathematics Festivals. The program began in 2010 as a single Math Teachers' Circle, and has expanded over the years to encompass these broader Math Circle and mathematical outreach experiences. Diana also serves as the Director of the National Association of Math Circles. As previously mentioned, this program of MSRI is designed to nurture the growth of Math Circles by seeding the creation of additional Math Circles in the USA; building a community of Math Circle leaders through which novice and existing leaders would be connected, encouraged, and inspired; providing high-quality resources that help Math Circles build and sustain effective programs; and documenting and disseminating the impact of Math Circle programs across the nation.

Our motivation in this manuscript is to detail some of the many benefits for faculty involved with a Math Circle. We will include our own personal stories, and we invite the reader to reflect on her own experiences or interests in mathematics outreach for connections to these themes. We focus on four areas where Math Circles promote (joyous) professional growth: understanding of teaching and learning mathematics; deepening connections at the department, university, and community levels; broadening leadership skills; and connecting to the larger mathematics community. We hope this inspires more faculty to become involved with Math Circles and sparks deeper research into some of the ideas presented here.

2 Promoting Active Learning

In 2012, one of the five overarching recommendations by the President's Council of Advisors on Science and Technology was to "catalyze widespread adoption of empirically validated teaching practices" [8]. In the wake of numerous studies and reports spanning decades, including a landmark meta-analysis of 225 studies published in the Proceedings of the National Academy of Sciences, having students actively engage with content in their classrooms has been identified as an effective means to increase positive student outcomes [4]. Active learning refers to the teaching practices employed by educators to promote this in-class active engagement. The Conference Board of the Mathematical Sciences (CBMS), an umbrella organization consisting of seventeen professional societies for the mathematical sciences, released a strong statement [1] of support for active learning with this central message:

> We call on institutions of higher education, mathematics departments and the mathematics faculty, public policy-makers, and funding agencies to invest time and resources to ensure that effective active learning is incorporated into post-secondary mathematics classrooms.

Despite this near-consensus support for active learning, a Notices article [3] and a six-part American Mathematical Society blog series [2] cited faculty adoption of active-learning strategies as a bottleneck in post-secondary mathematics teaching advancement. We argue that Math Circles provide a great opportunity for faculty to explore active-learning strategies.

Faculty are faced with many potential impediments to implementing active-learning strategies. Class size, grading support, contact time, learning outcomes for courses, student expectations, and instructor experience factor into each faculty members' decisions on how to structure courses and are at times beyond their control. There are also the pressures that come from fear of failure: if the students do not support the change of course teaching, then teaching evaluations may suffer. This can be especially stressful for pre-tenure faculty. Furthermore, the diversity of faculty environments due to these factors complicates the national dialogue [2]. It can be easy to think that what works at one institution or in one particular setting will not successfully transfer to another. However, we are aware of the support for active learning and have turned to our work with Math Circles to strengthen our development of a more student-centered classroom.

When running a Math Circle or facilitating a session, we find many of the aforementioned limiting factors are easily managed or simply nonexistent (grading, for example). Also, the Math Circle community has promoted a culture of active exploration, collaboration, and inquiry as hallmarks of successful sessions. Thus Math Circle leaders become immersed in the process of leading engaging sessions that involve and inspire all participants. Removing lecture, homework, exams, textbooks, and other common course components from the learning environment can be freeing but we have also found it to be intimidating. Thus Math Circles provide a space for faculty to practice a variety of active-learning strategies without fear of adverse repercussions, thereby allowing them to both build their skills and confidence.

For a typical mathematics course, structure and learning goals are centered around content, and problem-solving strategies must be inserted into this structure. Our work in Math Circles has allowed us to create (or borrow) lessons that are designed the other way around: start with problem-solving techniques and use mathematical settings to illustrate their uses, then present students with challenges from a variety of topics. What is gained from these sessions are authentic experiences for how problems are approached and solved in mathematics, making it easier to incorporate these ideas into more traditional classroom settings. Some of the materials from our Math Circle that we have used in college classrooms include puzzles designed to introduce approaches for problem-solving (e.g., "try an easier problem" or "work backwards"). These are particularly good prompts for introduction to proofs courses, where students must break out of procedural and algorithmic practices to explore the creative side of problem-solving. We have also drawn on the work of Joshua Zucker, a well-known Math Teachers' Circle leader, whose article "Be Less Helpful" addresses the nature of stepping out of the lecturing spotlight and allowing students to embrace the full experience of learning, including the struggle [10].

3   Connections Between Faculty, University, and Community

Mathematical outreach programs, and Math Circles in particular, lie at the intersection of the interests of universities, local communities, public policy-makers, and the mathematical community. Thus there is much potential for substantial connections between these stakeholders. However, when Math Circles are viewed as community service, they are often not valued as highly as scholarly research and only incidental to teaching in higher education. In "Making the Case for Professional Service," Ernest Lynton establishes a working definition of professional service, taking it to be work based on professional expertise. Lynton claims "professional service can and should be an important element in the definition of faculty roles and rewards— not only because of its societal and institutional benefits. It also can constitute scholarship of the highest order, equivalent in intellectual challenge, creativity, and importance to scholarly research and scholarly teaching" [6]. We absolutely agree. Math Circles are an exemplar of professional service in mathematics.

Math Circles satisfy Lynton's definition of professional service because they are outreach programs that rely heavily on the mathematical expertise of session leaders. As session leaders we must draw on our deep understanding of a wide range of mathematical ideas. We must understand entry points for these big ideas and be comfortable with deviations from the expected path of exploration. While it is often challenging (and keeps us on our toes!), there is nothing about leading a Math Circle that relies on rote knowledge or repetition. We draw on our expertise in crafting and leading these sessions in a way that represents a deep connection to the content and to the learners.

When arriving in a new town, faculty members may not be aware of existing tensions or bonds between the university and local community. In our personal experiences, there has been some positive and some negative aspects of the "town-gown" relationships. Thankfully, we have found that our towns are excited about the opportunity of having a Math Circle and the university appreciates the valuable link to the community. By inviting students and their families into the university, a Math Circle can literally open doors to the institution and show a commitment to the younger learners or teachers in the area. Some Math Circles travel to the local schools or libraries, again with the goal of breaking down barriers of access and bringing faculty expertise to non-university spaces. We have found that our professional and personal networks in the local community are also enriched by this work.

What we have found to be one consistent source of joy in working with Math Circles is the satisfaction of making strong connections with young learners, many of whom have never met a "mathematician" before. All three authors are first-generation college students and did not grow up knowing that people still worked on mathematics problems long past high school. We did not know that some mathematics problems look like puzzles or that sometimes some mathematics problems remain unsolved. It is all too common to hear people describe mathematics as completely objective, absolutely black and white, with knowing the right answer as the only currency of mathematics. Participants in a Math Circle not only meet mathematicians, but also they become them. They get to see the "play" behind real mathematics and appreciate that an answer is almost never as valuable to a good mathematics conversation as a question is. In turn, Math Circle leaders can learn from the participants some excellent strategies for being fearless, for asking "silly" questions, and for being honest when something is not fun or seems impossible.

4 Broadening Leadership Skills

Running a Math Circle means wearing many hats, including some that mathematics faculty seldom get to wear. One must connect with the community of interest, generally needing to make personal contact with teachers, community centers, or other organizations that can help with advertising the program to the target audience. There is a need for applications, flyers, and ideally at least a basic website. Mathematics faculty may or may not have skills for these various things. Often, they collaborate with others to accomplish them, and thus need to find volunteers and build relationships. In addition, as Math Circle leaders we have had to learn the ins and outs of hiring and training students, managing grant paperwork, submitting reports, and responding to press requests. All of these skills have prepared us to consider larger projects and for working with constituencies across our institutions and in our communities.

A large-scale program can involve dozens up to hundreds of K-12 participants, undergraduate and graduate student helpers, and a variety of faculty facilitators. The collaborative nature of Math Circles lends themselves naturally to partnerships between higher education institutions, K-12 schools, and business sponsors. For example, the Maize and Blue Math Circle in Michigan has partnered with Ford and DTE Energy Solution to support a free middle school Math Circle at the Dearborn STEM School. Learning how to build and manage these types of partnerships is a valuable skill for faculty seeking future leadership roles.

5 Connections and Identity in the Mathematics Community

Coming out of graduate school, a new faculty member is a part of a community built around an area of expertise. Over the years, new sub-communities are joined. Perhaps the faculty member is in the MAA's Project NExT program or joins a new faculty cohort at her hiring institution. What we have found unique about the Math Circle community is its ability to be vertical in nature, uniting K-12 students, teachers, undergraduate students, graduate students, mathematicians, and mathematically inclined parents or volunteers. Thus the Math Circle community in mathematics has far reach and interest in a variety of areas, not just to a sub-specialty of research.

The Math Circle community also functions with a refreshing lack of concern over tenure-track status or job titles. People from all sorts of backgrounds work with Math Circles, so there is great diversity among those involved, including long-term temporary faculty, industry professionals, government employees, and freelance or independent contractors. In fact, if you happen upon a group of Math Circle leaders at a conference or workshop, we recommend skipping the usual opener "Where are you at?" and opting instead for "Do you know the Math Circle Salute?" If you don't know it, watch James Tanton's video [7] or, better yet, help out at a Math Circle for a day and learn it! This fun bit of the Math Circle community hints at the playfulness of those involved; after all, it's hard to take yourself too seriously when your arms are twisted in knots in front of near strangers.

6 Our Stories: A Few Personal Notes on Math Circle Involvement

6.1 Sarah

My mathematical role models are those who have not only increased mathematical understanding but also contributed to broadening participation in mathematics. Whether through a Math Circle or some other activity that takes an authentic approach to exploring mathematics, the words of Francis Su can inspire us all: "The goal of broadly getting people to appreciate math is not at odds with bringing more people into deep mathematics. Connect with people in a deep way and you're going to draw more people into mathematics" [5]. Yes, this includes (in the case of our Math Circle) 4th and 5th graders. I am dedicated to increasing access and improving equity in mathematics, and being involved in Math Circles has helped enrich this part of my personal and professional mission. There have been other benefits as well, as I have been empowered to explore new lines of scholarly work and expand my mathematical network, thanks to my involvement in Math Circles.

6.2 Lance

Like other faculty members I have known, I became more interested in K–12 education as my daughter started elementary school. In 2014, when Sarah and I worked with the director of our daughter's school to start an extracurricular math program, we envisioned a small program running for a few years. I did not expect the overwhelming response from the community, for sessions to consistently have over fifteen participants, and I certainly did not expect to be working with Math Circle leaders across the country. I have seen first-hand how a Math Circle program can easily gain momentum and bring someone into a vibrant community of mathematicians.

My department has run an annual math event for high school students since the late 1950s. When I became director in 2010, the main event was a quiz bowl. After attending the Math Circles Mentorship and Partnership (MC-MAP) grant workshop in 2015 and participating in a Julia Robinson Math Festival (JRMF), I saw a better way to generate excitement about mathematics among the high school student and their teachers, while at the same time providing opportunities for our undergraduate students and faculty. I replaced the quiz bowl with a JRMF. We now hold several problem-solving sessions each year for our undergraduate students both for their own benefit and to prepare them to be table leaders for the JRMF and our Math Circle. Our faculty enjoy working on these problems as well, and I feel like my work with Math Circles has built connections across my campus that I would not have otherwise. So, in addition gaining new relationships to those in the Math Circle community, this work has added to many previous relationships with colleagues as well.

6.3 Diana

Being a Math Circle leader has affected every aspect of my career. In particular I credit almost all of the facilitation techniques that I use in my upper-level undergraduate mathematics classes to my work with Math Teachers' Circles. One of the initial leadership team members for my Math Teachers' Circle was a high-school mathematics teacher with extensive professional development expertise. Our collaboration resulted in me having the opportunity to observe and then practice implementing a wide variety of active-learning and group facilitation techniques with teachers. With feedback from this expert high-school teacher, my facilitation skills improved tremendously, and I proceeded to also implement them in my undergraduate mathematics classes. In addition, I also used these techniques (and more) in leading professional development workshops with other mathematics faculty.

7 Conclusion

Mathematicians deserve a chance to play, to grow as teachers without the inherent pressures of the evaluation system, to try new types of math problems outside their expertise, and to nurture new friendships and connections through projects they are passionate about. We have found Math Circles to be the perfect outlet for these areas of growth, giving us the opportunity to refine our professional skills in a fun and joyful environment. We encourage other faculty and math enthusiasts to consider starting or joining a Math Circle, not just because of all you will be giving to others involved, but also for all you will be receiving in turn.

We hope this manuscript serves as a call for more faculty to consider the ways that being involved in Math Circles is beneficial for them and very rewarding. We would also expect that our personal stories are common in the larger Math Circles community, a topic certainly worthy of continued research.

References

1. Braun, B. et. al. (2016). Active Learning in Post-Secondary Mathematics Education. Retrieved from www.cbmsweb.org/Statements/Active_Learning_Statement.pdf.
2. Braun, B., Bremser, P, Duval, A., Lockwood, E. and White, D. (September-November 2015). Active Learning in Mathematics, Parts I-VI [blogpost] American Mathematical Society Blog on Teaching and Learning in Mathematics. Retrieved from blogs.ams.org/matheducation/category/active-learning-in-mathematics-series-2015/
3. Braun, B., Bremser, P, Duval, A., Lockwood, E. and White, D.. What Does Active Learning Mean For Mathematicians?, Notices of the American Mathematical Society, February 2017, Volume 64, Number 2, pp. 124–129.
4. Freeman, S., Eddy, S.L., McDonough, M., Smith, M.K., Okoroafor, N., Jordt, H., and Wenderoth, M.P. (2014). Active learning increases student performance in science, engineering, and mathematics. Proceedings of the National Academy of Sciences, 111(23), 8410?8415. Retrieved from www.pnas.org/content/111/23/8410
5. Hartnett, K. (February 2017). To Live Your Best Life, Do Mathematics. Quanta Magazine. Retrieved from www.quantamagazine.org/math-and-the-best-life-an-interview-with-francis-su-20170202/
6. Lynton, E. A (1995). Making the case for professional service. Washington, DC: American Association for Higher Education Forum on Faculty Roles and Rewards.
7. MathCircles. The Math Salute. YouTube video. Dec 22, 2010. www.youtube.com/watch?v=43-X3mEY1Xg.
8. PCAST STEM Undergraduate Working Group (2012). Engage to Excel: Producing One Million Additional College Graduates with Degrees in Science, Technology, Engineering, and Mathematics. Eds Gates SJ, Jr, Handelsman J, Lepage GP, Mirkin C (Office of the President, Washington).
9. Wiegers, B. and White, D. (2016). The establishment and growth of Math Circles in America. Proceedings of the Canadian Society for History and Philosophy of Mathematics/La Société Canadienne d'Histoire et de Philosophie des Mathématiques, pp. 237–248.
10. Zucker, J. (2012). Be Less Helpful: Embracing Perplexity to Create a Productive Classroom. MTCircular. pp. 5–7. Retrieved from www.mathteacherscircle.org/assets/legacy/newsletter/MTCircularAutumn2012.pdf.

The Women in Mathematics Symposia: An Organic Extension of the EDGE Program

Amy Buchmann, Yen Duong, and Ami Radunskaya

Abstract The Women in Mathematics Symposia are a collection of annual regional mathematics conferences for women mathematicians which have all been organized by EDGE affiliates. To the best of our knowledge, they exist in California, the Midwest, Texas, the Carolinas, and Israel. We explore the history and original motivations of the WiMSoCal organizers and how WIMS spread and adapted to other regions. We include notes on participation, diversity and inclusion, and organization from the conferences. We include quotes from a survey about the conference's effectiveness, data about the attendees, and reflections from attendees on the impact it has had on their research and careers.

1 Introduction

The Women in Mathematics Symposia (WIMS) began in Southern California in 2009 and have affected over a thousand women in mathematics across the world in the past decade. We three decided to write this chapter as both a record of the past and a blueprint for the future. In it, we have past participants and organizers reflecting on what worked and what didn't in past WIMS. We hope this chapter encourages some readers to organize their own vertically integrated conferences for local women in mathematics. Like EDGE, the main goal of WIMS is to support

A. Buchmann (✉)
University of San Diego, San Diego, CA, USA
e-mail: abuchmann@sandiego.edu

Y. Duong
Freelance Writer, Charlotte, NC, USA
e-mail: yenergy@gmail.com

A. Radunskaya
Pomona College, Claremont, CA, USA
e-mail: ami.radunskaya@pomona.edu

© The Author(s) and the Association for Women in Mathematics 2019
S. D'Agostino et al. (eds.), *A Celebration of the EDGE Program's Impact on the Mathematics Community and Beyond*, Association for Women in Mathematics Series 18, https://doi.org/10.1007/978-3-030-19486-4_10

women mathematicians. We want this chapter to do the same by motivating and inspiring would-be organizers and participants.

WIMS have been held in several cities in Southern California, as well as in the Chicago and Midwest area, Texas, the Carolinas, and Israel. Of over 1000 participants, 40 women have organized at least one WIMS. These 40 include many graduate students and early career faculty, and so WIMS have offered an opportunity for professional development through conference organization.

The WIMS symposia form pieces of a mosaic of mathematics conferences across the country that focus on woman-identified mathematicians. This mosaic includes but is not limited to:

- Association for Women in Mathematics (AWM) Research Symposia, held every other year in a different location across the country
- Nebraska Conference for Undergraduate Women in Mathematics
- Graduate Research Opportunities for Women conferences for undergraduate women that originated at Northwestern University in 2015
- Southeastern Conference for Undergraduate Women in Mathematics
- Women in Mathematics in New England conferences for undergraduate women
- Woman and Mathematics program at the Institute for Advanced Studies, which is a one- or two-week long program targeting undergraduate and graduate students.

The WIMS symposia differ from these events in their goals, targeted participants, and local focus.

Broadly speaking, the goal of WIMS is to support all woman-identified mathematicians. While undergraduates are often encouraged to attend, the focus of WIMS is not to convince undergraduates to pursue graduate studies in mathematics. "Local" is a key adjective in defining the WIMS objectives: in contrast to the national AWM Research Symposia, we hope to bring together mathematicians within driving distance of each other. There were four particular goals of the original WIMS; different conferences have adapted and changed these goals to fit their regions and cultures.

1. Strengthen the network of women mathematicians in the region.
2. Facilitate tiered mentoring between junior and senior mathematicians.
3. Highlight women's contributions to the mathematical community.
4. Encourage new collaborations.

We begin this chapter with some history and facts about WIMS. Next are results from a survey of past participants,[1] a program description with schedule suggestions, and notes about breakout sessions and other information that may be useful to organizers. We conclude the chapter with tips and tricks for readers who want to start their own WIMS.

Throughout the chapter we include quotes from participants and organizers which we solicited via an online survey sent in spring 2018.

---

[1] Informed consent: prior to completing the survey, participants were informed that quotes from the survey could be included in this chapter.

2   History of WIMS

In 2005, EDGE directors Sylvia Bozeman and Rhonda Hughes came up with the idea of formal "EDGE clusters." These regional groups would nurture the networks formed in the EDGE summer session, and enlarge the network by including other women mathematicians at various stages in their careers. The regional clusters resulted from a 3-year pilot project supported by an NSF ADVANCE grant.

The project created a network among a small group of women in close geographical proximity composed of senior and junior faculty and graduate students. Senior women mentored junior women, while those junior faculty mentored graduate students. The network was formed to help the younger groups advance toward their professional goals with advice from senior faculty. As part of this pilot project, six clusters were formed: the Georgia cluster, the Mid-Atlantic cluster, the Indiana cluster, the North Carolina cluster, the Iowa cluster, and the Southern California cluster.

In October 2005, the Southern California cluster met for the first time, with dinner after a mathematics talk on "Classifying Division Algebras" by Ulrica Wilson, who was at the time a visiting assistant professor at Claremont McKenna College and later an EDGE co-director. The organizers of the event were Ami Radunskaya (Pomona College) and Cymra Haskell (University of Southern California). The talk was part of a series that Cymra organized at USC with the support of Women in Science and Engineering. The dinner meeting included six faculty, five graduate students, and one undergraduate. The cluster grew and met regularly for advice, stories, and social outings.

In June 2007, at a barbecue at Cymra's house, over a dozen women discussed the future vision of the group. What did the participants want from the network? A majority of the cluster landed on: "We want the opportunity to talk about *mathematics* with each other". And so, the Women in Math in Southern California (WiMSoCal) Symposium was born.

The first WiMSoCal Symposium took place at Loyola Marymount University in (LMU), co-organized by Alissa Crans (LMU), Cymra, and Ami. The symposium had 34 participants: 4 undergraduates, 20 graduate students, and 10 faculty members from 9 local institutions. The eight talks included three talks by graduate students. Feedback after the symposium indicated most participants wanted a similar event once a semester, with the location rotating around Southern California. The rest of the feedback led directly to some of the main features of WIMS. An incomplete list of requests by participants follows.

- Vertical mentoring in the form of organized discussions: one for undergraduates by graduate students, and one for graduate students by junior faculty
- Travel reimbursement funding
- Discussion about issues for women in mathematics
- Parallel sessions of shorter talks organized by research interest
- Keynote talks.

Since 2009, ten more WiMSoCal symposia have occurred, and the number of participants has grown to over one hundred. It has been gratifying for the original organizers to see the idea of WiMSoCal replicated across the country after the EDGE regional clusters led directly to WIMS.

3 Why WIMS?

Of those who enter graduate school in math-intensive fields, more women than men drop out or change fields, and of those who complete doctorates, fewer women apply for tenure track positions. Women drop out of scientific careers - especially math and physical sciences - after entering them as assistant professors at higher rates than men, and this remains true as women advance through the ranks [4].

Events for women in mathematics fight a battle to change the bleakness of the quote above. Between 2007 and 2016, the proportion of math Ph.D. recipients who were women fluctuated between 27% and 37%, as reported by the *AMS Notices*. Over that same time period, women made up only 20% of the Invited Hour Address Speakers at AMS meetings. Including at least one woman organizer for an AMS special session in 2017 increased the percentage of women speakers slightly from 18–31% to 25–37% [1]. The ambiguity comes from speakers identified with initials or non-gender-specific names.

The AMS does not keep track of demographic information of members, but the Society for Industrial and Applied Mathematics (SIAM) does. Of the 14,638 members at the end of 2017, 2517 identified themselves as women, and 1704 did not enter their gender into the SIAM database. So the percentage of women SIAM members was between 17% and 29%.

To increase the participation of women in mathematics, the 2006 InterAcademy Council report "Women for Science" [9] called for increasing the visibility of women scientists, providing mentoring and networking opportunities to combat isolation, and offering resources for launching careers. By highlighting women mathematicians as keynote speakers, uniting women across a specific geographic area, and including lively and relevant career and personal discussions, WIMS meets each of these goals.

One of the greatest strengths of WIMS is its status as a small regional conference. Women don't need to travel far to attend. It's easier to meet new people when there aren't too many of us. Many of us also share the experience of, at some point, counting ourselves as the only woman in a room of mathematicians. After connecting with conversations at WIMS, participants often see each other again at larger regional events, strengthening the network.

We asked participants and organizers why they decided to become part of the WIMS community. Their responses reminded us of some of the benefits we see in EDGE. We include some of them here for future organizers to use in grant applications or proposals.

Several women mentioned the novelty and support of a women-only conference:

> The first time, I was curious to see what a math conference with only women would be like and it was conveniently located so I attended. I enjoyed the experience so much that I continued to attend until I left the midwest. I found the social dynamic to be quite different (much more relaxed and welcoming) than many other conferences I have attended.

> I am really thankful for WIMSoCaL. Having known so many other women in the [Southern California] area from attending many years of WIMSoCaL meetings always brings me a strong sense of comfort and confidence whenever I attend local meetings. WIMSoCaL created a strong network of women that help support each other academically and personally. I have to attribute the success I have today to the wonderful work of the WIMSoCaL organizers.

> It was a great chance to network with other women in mathematics and give a talk in a supportive environment. I gave my first talk on my research at Midwest WIMS. In terms of niche, WIMS does a great job of highlighting the research of women and being a supportive environment. It also is a great place to hear about research outside of your comfort zone.

Organizers can work to create that "supportive environment" by including the phrase on promotional materials, websites, and handouts. They can also encourage expository or beginning research talks to support graduate students.

Breakout sessions or small group discussions are a capstone of WIMS, described in detail in Sect. 6. Every person who attends WIMS comes with their own unique set of experiences and reflections, and should feel valued within their small group.

> Even before the conferences I knew this is something worth doing, but only during [it] did I understand the psychological and intellectual impact of intense learning and interacting in such a supportive and non-competitive atmosphere.

> Midwest WIMS had small group discussions on a range of topics and concerns. It was really important to get advice from mathematicians who had similar goals and experiences.

> I like that the conversation topics at breakout sessions at Midwest WIMS 1–3 were so broad–not everyone wants to talk about work/life balance or finding a job–but there are opportunities to find a conversation that is helpful no matter what career and life stage you are at. It's nice to share experiences, which are sometimes unexpected (in our algebra session we had a lengthy discussion about clothing between talks).

Women have commented that they attended WIMS to have the opportunity to present their research in a welcoming environment. One speaker said she "thought WIMS would be a less stressful environment in which to do this because it was mostly women." Another also remarked on the uniqueness of the WIMS environment:

> I never realized how different the intellectual atmosphere could be in a room filled mostly female mathematicians. When I spoke, I felt immediately valued as opposed [to] apprehensive about saying something of value.

The above quotes and discussion may be helpful for people who have already committed to organizing a WIMS, but there are plenty of women who may be considering it and wary of the time commitment. As graduate student organizers, Amy and Yen believe in the lasting professional and personal benefits from creating and running one's first conference. At some point every person learns about bureaucracy and politics and fundraising, but the stakes are pleasantly lower as a graduate student and you have the benefit of faculty advisors for advice and help. Faculty of course can and should organize WIMS, but we encourage them to include students if possible.

> As a graduate student [WIMS] was one of the first opportunities I had to be involved in decision making processes. As I transition to a tenure-track role I feel that the skills I developed in organizing and planning such an event are invaluable, as I plan sessions at conferences and serve on department and university committees. (On a personal note, there are a lot of similarities to planning a wedding!)

Benefits of organizing a WIMS are not limited to graduate students. Because they contact the participants and different schools, organizers can form even stronger networks than participants do. They can also increase their own visibility within their departments and universities.

> I got to know a lot more people because I was the organizer. I exchanged emails with almost everyone who attended. My dean was also really excited that I did it, which definitely didn't hurt my career.

This last quote summarizes how far we have come as women in mathematics, and encourages us to imagine how far we can go.

> I think it is very important that we continue to bring together women, at various stages of their careers, so that they can showcase their research, and have networking opportunities. Although almost all conferences nowadays make an effort to increase the participation of women, the reality is that there is still a large disproportion in gender representation in most of them. It was refreshing to see so many women gathering together, all presenting very high quality work. This would have been almost unthinkable when I started my career.

4 Diversity and Inclusion

While almost all WIMS participants share the experience of being a woman in math, each has a different perspective to offer during the invaluable breakout sessions and discussions. This section discusses different types of diversity that organizers can seek out. We also discuss strategies for creating an inclusive environment.

We strongly encourage WIMS organizers to spend a few minutes during their organizing process to reflect on who they want to serve with their conference, and include language on their promotional materials to that end. For instance, the Midwest WIMS 2017 website emphasizes that "Graduate students, postdoctoral scholars, and researchers at small institutions are especially encouraged to apply" for funding.

Organizers are the face of the conference to attendees. Organizing committees may want to think about racial diversity among themselves, as one participant noted:

> I'd like to see more diversity within this community. More people from underrepresented communities giving talks and holding leading roles would be very nice.

4.1 Career Stage

One question that often arises: what is the target career stage of a WIMS conference? Certainly faculty and postdocs are welcome to present their research, network with each other, and look for collaborators. But we argue that as one goal is to support up-and-coming women mathematicians, WIMS is an invaluable resource for the early graduate student who hasn't yet given a research talk at a conference. We suggest that organizers explicitly encourage graduate students to speak, either on original research or expository topics. Organizers have taken different stands on including undergraduates or not.

If organizers want undergraduates to benefit from the conference, they may consider including a poster session or a parallel session consisting entirely of undergraduate talks. If they want graduate students to benefit, they may explicitly encourage talks by beginning researchers or even expository talks, and advertise the supportive environment. Such a conference could still include invited talks to set a research or teaching-focused motif for the day without discouraging less experienced conference attendees from participating. Again, we encourage organizers to include graduate students in the organizing committee who can speak to graduate student concerns. For instance, one person who attended one WIMS in 2015 and another in 2017 wrote:

> I liked that at the 2015 WIMS, people from smaller schools with fewer resources and connections had more opportunities to speak. For the 2017 WIMS, there were way fewer opportunities for people to speak, and there seemed to be more of a focus on talks from people who were already well established. Those are not the people who need more opportunities to talk.

For early career faculty, including a variety of questions about school and career transitions in the breakout sessions can make them feel valuable as mentors for students, while they benefit with knowledge from tenured faculty. Faculty may also enjoy open problem sessions, but after our experiences with these, we urge organizers to also include a parallel session for students, perhaps to do with professional development.

4.2 University Size and Focus

While WIMSoCal began as a research-focused extension of an existing social network, these new conferences concurrently offer a serious research environment and create a new regional network of researchers, educators, and learners.

We urge research-focused organizers to remember teaching-oriented schools and faculty, who have different relationships with students and offer unique perspectives as mentors and examples for graduate and undergraduate students. They often have smaller faculty and may be the only woman in their department, and hence could benefit the most from a WIMS.

On the other hand, one organizer cautioned future organizers:

> Please be sure not to reinforce stereotypes about women. When a conference for women is all about how to get a teaching job, then what message are we sending? It is great to talk about such issues, but please be balanced.

It is not the case that those younger than us "should" follow our own paths, but instead that we should show them a variety of paths. For finding people in different fields, organizers suggested seeking out other organizations and using existing networks.

> As much as possible, include women from non-doctoral institutions. Reach out to people who have done Project NExT to find recent PhDs in your area. They may be the only woman in their department or looking for collaborators. They may not have large budgets to travel to other conferences so regional conferences are especially helpful. It is also useful to include such participants because many graduate students will not end up at research universities. These women can help mentor these students. They may even want to help organize the conference!

> Use your own network of professional connections to invite a diverse group of speakers, in terms of research, professional affiliation, etc. Advertise the conference widely.

4.3   Vertical Integration

Mixing undergraduates, faculty, and people at all career levels creates vertical integration. Organizing several vertically integrated breakout sessions can be a tedious task for organizers, but ultimately adds value. Explicitly, this requires an organizer to assign participants to different discussion groups while keeping track of the career levels represented in each group.

In vertically integrated breakout sessions, younger graduate students can chat with more experienced ones, postdocs can chat with faculty, and tenured faculty can share their experiences with those earlier on the career ladder. Mentoring junior participants gives more senior ones a sense of accomplishment which can help propel them through the next tricky research or teaching problem [11]. WIMS creates an environment to ask open-ended questions and receive candid answers about the commitments and experiences of being a professional mathematician.

4.4   Parenthood

Breakout sessions lead to insightful connections on topics which one may not encounter at other conferences. For instance, Yen, who had two children during graduate school, first thought about the timing for her children when attending WIMSoCal in 2011, as motherhood came up in conversation during the breakout session. Other participants also remarked on this topic:

> At Chicago, it was really useful to hear stories about "math and motherhood" that you can't really learn about except in person. At the Texas WIMS, I thought that participants at all levels (from undergrad to faculty) really got a chance to connect and talk. At both conferences, I really got a lot out of meeting and talking with other women mathematicians – and presenting my research to a broad audience was really valuable.

On the topic of motherhood, late graduate school and early career faculty may be the most in need of WIMS-style support networks [7]. Mentoring and fostering a sense of "belonging" has been shown to be critical at this career stage [2, 5].

> The tenure structure in academe demands that women having children make their greatest intellectual contributions contemporaneously with their greatest physical and emotional achievements, a feat not expected of men. When women opt out of full-time careers to have and rear children this is a choice - constrained by biology - that men are not required to make. [4]

One word of caution, which appears as common sense to some: avoid questioning women on their reproductive choices. Creating a welcoming environment in which to discuss topics like parenting or caring for older family members does not entitle people to the sometimes-painful details of others' lives.

Seek out money for childcare grants—organizers often hear "no" but sometimes hear "yes" on this issue. Secure a lactation room, which requires a chair, an electric outlet, a small table, and a door with a lock, before the conference and let participants know that it is available [3].

4.5   Gender

We want to include a note about two other topics: transgender people, people of minority genders, and men. In the spirit of inclusion, we expect most WIMS organizers are open to people of minority genders and transgender people, but may have overlooked them. One can include specific, inclusive language on promotional materials and websites—the cost is low to organizers, and the potential benefits for participants are high. For example, the WiMSoCal 2018 website says, "All are welcome to register or give a talk/poster, regardless of gender."

If the WIMS is listed as "in cooperation with the AWM" (see Sect. 7), then organizers could include part of the AWM "Welcoming Environment" statement on their website:

> In pursuit of that ideal, the AWM is committed to the promotion of equality of opportunity and treatment for all AWM members and participants in AWM-sponsored events, regardless of gender, gender identity or expression, race, color, national or ethnic origin, religion or religious belief, age, marital status, sexual orientation, immigration status, disabilities, veteran status, or any other reason not related to scientific merit.

As for men, different organizers have taken different tactics. If you receive federal funding, you cannot include discriminatory language excluding men. Men who are interested in a WIMS are often the men we want as allies in departmental politics and collegial relationships. Though many organizers and participants noted that having no men in a room changed the dynamic considerably, we endorse inclusivity.

One tactic we do not endorse: putting men, transgender people, and people of other minority genders in a separate room for breakout sessions, etc. This is an alienating practice—an exception to this is if you have enough men to form their own breakout session. They may benefit from discussing gender issues together.

5 Measurable Impact

To inform our understanding of the magnitude and impact of WIMS, we surveyed past participants and organizers of WIMS—we received 127 responses from the estimated 1000 people who have attended any WIMS, many of whom attended multiple years and locations. Although we had a low response rate, we still believe this data is valuable and we share it with the understanding that it reflects only a small percentage of WIMS participants.

Figures 1 and 2 show how WIMS participants advanced in their careers. Undergraduates became graduate students, graduate students became tenure track faculty, postdocs, and lecturers, etc. Of the survey respondents who attended WIMS, nearly half were students when they attended, and 37.6% were tenured or tenure-track faculty. At the time of the survey, roughly a third were students, just over 40% were tenure or tenure track faculty, and the remaining quarter fit into the postdoc/lecturer/other category.

In particular, we hope future organizers notice the diversity of career stages at WIMS, and plan their own WIMS with inclusion in mind, as discussed in Sect. 4. Students in particular may be interested in non-academic jobs discussed in a breakout session or panel, though plenty of postdocs and professors have made career switches. Some jobs that fit into the "other" category: tour guide, inventory

Fig. 1 Career stage at time of attending WIMS

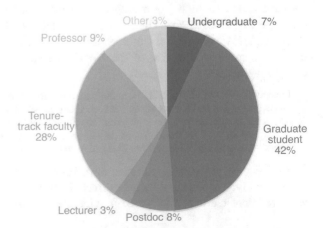

Fig. 2 Career stage when responding to survey

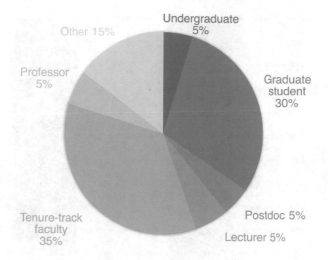

auditor, project manager, researchers at Microsoft, the US Navy, Duke Medical Center, and Georgia Tech, software developer and journalist.

WIMS aims to strengthen regional networks of women. One benefit of a strong network is increased collaboration and more innovative, fruitful mathematical ideas. Of the 127 respondents, eleven people said they had collaborated with someone they had met at WIMS.

One person met someone at a conference at one university, and then later in her career ended up at that university. Knowing someone in the department ahead of time certainly did not hurt her job prospects. Another person met a graduate student and discussed the graduate student's advisor, who became her postdoc supervisor.

Several people said that they later saw WIMS participants at other conferences. Approximately 40% of respondents said they are still in touch with someone they met at WIMS.

WIMS participants almost universally enjoy and recommend WIMS. Out of the respondents, over 90% said they would recommend WIMS to a colleague. Over 80% agreed or strongly agreed that WIMS benefited them personally, while over 70% agreed or strongly agreed that it benefited them professionally (see Fig. 3).

However, of the 127 respondents, 13 reported that WIMS in some way did not meet their expectations. Inappropriate research topics were an overarching theme of the negative feedback—either parallel sessions were too specialized or the attendee had no peers with whom to discuss research. It's helpful to have broad research areas represented in sessions, but we recommend that organizers stress the uniqueness of WIMS as a way for women mathematicians to meet, not necessarily as just another mathematical research conference, to temper these expectations.

> It was always an issue whether there were going to be other people in my field or not. It never seemed worthwhile to go if I was going to be the only one.

> [The] conference was tailored to areas of math that were too specialized, so I didn't learn anything

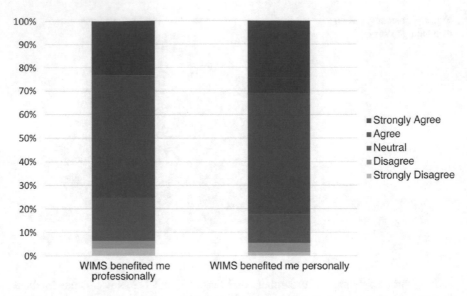

Fig. 3 A majority of respondents indicated that WIMS benefited them professionally and personally

> I felt like the talks were restricted to topology/geometry and analysis, rather than having a larger variety of topics.

> The WIMS in 2015 was at a SLAC and so had a more varied topics and attendees, since the other ones have been held at Research Universities, the focus has been on research, which is great, but as I have not had as much time to keep up, unless the speakers are in algebraic topology, it is not so helpful.

> Some lectures I heard in WIMS were very good, but otherwise I felt there is not much that connects the participants (that were from different areas and different stages of their career).

6   Components of WIMS

Most WIMS have taken place over one day, which may lessen the burden of traveling to a conference for women with family responsibilities. We include two sample schedules from WiMSoCal 2014 and Midwest WIMS 2014 (Table 1) to show the variety of sessions that have been included in the single day meetings.

Below we discuss some of the sessions that have been included in WIMS and how they support the goals of WIMS.

Parallel Sessions Parallel sessions allow more women to present their research, which is especially important for graduate students and early career faculty.

Parallel sessions are often divided into several mathematical subfields. This gives women the opportunity to learn about research in their subfield by women

Table 1 Example schedules from Midwest WIMS 2014 held at the University of Notre Dame (left) and Twin WIMSoCal 2014 held at the University of San Diego (right)

Midwest WIMS 2014		Twin WIMSoCal 2014	
8:30 AM	Registration and breakfast	8:30 AM	Coffee/Registration
9:15 AM	Opening remarks	9:00 AM	Opening remarks
9:30 AM	Parallel sessions	9:10 AM	Parallel sessions
11:30 AM	Coffee break	10:15 AM	Small group discussions
12:00 PM	Problem session	11:00 AM	Coffee break
1:00 PM	Lunch with breakout sessions	11:30 AM	Parallel sessions
2:30 PM	Plenary speaker	12:30 PM	Lunch and invited talk
3:30 PM	Poster session	1:15 PM	Panel discussion
4:30 PM	Parallel sessions	2:00 PM	Parallel sessions
5:30 PM	Evening reception	3:00 PM	Coffee break
		3:30 PM	Invited talk
		4:15 PM	Social event

in their area. Many fields have been represented in WIMS meetings including Algebra and Combinatorics, Dynamical Systems, Geometry and Topology, Logic, Mathematical Biology, Partial Differential Equations, Numerical Methods and Modeling, Statistics, and Knot Theory.

Poster Session Poster sessions can be included during coffee breaks to enable more junior mathematicians to present their work. This is especially valuable for undergraduate and graduate students who may have research results to present but are not yet comfortable with giving a talk. Poster sessions allow women to have one-on-one discussions about their research where they can give and get feedback.

Problem Session A problem session may be included to support the goal of encouraging new, regional collaborations. Women are encouraged to submit open problems before the conference and are given a few minutes to discuss the problem and solicit collaborators.

Invited Talk/Plenary Speaker A plenary talk given by an invited speaker showcases some of the work of women mathematicians in the area and entices participants to attend a new WIMS. We all know women who inspire us—inviting one of these to give a plenary talk is an exciting way to honor them. Publicizing the participation of a well-known, engaging mathematician will be a draw, since participants will have ample opportunities to talk to the plenary speaker at informal events throughout the conference.

Panel Discussion Panel discussions can cover a wide range of topics, from the ever-popular Work-Life Balance which can include caregiving for elderly and young people to Career Options and Applying to Grad School/for Jobs. We strongly endorse moderators with prepared questions, as well as plenty of time for audience questions and discussion. Consider the size of WIMS when structuring a panel—generally two or three panelists will be sufficient for these smaller conferences.

Breakout Groups Navigating the early stages of a career as a mathematician can be challenging for anyone, and there are some challenges particular to women. To facilitate mentoring, several WIMS meetings have included time for small group discussions. These discussions can be scheduled in between other sessions or during lunch. At Midwest WIMS 2014, topics were selected ahead of time, and a woman was invited for each discussion to facilitate the conversation. The discussions were held over lunch with a discussion topic for each table. Participants were able to move to different tables during the lunch.

Discussion topics have included: networking/mentoring, academic job market, research oriented topics, life in graduate school, success in academic life, getting tenure, and work-life balance. Small group discussions may be more successful at larger, more diverse events, while smaller events or those with many people in the same career stage may prefer panelists.

Group Activities In lieu of discussion groups as above, some WIMS set up small group activities that encourage participants to get to know each other. Speaking up can be challenging given extant power dynamics; for example, undergraduates may not feel comfortable sharing their thoughts when senior mathematicians are at the table. Some activities that we have tried with some success include:

- Math Haiku. Each group is given a set of words to create a haiku. The words in the given "pile" are intended to provoke conversation. For example, words such as "imposter," "gender," "identity," and "outlier" might be included, as well as a set of mathematical terms, and some words evoking emotion, such as "begrudge" and "isolated." Here is an example from WiMSoCal 2014, with words from the given collection indicated in **bold**:

  > Blá, an **outlier**
  > **Isolated** in a **ring**
  > **Continuity**
  >   **Begrudg**ingly   She
  > Feels **closed** in **chaotic sand**
  > wants **Transformation**
  >   Seeks connected **graph**
  > **Iterating** to stable
  > **Equilibrium**
  >   She **generate**s a
  > **Fundamental** network of
  > **Passionate mentor**s

- Proofs without words. Each group is given a choice of theorems or properties to illustrate without words. They are allowed to use pictures, videos, or any other medium. Example prompts: the Heine Borel Theorem or the associative property of multiplication.
- Knot tiling activity. Each group is given a set of knot tiles, and they use these to answer a set of questions about knots, and then create a "knot mosaic" [10].

Results of these activities can be displayed at the end of the day during the concluding reception. We have found that these activities generate conversation, bring people from different backgrounds together, and provide fodder for conversation at the end of the day. For a more detailed description, see the AWM Newsletters from March 2016 and May 2017 [6, 8].

Lunch and Coffee Breaks It's important to provide opportunities for informal interactions at the conference venue. We recommend a coffee/continental breakfast at the start of the day, coffee breaks mid-morning and mid-afternoon, and lunch in a setting conducive to meeting new people and chatting. When food is available on site, more formal activities can be offered during lunch, such as the panel discussions or breakout groups listed above, or group walks. Whether or not there are formal activities programs, one lesson we learned at EDGE is that eating together means coming together.

One person bemoaned a lack of interaction time between talks, which can be planned for with strategic breaks.

> I also felt that there weren't many instances to talk to those at the conference since the talks were back to back.

Evening Reception A reception or social event allows time for informal discussions to strengthen the connections made during the day, and serves as a great place to announce the hosts for the following year's WIMS. It's also a good time to collect surveys. One participant's remark about the closing reception sums it up:

> It was a great chance to meet active female researchers in the field and to meet role models who had advice on how I could make it too as a women in math!

7   Suggestions for Organizing

Several organizers commented that organizing WIMS was a large job and was more time consuming than anticipated. To ease the burden of planning and facilitating the meeting we provide some suggestions below.

7.1   Steering Committee

To keep WIMS an annual event and to avoid reinventing the wheel, it can be helpful to have a steering committee made up of organizers from the current and previous WIMS meetings. Past organizers can share planning information such as submitted grants, conference planning timeline, and catering orders as well as lessons learned along the way.

The steering committee should also be responsible for establishing the steering committee for the following year. Past organizers have suggested setting aside time during the conference to identify a date and location for the next meeting with the steering committee.

7.2 Planning Ahead

Organizing a meeting is a large commitment with several tasks that need to be done months in advance. As one organizer suggested, "start planning early and ask for help." Creating a sample planning calendar like the one shown in Table 2 can help establish deadlines and keep the organization of the meeting on track.

7.3 Securing Funding

Ideally, organizers secure enough funding to support travel and lodging for all who wish to participate. Travel funding makes the meeting more accessible to everyone, especially early career mathematicians such as graduate students and postdocs, as well as faculty with limited or no institutional travel funding. A sample budget for a meeting with travel support is shown in Table 3. If organizing a

Table 2 Sample organization calendar

Approximate time	Tasks
12 months out	• Determine which university will host WIMS
	• Request grant from previous year
6–11 months out[a]	• Submit grant to secure funding
	• Get departmental support
	• Get other support as needed
	• Invite keynote speakers
	• Advertise the meeting
3–6 months out	• Send out email announcement
	• Invite parallel session organizers
	• Sign up for "in cooperation with AWM" status
	• Set up webpage
	• Open registration and abstract submission
	• Establish funding application procedure
3 weeks out	• Finalize catering
	• Finalize schedule
	• Travel funding decisions and notification
	• Hotel information
	• Finalize rooms, session chairs, breakout sessions, panel
	• Organize local help (student volunteers?)
1–2 Days out	Print any handouts
Day of	Enjoy the conference!

[a]Depending on the size and budget of the conference. If a grant is being submitted to fund participant travel, organizers should err on the side of caution, and have these items done early to submit the grant in a timely manner

Table 3 Sample budget for large meeting with lodging and travel funding

Item	Budget
Catering	$8000
Invited speaker travel and lodging	$3000
Participant travel and lodging	$10,000
Mini-grants	$2000
Other	$200

Table 4 Sample budget for a small meeting with no funding for travel and lodging

Item	Budget
Light breakfast, coffee breaks, and refreshments	$400
Lunch	$1000
Paper materials	$100

meeting for an existing WIMS region, budgets of meetings held in that particular region will give a better idea of an appropriate budget, as regions vary in size and transportation options. Though travel funding is preferred for large geographical regions, many WIMS meetings have been successfully organized with modest budgets. An example of a minimal budget is shown in Table 4.

7.4 Funding Opportunities

Funding from Host Institutions Many of the WIMS meetings have been made possible with modest amounts of funding secured from the host institutions. The host department, college, and/or institution may have funds available for conferences organized at that institution.

National Funding Agencies Several WIMS meetings have been funded with grants from national funding agencies. Include in the grant: conference goals, any confirmed funding from the hosting institution, a sample schedule including any confirmed or invited speakers, and a list of schools to invite. It is especially helpful to talk to past organizers who have been successful in securing grants from national funding agencies.

7.5 Advertising

A great way to start advertising WIMS is by emailing all math departments within a certain driving distance or geographic region. Include both research and teaching oriented departments. Creating a thorough list of such universities is time-consuming, but once created, such a list can be passed down to future organizers and updated each year.

A web page is necessary to convey logistical information about the conference. Some meetings have also been advertised on social media. Many national associations advertise meetings on their calendars. Some associations to consider are the Association for Women in Mathematics, The American Mathematical Society, and the Mathematical Association of America.

To obtain "in cooperation with the AWM" status, visit the AWM website and enter the conference information. Then include the AWM's non-discrimination statement on the website and conference materials. This is a convenient way to avoid writing an original nondiscrimination statement.

7.6   Other Considerations

7.6.1   Supporting Parents

It can be especially challenging for women with small children to attend conferences. Organizers can help by identifying a mother's room on campus and investigating childcare options [3].

7.6.2   It Takes a Village

There are many opportunities for volunteers during the meetings. Women can be invited to chair parallel sessions, and volunteers can help with registration and other logistics. Undergraduate students can organize their own poster session, or brainstorm together about ice-breakers. Social media aficionadas can organize a blog or hashtag, departments and administrators can demonstrate support by hosting one of the social events, and university higher-ups can give some friendly words of welcome.

8   Conclusions

We hope this chapter encourages you, the reader, to organize your own regional women in mathematics conference in the same vein as WIMS. Through the words of participants and organizers, we demonstrated the purpose and impact of WIMS in hopes of motivating you to make your own. We combined our years of experience with WIMS to offer insights on scheduling, planning, diversifying, and funding such a conference. We believe that WIMS, which has affected hundreds of women across the globe, demonstrates the long-term value of EDGE as a place to cultivate future women leaders and organizers of the mathematical community.

Acknowledgements Thank you to all WIMS participants and organizers who filled out our survey. We also want to thank all the host departments and institutions for their support, and to Sylvia, Rhonda, and the NSF for the ADVANCE grant that supported the first EDGE clusters.

References

1. Statistics on women mathematicians. *Notices of the AMS*, 9:994, 2017.
2. S. Bryant and K. Pershell. The EDGE program: Mentoring women pursuing careers in the mathematical sciences. *University of New Mexico 8th International Annual Mentoring Conference Proceedings*, 2015.
3. R. M. Calisi and a Working Group of Mothers in Science. Opinion: How to tackle the childcare–conference conundrum. *Proceedings of the National Academy of Sciences*, 2018.
4. S. Ceci and W. Williams. Sex differences in math-intensive fields. *Curr Dir Psychol Sc*, 19:275–279., 2010.
5. C. Good, A. Rattan, and C. Dweck. Why do women opt out? Sense of belonging and women's representation of mathematics. *Journal of Personality and Social Psychology*, 102:700–717, 2012.
6. C. Haskell. Wimsocal 2017. *AWM Newsletter*, 47:16–17, 2017.
7. Institute of Medicine and National Academy of Sciences and National Academy of Engineering. *Enhancing the Postdoctoral Experience for Scientists and Engineers: A Guide for Postdoctoral Scholars, Advisers, Institutions, Funding Organizations, and Disciplinary Societies*. The National Academies Press, Washington, DC, 2000.
8. G. Karaali and A. Radunskaya. Collaboration and creativity in southern California: an offering. *AWM Newsletter*, 46:30–32, 2017.
9. J. Sengers and M. Sharma. Women for science. *InterAcademy Council*, 2006.
10. F. Tabing. Topological images with modular block print tiles. In *Mathematics, Music, Art, Architecture, Education, Culture*. Bridges Waterloo 2017, Tessellations Publishing, 2017.
11. R. T. Taylor and L. Karcinski. Science and mathematics mentees and mentors: Who benefits the most? *Mentoring & Tutoring: Partnership in Learning*, 24(3):213–227, 2016.

The Career Mentoring Workshop: A Second-Generation EDGE Program

Rachelle C. DeCoste

Abstract The EDGE program, originally founded to provide support for women entering doctoral programs in the mathematical sciences, has had a dramatic, positive impact on the mathematics community well beyond its program participants. Many of the women who have participated in the EDGE summer sessions have not only successfully earned doctorates in the mathematical sciences, but have subsequently assumed leadership roles in new outreach efforts aimed at diversifying the United States mathematics community. In this paper, we examine in more detail the impact of the author's—an EDGE alumna and current Associate Professor of Mathematics—efforts to diversify the mathematics community by way of founding the Career Mentoring Workshop (CaMeW) for women completing their math doctorates. Lessons learned, including challenges and successes, will be shared for others who may consider initiating similar outreach efforts. CaMeW is an example of a "second-generation EDGE program," that is a program founded by an EDGE alumna who is actively working to diversify the math community.

1 Introduction

In the summer of 1998, I participated in the first EDGE program. It was held at Bryn Mawr College and I was one of the eight women participants who were about to enter their first year of graduate school in mathematics. In the 20 years that have followed, I have completed a PhD, held a National Research Council Postdoctoral Fellowship, and earned tenure and promotion at a small selective liberal arts college—my dream job. I regularly publish research articles on the geometry of nilmanifolds arising from Lie groups; I teach undergraduate mathematics classes at all levels; and I mentor students, some of whom go on to graduate school in mathematics and other disciplines. However, I am not just a typically successful

R. C. DeCoste (✉)
Wheaton College, Norton, MA, USA
e-mail: decoste_rachelle@wheatoncollege.edu

© The Author(s) and the Association for Women in Mathematics 2019
S. D'Agostino et al. (eds.), *A Celebration of the EDGE Program's Impact on the Mathematics Community and Beyond*, Association for Women in Mathematics Series 18, https://doi.org/10.1007/978-3-030-19486-4_11

academic mathematician; I am an EDGE mathematician. The impact of the EDGE program is most heavily seen through the work I have done as a member of the mathematics community in the 15 years, since I completed my PhD. EDGE mathematicians are deeply committed to work diversifying mathematics.

I am the founder and director of the Career Mentoring Workshop (CaMeW), a program that has mentored 95 women finishing their PhDs over the course of seven workshops. This program helps women secure first postdoctoral positions—either inside or outside of academe—and aims to ensure that they feel supported moving forward.

2   CaMeW Rationale and Structure

CaMeW seeks to create a supportive community for women mathematicians in pursuit of fulfilling professional careers. This is important for the individuals, but also for the mathematics community. Women faculty enhance the persistence of women students, as supported in Dasgupta [2] and Herrmann [7].

The Career Mentoring Workshop began as a hallway conversation with another EDGE community member at the 2007 Joint Mathematics Meeting (JMM). At the time, I was a National Research Council Fellow with a joint appointment at the United States Military Academy at West Point and the United States Army Soldier Research Center. This EDGE alumna was completing her PhD, searching for her first postdoctoral position, and experiencing self-doubt about her ability and potential to secure a faculty position. She was surprised to hear that I had felt similarly when I had sought my first position. Our common experience was a classic example of imposter syndrome, a well-studied phenomenon first introduced by Clance and Imes [1] that tends to impact even the most highly successful professionals.

This JMM conversation started a chain reaction. I could no longer ignore the palpable consequences that doubt was having on women's potential to access careers they not only wanted, but were well qualified to fulfill. I decided to establish a space where women could have open, honest conversations about the job search process. Indeed, according to Hill [8], women in STEM (Science, Technology, Engineering and Mathematics) fields are more susceptible to feeling like they do not belong in academia, thus they are posited to benefit directly and strongly from mentoring by other women. According to Fridkis-Hareli [3], the majority of women in science who have been successful in their careers have been mentored, whether formally or informally.

That fateful hallway conversation led to the first CaMeW held in the summer of 2007 at the United States Military Academy at West Point. The workshop was supported by a Mathematical Association of America (MAA)/Tensor Grant for Women and the Department of Mathematical Sciences at West Point. Ten women were invited to participate in the workshop from a pool of twenty-five applicants. Since then, funding has been secured for the workshop to run in 2007–2010,

2012, 2016, and 2018, with financial support from the MAA/Tensor Program, the Department of Mathematical Sciences at West Point, Wheaton College (MA), the NSA, the EDGE Foundation, the Department of Mathematics and Statistics at Mount Holyoke College, and the Summer Mathematics Program at Carleton. Each CaMeW session has hosted 12–15 graduate student participants. Even though transportation for faculty mentors was funded, faculty did not receive stipends for work preparing for or participating in CaMeW in the first five workshops. These faculty volunteers demonstrated a deep commitment to supporting other women mathematicians. June [10] describes tension between expecting professionals to mentor the next generation and to perform service to the community and the fact that much unpaid service is performed by women and other underrepresented minorities. Moving forward, CaMeW expects to compensate faculty for their participation in the program, though we recognize that the rate of compensation is well below what their professional expertise should dictate. Throughout the iterations of CaMeW, several of the faculty mentors have been members of the EDGE community.

The 3-day format of CaMeW has remained consistent throughout the years. While a longer workshop would allow for more reflection time and community building, a 3-day commitment is as much time as most participants can afford away from their doctoral theses. Additionally, in the early years of the program, asking faculty to volunteer more than 3 days of their time without compensation seemed like too large a request. The workshops have been held in the Mathematics Departments at West Point, Wheaton College, and most recently, at Mount Holyoke College. The suggestion to move the workshop to a national mathematics meeting to save money has been rejected because of the privacy afforded by these more intimate venues. To hold CaMeW at MathFest, for example, would not allow for the same level of comfort that leads to deep, honest conversations. When one's classmates, advisor, or potential employer might be just around the corner, it is harder to feel safe with the level of vulnerability necessary to achieve true connections between our participants and faculty.

The goals of the workshop are multiple. The immediate objectives are to educate women about possible career paths and to help them establish, or build upon, existing peer and mentor networks. Toward this end, several of the sessions consist of faculty presenting material on topics such as possible postdoctoral positions, the steps of the job search process, types and expectations of interviews, and negotiating strategies. Each participant is asked to prepare and submit job search materials (cover letter, curriculum vitae, research statement, and teaching statement) prior to the workshop. Two different faculty members read each participant's package and then meet with the individuals to provide personalized feedback during the workshop. This intentional mechanism allows the participants to have solid material for their job search prior to the mathematics academic job search season, which typically begins in early fall. In addition to preparing practical job materials, participants also prepare and present a short talk about their dissertation research in small group sessions where they receive individual written and verbal feedback. The supportive atmosphere allows the participants to gain valuable personal feedback without the negative consequences of a stressful environment. These practice presentations help prepare the participants for talks at the JMM or for job interviews. Feedback on the talks focuses on both content and delivery, with the intent to help each participant improve her ability to give an effective talk. In addition to the formal interactions and sessions, time is set aside during CaMeW for each participant to reflect on her stated goals and their outcomes both at the beginning and at the end of the workshop.

As stated in the beginning of this essay, the overarching goal of CaMeW is to help women find first postdoctoral positions that help them establish life-long careers in mathematics. We believe that a first step in this process is to ask women participants to consider the kind of a career they want to pursue.

CaMeW is often the first place where participants are asked to reflect on the path they want for their future. Indeed, they usually know what their PhD advisors envision, but they are often surprised when they realize that they have never questioned whether their advisor's goals align with their own. In fact, over the years, we have helped several participants navigate difficult conversations and relationships with advisors. We encourage participants to think about their long-term career goals, by thinking individually about their desired day-to-day experiences, as well as the more encompassing characteristics of their desired future work life. We ask them to consider the balance between research and teaching, both short term and long term. We challenge them to prioritize certain aspects of their future life from a long list that includes job-related details—these include number of classes taught each semester, sabbatical support, travel support, types of courses taught, class size, size of the faculty, etc. We also validate the important consideration of the quality of life when thinking about their future, and we encourage them to examine schools' idiosyncrasies that are not directly work-related, such as geographic location, urban vs rural locations, availability of social support, etc. We strongly believe that thinking about priorities leads to a fulfilling work–life balance and is essential to establishing a successful career in mathematics. We use this exercise to help participants see that they are not a homogeneous group of 15 graduate students vying for the same set of jobs. We embolden each participant to seek a job that aligns with her individuality and with her own personal objectives. We openly and strongly affirm that each list is valid and that each choice is legitimate.

After we ask participants to think about their ideal jobs and we give feedback on their application materials, we hold an interactive workshop session on interviews, during which we engage in mock interviews. The participants often find this awkward at first, but we ask them to pretend that they are interviewing for their self-defined "perfect job." This experience gives the applicants an opportunity to explain to someone else why they want this particular job and how they are qualified for that position.

We ask questions that they might be asked on an initial interview, again offering the opportunity for them to practice responses in a situation where there is no risk. Participants also brainstorm lists of questions they think might arise and questions they should ask during the interview process. We find that many participants do not know that they are expected to ask questions during an interview. The interactive workshop session opens the door for us to remind them that interviews are dialogues where they should gather information to understand whether the position fits their personal and professional objectives, as discussed above. Through these sessions, our participants become better communicators and confident speakers who can advocate for themselves.

3   CaMeW Follow-Up

After the participants leave CaMeW the core faculty routinely review second drafts of application material, discuss interviews, and provide advice on negotiations as the participants move through the job search during the following year. We also arrange an informal gathering at the JMM each year. In this meeting, we invite participants from all years of CaMeW; this allows them to share their job search experiences, while interacting with others who have been through the process, providing additional role models and success stories. As they continue in their career trajectories, we offer discussions about changing jobs, balancing work and family life, and becoming more established in their professional settings. Today, the community of women who continue to support each other as CaMeW alumnae is impressive. They continue to mentor and advise each other, in person or online through social media.

4   CaMeW Staffing

The workshop is faculty-intensive. For instance, over the course of 3 days in 2018, 15 participants interacted with 16 faculty and other professional mathematicians. For each iteration of CaMeW, six core faculty are present for the entirety of the workshop: the core faculty are there for every session, every meal, and even for breaks. For the middle day of the workshop, the core faculty are joined by additional faculty from nearby schools who participate in the feedback sessions and later host a panel discussion for the participants. A benefit to holding CaMeW at schools located in the Northeast has been the close proximity of many other colleges and universities. It is necessary to include additional faculty to provide hands-on feedback for the participants, but also to increase the number of perspectives shared during the workshop, which is necessary to support our goal of allowing participants to interact with a range of successful women in mathematics. We strive to represent racial and ethnic diversity for the faculty and we also aim to invite faculty mentors from a diverse range of schools, from traditional liberal arts colleges, to comprehensive universities and R1 institutions. We seek to include faculty who are at various points in their careers. For instance, we rely on more junior faculty to discuss current trends in the job search in mathematics, as we know these trends change over time. We ask mid-career faculty who have achieved tenure and who may have changed jobs to offer their perspective on the often-elusive life–work balance. We also make sure to provide the perspectives of more senior faculty who have mentored junior colleagues for years and who can share lessons learned from their seasoned careers. The presence of senior faculty also benefits the junior faculty as it allows for strong connections and mentoring relationships to form among these faculty members. It should be noted that in the initial years of CaMeW, there were men included among the faculty members. However, in later years, the faculty have been all women, based on the belief that there are many male mentors in the mathematics community. With only 20% of the women among full-time faculty at R1 institutions, according to AMS data (2018), participants may be lacking in both female role models and mentors.

Faculty who participate for 1 day are fully engaged through their individual meetings to review application materials, through the feedback they give during the math presentations, and through their panel session. They are also committed to the success of the workshop through all the informal interactions that occur during breaks and meals. Topics during the panel usually range from practical questions about what makes a job application stand out, to personal experiences tied to the challenges and rewards of life as a woman in mathematics. We end our one full day together with a keynote talk by a woman mathematician who discusses her own journey through mathematics. The most successful talks have been the ones where speakers have shown their own personalities, honestly shared stories of successes and challenges, and have humanized the life of a successful woman in mathematics. Keynote speakers have included Ruth Haas, cofounder of the Smith College Center for Women in Mathematics; Rhonda Hughes, Bryn Mawr College, cofounder of the EDGE program; Suzanne Weekes, Worcester Polytechnic Institute, cofounder of Preparation for Industrial Careers in Mathematical Sciences; Liz McMahon, Lafayette College; Catherine Roberts, College of the Holy Cross and current AMS Executive Director; Kathi Crow, Salem State University.

Given all these reasons and goals, it is evident that one of the challenges for each CaMeW has been to find a well-balanced cohort of faculty who can represent a wide range of women mathematicians. Each individual must be comfortable with the vulnerability inherent in sharing her personal story, including the obstacles and the failures faced along the way. To stay true to the original intent of the workshop, we must provide a space for women to feel safe in sharing their struggles honestly. In our experiences, the most effective faculty mentors are those who do not pretend that their success is void of challenges, but the ones who share their full stories along with some effective strategies for overcoming the difficulties that can arise in the life of a woman mathematician. In addition to ensuring that each individual faculty member will contribute to the overall goals of the workshop, we strive to ensure the group is both diverse in its composition and cohesive in the overall message, which is to validate and support each individual's choices of professional path.

5 Outcomes of CaMeW

Over the seven iterations of CaMeW, we have mentored 95 women from 54 different graduate programs. Following the completion of their doctorates, participants of CaMeW have gone on to successful careers in academia and in industry. Their first postdoctoral positions include research postdocs, teaching postdocs, tenure-track positions, visiting positions, industry positions, and full-time parenting. At CaMeW, we discuss that many academic job paths are not linear and that many people change jobs during their careers, including several of our faculty mentors. We provide women with tools to persist in their chosen careers and we want them to recognize that if they are in academic positions that do not fulfill them—for either professional or personal reasons—they should know that the issue may not lie with them not belonging in academia. We also recognize that sometimes women DO choose to leave academia and move to industry; other times women choose to remain at home after having children. These are all valid career choices as long as they are ones that the women themselves feel empowered to make, rather than feeling they must leave mathematics because they do not belong.

Following CaMeW, participants are able to finish preparing materials that are tailored to specific jobs and discuss their preparation for these positions. During interviews, CaMeW participants are able to articulate their own interests and understand how they fit various positions. Overall, we believe that participants are more confident in their ability to navigate the job search and that this contributes to the successful attainment of their first postdoctoral position. Taking away the unknown and the feeling of isolation provides a level of comfort that allows them to focus on finding a job that is right for them instead of wondering if they belong anywhere in mathematics. With our encouragement, CaMeW participants pass on the knowledge and experiences they gain from their participation in CaMeW with their peers. We also encourage them to apply for participation in other professional development opportunities, such as the Association for Women in Mathematics Graduate Poster Session at the JMM and MAA's Project NExT.

6 CaMeW Challenges

We are aware of the possibility that participants may arrive at CaMeW with the notion that they are in "competition" with each other on the job market. We ask that they list their own personal objectives, which allows them to see that the overlap of desired jobs is small. We ask participants to set aside this potential competitive angle so that we may all support each other's individual goals. During each iteration, we have succeeded in creating a genuinely supportive environment in which participants establish lasting relationships.

Funding remains a challenge for the sustainability and continuation of CaMeW. The lack of funding has led to years when CaMeW could not run. The current iterations of the program, including funding for faculty, cost on average just over $17,000. Though this small investment (just over $1000 per participant) effectively supports diversifying mathematics, regular funding remains elusive. The future of CaMeW is in doubt.

7   The Future of CaMeW

Why should programs like the EDGE program and CaMeW continue to exist? Do women no longer need targeted programs to support their career-long participation in the mathematics community? One needs look no further than the recent report about sexual harassment in STEM fields edited by Johnson et al. [9] to see that all is not equal in academia. Women remain underrepresented in the professoriate, as documented in the AMS surveys of graduate students and departmental profiles by Golbeck et al. [4], and are still internalizing these community-wide issues as personal ones. The AMS recently participated in the National Science Foundation-funded STEM Inclusion Study. While published results are not available, some indications of the results were written about by Helen G. Grundman, AMS Director of Education and Diversity, in the inclusion/exclusion blog of the AMS (November 29, 2018) [6]. As she indicates, women are significantly more likely than men to agree that they have to work harder to be perceived as a legitimate professional.

In the same Inclusion Study, women members of the AMS report significantly higher frequencies of being harassed, verbally or in writing on the job in the last year than men. As a community, we need to do more to support the persistence of women in mathematics at all levels. According to NSF data [11], women earned 41.7% of the undergraduate degrees in mathematics and statistics in the United States in 2014. Further, according to the Fall 2016 Departmental Profile Report in the 2016 Annual Survey of the Mathematical Sciences in the US reported in the September 2018 Notices by Golbeck et al. [5], women account for 31% of all full-time faculty in the mathematical sciences. This mirrors the percentage of women earning PhDs in the United States, which was 30% in 2016 (31% in 2015) according to the AMS Survey of new PhDs, again reported by Golbeck et al. [5]. However, when one delves deeper into the employment data, one sees that women hold only 14% of full-time tenured and 26% of full-time tenure-eligible positions in doctoral mathematics departments. Though women are persisting in some areas of academia, they are still largely underrepresented at graduate degree granting institutions.

8   CaMeW Conclusions and the Larger EDGE Community

In my career, I have intentionally surrounded myself with women who are deeply committed to supporting each other and the larger mathematics community. All mathematicians should assist with inclusivity efforts in the mathematics community.

CaMeW is one program that has been successful in creating an inclusive community of women who support each other as individuals. In Chapter 5, I identify many other programs created or supported by members of the EDGE community. EDGE community members are creating, directing, and leading initiatives aimed at diversifying mathematics beyond—but because of—EDGE: they are second-generation EDGE activities.

Acknowledgments The author would like to thank the editors of this volume for their endless patience and encouragement. Also, sincere gratitude is owed to Touba Ghadessi, PhD who is a kind, generous and supportive colleague and friend.

References

1. Clance, P. R., & Imes, S. A. (1978) The imposter phenomenon in high achieving women: Dynamics and therapeutic intervention. *Psychotherapy: Theory, Research & Practice, 15*(3): 241-247
2. Dasgupta, N, Stout, J.G. (2014) Girls and Women in Science, Technology, Engineering, and Mathematics: STEMing the Tide and Broadening Participation in STEM Careers. Policy Insights from the Behavioral and Brain Sciences 1(1): 21-29
3. Fridkis-Hareli, M. (2011) A mentoring program for women scientists meets a pressing need. Nature Biotechnology 29(3): 287-288
4. Golbeck, A.L., Barr, T.H, Rose, C.A. (2018) Report on the 2015-2016 New Doctoral Recipients. Notices of the AMS, March 2018: 771-785
5. Golbeck, A.L, Barr, T.A., Rose, C.A. (2018) Fall 2016 Departmental Profile Report. Notices of the AMS, September 2018: 949-959
6. Grundman, H.J. (2018) The STEM Inclusion Study: What we've learned so far. In: The AMS inclusion/exclusion blog. American Mathematical Society. https://blogs.ams.org/inclusionexclusion/2018/11/29/the-stem-inclusion-study-what-weve-learned-so-far/
7. Herrmann, S.D., Adelman, R.M., Bodford, J.E., Graudejus, O, Okun, M.A., Kwan, V.S.Y. (2016) The Effects of a Female Role Model on Academic Performance and Persistence of Women in STEM Courses. Basic and Applied Social Psychology, 38(5): 258-268
8. Hill, C., Corbett, C., St. Rose, A. (2010) Why So Few? Women in Science, Technology, Engineering and Mathematics. AAUW, Washington, DC.
9. Johnson, P.A., Widnall, S.E., Benya, F.F. (eds) (2018) *Sexual Harassment of Women: Climate, Culture, and Consequences in Academic Sciences, Engineering, and Medicine*. National Academies of Sciences, Engineering, and Medicine. Washington, DC.
10. June, A.W. (2015) The Invisible Labor of Minority Professors. The Chronicle of Higher Education, 62(11)
11. National Science Foundation, National Center for Science and Engineering Statistics. (2017) *Women, Minorities, and Persons with Disabilities in Science and Engineering*. Special Report NSF 17-310. Arlington, VA.

Emotional Labor in Mathematics: Reflections on Mathematical Communities, Mentoring Structures, and EDGE

Gizem Karaali

> *The Bureau of Labor Statistics puts both "diplomat" and "mathematician" in the "professional" category, yet the emotional labor of a diplomat is crucial to his work whereas that of a mathematician is not.*
>
> Arlie Russell Hochschild
> *The Managed Heart [21, p. 148].*

Abstract Terms such as "affective labor" and "emotional labor" pepper feminist critiques of the workplace. Though there are theoretical nuances between the two phrases, both kinds of labor involve the management of emotions; some acts associated with these constructs involve caring, listening, comforting, reassuring, and smiling. In this article I explore the different ways academic mathematicians are called to provide emotional labor in the discipline, thereby illuminating a rarely visible component of a mathematical life in the academy. Underlying this work is my contention that a conceptualization of labor involved in managing emotions is of value to the project of understanding the character, values, and boundaries of such a life. In order to investigate the various dimensions of emotional labor in the context of academic mathematics, I extend the basic framework of Morris and Feldman (Acad. Manag. Rev. 21:986–1010, 1996) and then apply this extended framework to the mathematical sciences. Other researchers have mainly focused on the negative effects of emotional labor on a laborer's physical, emotional, and mental health, and several examples in this article align with this framing. However, at the end of the article, I argue that mathematical communities and mentoring structures such as EDGE help diminish some of the negative aspects of emotional labor while also accentuating the positives.

G. Karaali (✉)
Pomona College, Department of Mathematics, Claremont, CA, USA
e-mail: gizem.karaali@pomona.edu

© The Author(s) and the Association for Women in Mathematics 2019
S. D'Agostino et al. (eds.), *A Celebration of the EDGE Program's Impact on the Mathematics Community and Beyond*, Association for Women in Mathematics Series 18, https://doi.org/10.1007/978-3-030-19486-4_12

1   Introduction

Sociologist Arlie Russell Hochschild launched the term "emotional labor" into
the mainstream with her 1983 book *The Managed Heart: Commercialization of
Human Feeling*. Inspired by Hochschild, and the large literature on emotional labor
following her seminal work (see, for instance, [39] for a careful review of this work
up to the end of the twentieth century, or [16] for a more recent volume), in this
paper I will define emotional labor as any labor that involves the management of
emotions, of the self or of others. For some nuances that might help the reader
engage with contemporary literature on this theme, see Sect. 2.

No matter how one defines emotional labor, according to Hochschild, the job
of a mathematician is quite independent of this kind of labor (as per the epigraphed
quote). Indeed many mathematicians enter this profession with similar illusions. For
some it is even an appealing factor that the human contact required in many other
professions would not be relevant. For others, emotional labor is a non-issue; they
at some point decide they love and cannot live without mathematics, or they decide
math is something they can do well enough to feed themselves and their loved ones;
in either case, the emotional dimensions of labor do not come into play in their
internal negotiations about future career plans.

However many academic mathematicians soon find that their job entails emotional labor even if it is not part of the explicit job description.[1] In this article I
investigate the different ways mathematicians are called upon to provide emotional
labor in the discipline. Even though many scholars have explored the features of
emotional labor in academia, literature does not engage with the specific context
and experiences of those in the mathematical sciences. In this article I probe the
construct of emotional labor in the context of academic mathematics in order to
shed light on this oft-neglected dimension of work in the discipline, and to highlight
some aspects of it that might otherwise be missed. Thus this article may be viewed as
a case study of sorts for emotional labor scholarship on the one hand and a reflection
exercise for mathematicians on the other.

In Sect. 2, I review the literature on emotional and affective labor, and building
on prior work, I tease out a nine-dimensional framework that undergirds the
discussion in the rest of the paper. In Sect. 3, I apply this framework to the three
contexts of teaching, academic service, and academic research. In this section I also
begin to answer the related question: "Who hears the call to emotional labor?" In
Sect. 4, I focus on the mathematical context and identify the kinds of emotional
labor mathematicians are called to do, adapting the framework of Sect. 2 and the
examples of Sect. 3 into the mathematical sciences, and supplementing them with

---

[1]Here for reference is the official job description the United States government provides:
mathematicians "conduct research in fundamental mathematics or in application of mathematical techniques to science, management, and other fields," see https://www.bls.gov/oes/current/
oes152021.htm, last accessed on December 20, 2018.

new ideas as needed. In Sect. 5, I zero in on a specific category of work demanding emotional labor. Thus in this section I explore whether, and if so how, mathematical communities and mentoring structures (such as EDGE) may help diminish some of the negative aspects of emotional labor while also accentuating the positives. I also address more fully the "who" question posed earlier. Section 6 wraps up this article, with a few brief remarks.

2 A Framework for Emotional Labor

Theorists of labor have explored various types of work through the last century. Manual/physical labor, that is, work that engages the human body in the production of commodities, has been central to most labor movements of the twentieth century. Intellectual/cognitive labor is of interest to many living in today's "post-industrial" "knowledge economy" environment. This paper is about a third type of labor, which as Hochschild put it bluntly has been "seldom recognized by those who tell us what labor is" [21, p. 197], also see [18]. Indeed emotional labor is often viewed to be feminine and thus "less-than."

In this paper we take any labor that involves managing the emotions of the worker or of those they interact with to be emotional labor. Some acts associated with this kind of labor include, but are not limited to, caring, listening, comforting, reassuring, and smiling. However there are two distinct components here: the self-management component remains internal, while the outward management of the emotions of the other (the client, the patient, the passenger, or, in the classroom setting, the student) is often more explicitly delineated and externally monitored by the employer. Even though both kinds of emotion-related work were labeled emotional labor by Hochschild in her [21], today these two are typically analyzed under different terms.

In most contemporary scholarship, work that entails the monitoring and managing of the emotions of the laborer is called "emotional labor"; see, for example, [32], as well as the many articles in [38]. Work that entails the creation or management of emotions in a designated other (or a designated group of others) is called "affective labor", after [19]. These terms and the scholarship engaging with them are not uncontroversial; see, for instance, [35] for a critique of how the phrase "affective labor" might be used to create a gendered hierarchy of labor. Nonetheless, the study of emotions in the workplace in general has led to productive ways of thinking about the well-being of workers and more broadly about organizational behavior [14]. Furthermore, understanding how mental, manual, and emotional labor come together in today's work may lead to "a potentially more comprehensive understanding of nature and social life" [18]. Thus, it is my contention that a conceptualization of labor involved in managing emotions is of value to the project of understanding the character, values, and boundaries of a mathematical life in the academy. In the following, therefore, I will use the term "emotional labor" to capture both types of emotion work, pointing out explicitly the distinct aspects of different types when needed.

A formal framework to conceptualize the type of emotional labor involving the management of the emotions of the laborer is presented in [32], where four dimensions are proposed:

(a)$_{self}$ frequency of appropriate emotional display,
(b)$_{self}$ attentiveness to required display rules,
(c)$_{self}$ variety of emotions required to be displayed, and
(d)$_{self}$ emotional dissonance generated as the result of having to express organizationally desired emotions not genuinely felt.

The authors then argue that "although some dimensions of emotional labor (e.g., variety of emotions that are displayed) are likely to be associated with higher emotional exhaustion, it is mainly emotional dissonance that is likely to lead to lower job satisfaction." Some of what follows will have resonances with this perspective; see in particular Sect. 5.

A comprehensive framework for an analysis of emotional labor in the context of academic mathematics will also need to account for the dimensions of the type of emotion work that involves the creation and management of desired emotions in designated others. Analogous to the four dimensions above, I propose the following:

(a)$_{others}$ frequency of instances of management of the emotions of designated others,
(b)$_{others}$ attentiveness to designated others' current emotions,
(c)$_{others}$ variety of emotions one is required to engender or sustain in designated others, and
(d)$_{others}$ emotional burden generated as the result of having to focus on designated others' emotions at the expense of other priorities or personal values.

To the above eight dimensions, I will add a ninth that does not require the management of displayed emotion and yet is very much related to internal self-directed emotion management:

(e)$_{self}$ internal self-management of emotion required to continue to perform effectively in the job.

This dimension of emotion work, related in various ways to the mental health of the laborer, is typically not monitored by the employer and yet is absolutely crucial to the employee's performance and sustained effectiveness. See [48] for a review of the psychosocial consequences of the self-monitoring of emotions, and [34] for recent work that points toward the various significant effects of self-focused emotion management on the mental health and well-being of workers.

3 Emotional Labor in the Academy

There are many ways of doing emotional labor in the academy. See [41] for a "contemporary account of what it means to experience and feel academia, as a privilege, risk, entitlement, or failure" [page 1]. Bellas [5] offers a critique of the rewards system within the academy that values the "masculine" aspects of the job (research and administration) over the "feminine" aspects (teaching and service). However she also points out that emotional labor plays a significant role in all these areas of academic work.

In the three subsections of this section I explore some examples of the emotional labor involved in the trifecta of teaching, research, and service in terms of the nine dimensions described above. This will be a general exploration; I will leave the analysis of the specific context of the academic mathematician to the next section.

In the examples of this section (and the next), I intentionally focus almost exclusively on the burdens that may accrue from the emotional dimensions of teaching, service, and research. I do not deny that academics also get a lot of emotional satisfaction from their work, and that for many, some of these aspects serve as the highlight of their careers and sustain them for years. In fact in Sect. 5, I will come back to teaching and service in particular, and zero in on some of the positives of the emotional labor involved in these two aspects of an academic career.

3.1 Emotional Dimensions of Teaching

Teaching involves a significant amount of emotional work. Cavanagh in [10] explores in depth the emotional dimensions of teaching and learning; see especially pages 102–108 where the emotional labor of teaching is explicitly discussed. Näring et al. in [33] explore specifically the relationship between the emotion work of teachers and their emotional exhaustion. For this paper, however, it will be sufficient to specifically point out examples of emotional labor related to teaching that can be described in terms of the nine dimensions of Sect. 2.

During the academic year, professors typically meet their students in the classroom a few times a week ((a)$_{self}$). During these regular sessions, they often aim to display effortless expertise, enthusiasm for the discipline, and joy of teaching ((c)$_{self}$). These impressions are not always easy to sustain, and for those instructors who are traditionally underrepresented in the professoriate, they may be somewhat difficult to sustain simultaneously. ("She is so bubbly enthusiastic about her topic! She must not really know what she is talking about.") Thus the professor must often pay close attention to carefully balancing the displayed emotions ((b)$_{self}$). This balancing act is often difficult. Too much enthusiasm might be counterproductive. One must be perceived as professional and yet friendly, charismatic and yet not too distant, and so on. This delicate performance aspect of teaching might be additionally difficult for the typical introvert academic, who might be exhausted by putting on a show for the students every other day.

During office hours, professors regularly interact with students and need to manage their feelings $((a)_{others})$. They need to attend to feelings of helplessness, distrust, and apathy, and find ways of supporting students' confidence, interest, and enthusiasm $((b)_{others}, (c)_{others})$. They may need to turn on this others'-feelings focus, day after day, even when they themselves are not feeling emotionally healthy or when they have other needs of their own $((d)_{others})$. Beyond the standard requirements of office hour performance, occasionally professors find themselves in the roles of therapist, mother, experienced older brother, or wise elder, where they are expected to help students process and manage their emotions about various life issues.

In academia a disproportionate amount of teaching-related emotional work falls on the shoulders of female professors; in particular [13] highlights the extra burden on female faculty. Add to this the gender bias in student evaluations of teaching, which have been explored at least since the publication of [4]. Such extra work demands and special requests for favors as well as concerns about student evaluations certainly contribute to the burden of emotional work related to teaching, in particular in the dimensions of $(b)_{self}, (d)_{self}, (e)_{self}, (b)_{others}, (d)_{others}$.

It should be noted at this point that several statements made in this paper about the added burden of gender on female faculty in general and female mathematicians in particular may apply also to other minoritized or underrepresented groups. Most of the scholarship I have come across on the topic of emotional labor works with gender as the main variable. Scholarship that works with other variables such as race, ethnicity, and socioeconomic status in the context of the academy does not always engage explicitly with terms such as "emotional labor" or "affective labor." This of course does not mean that there is no extra burden on other minoritized groups in the academy. See, for instance, [44] for a discussion of the types of added stress African American faculty live with, and [27] for an investigation into how underrepresented faculty may be managing their emotions. What is called "emotional drain" in [45] experienced by women of color in the academy is also very much in the same territory.

3.2 Emotional Dimensions of Academic Service

There is evidence that supports the claim that women do significantly more academic service than men; research also seems to imply that the main source of discrepancy is in the amount of internal service [17]. Furthermore Bellas [5] argues that there may be "greater demands for emotional labor on women as women attempt to convey, justify, and legitimize their contributions and, indeed, their presence." Noting how women's ways of leading and communication are often devalued, she concludes her section on emotional labor and academic service with "All this may place additional demands for emotional labor on faculty women, especially women in the lower ranks and women of color, who may suffer devaluation in interactive as well as financial contexts." Once again readers may assume that the above will apply to other minoritized groups.

Two significant components of internal service are student advising and committee work [5]. Emotional labor related to student advising is much akin to the emotion work a professor does during office hours, even though often the others'-feelings focus is even more dominant in advising. Advising professors must remain in tune with their students' emotional needs and general emotional condition in any advising session $((b)_{others})$. Furthermore, they should preserve their "caring professional" presentation $((b)_{self})$.

In faculty committees, which involve work on the emotions of both self and others, professors often meet regularly to discuss matters that are either too small or too large to be resolved by said committees. The frequency of these meetings $((a)_{self}, (a)_{others})$, especially if in inverse proportion to their actual effectiveness, may lead to burnout and sometimes apathy. However the main emotional dimensions of this kind of academic work involve interpersonal relations between colleagues. One should, for instance, make sure to help others feel good about themselves or at least not offend their sensibilities too much $((b)_{others}, (c)_{others})$. One should also attend to seeming interested and competent $((b)_{self}, (c)_{self})$. Faculty can feel emotionally burnt out if they find themselves assigned to committees whose work does not interest or challenge them, or if they find themselves on committees where their contributions are not valued. If they want to be good team players, they still feel the pressure to seem interested or at least act as if they care, all the while not feeling that way at all $((d)_{self})$. If on top of all this, their fellow committee members are high-maintenance folks who need extra emotion management, those professors who feel obliged to perform said management may be additionally burdened $((d)_{others})$.

Undeniably some faculty find academic service outlets that are emotionally very fulfilling for them. Here, as mentioned in the preamble to Sect. 3, I focused exclusively on the negative dimensions of emotion work related to service. See Sect. 5 for an exploration of the positive features to complement this discussion.

3.3   Emotional Dimensions of Academic Research

Bellas [5] focuses on how emotional work relates to academic research in a section titled "Emotional Labor and Research." In the following I mainly follow her examples, coding them according to the framework described earlier in Sect. 2.

Certain types of research, in particular social science research which involves issues of personal relevance to the researcher, challenge the researcher to remain neutral and objective, or at least conscious of their biases; this might be emotionally challenging $((e)_{self})$. This, together with the expectation of neutrality in the presentation of the final product of the work, might lead to emotional dissonance $((d)_{self})$. In quantitative or empirical research work, one might still find emotional labor lurking in the background. If research involves interview or experiment participants, then the researcher may need to manage the emotions of said participants $((b)_{others}-(d)_{others})$. In all work, researchers need to remain vigilant against wishful thinking and overly optimistic interpretations of experimental results and other data $((e)_{self})$. If the research involves other researchers, such as training assistants, then the researcher once again will need to manage emotions of others, and this time probably at regular intervals $((a)_{others}-(d)_{others})$.

Researchers need to attend to how the audience may view them during a conference talk, making sure to present a professional, competent, and yet interesting persona, though they may not really feel that way $((b)_{self}, (c)_{self})$. When submitting work for publication or proposals for grants or presentations, they should present themselves as competent and confident $((b)_{self}, (c)_{self})$. If said work is rejected in a dismissive or rude manner, they must pretend to be mature and generous and graciously take the given feedback $((d)_{self})$. This type of effort might also be needed to handle certain audience members during conference presentations.

4 Emotional Labor in Mathematics

When I became a mathematician, I knew that my job would not involve much manual/physical labor.[2] If I did think in terms of labor economics at all, I simply assumed that I would be part of today's knowledge economy, where I would be contributing to the production and dissemination of mathematical knowledge. But what was not obvious to my naive young self is that both research and education are a part of what is called the "service-providing sector"; see, for instance, the classification offered by the United States Bureau of Labor Statistics [46]. And this sector is today, possibly even more so than it had been during the writing of [21], the largest sector that demands emotional labor from its participants.

In Sects. 3.1–3.3, I explored specific ways in which the three main parts of an academic career (teaching, research, and service) might involve emotional labor. In this section I focus more explicitly on the context of the mathematical sciences. Note that in this section, too, my emphasis will be almost exclusively on the negative burden of these aspects of academic work.

Perhaps it is natural that teaching mathematics is intrinsically emotional [7, 50]. Students come into our classrooms with many emotions about mathematics. Some have math anxiety, some have self-doubt, some have a confidence level which may not serve them well in their next course. Students also bring along non-mathematical emotions, which contribute in all sorts of ways to how they engage with our content and pedagogy. If they just broke up with a partner, or if they have a sick relative, or

---

[2]This is not to claim that mathematicians are disembodied workers. When I had a minor shoulder injury and had visions of not being able to use the chalk board for several weeks, or during that stressful time when I lost my voice unexpectedly, I very clearly noted the physical aspects of my role in the academy.

if they are anxious about paying the next month's rent, their classroom participation as well as their learning will be impacted. Thus it makes sense that emotional labor, in particular others'-feelings oriented work along the dimensions of ((b)$_{others}$–(d)$_{others}$), makes up a significant portion of the work of a teaching mathematician.

Most mathematics professors can identify several familiar aspects of teaching described in Sect. 3.1, if not in their own experiences, then in some of their colleagues'. In particular many departments have that one professor whose office hours tend to turn into what seem like therapy sessions from outside the door. This professor, more often than not, belongs to a minoritized group in the discipline; it can be a woman or a person of color, for instance. It is clear that the amount of teaching-related emotional labor an individual professor takes on is not independent of the identity of that said professor. In particular, I have already mentioned in Sect. 3.1 that the distribution of emotional labor related to teaching does not seem to be gender-neutral; there is no reason to expect that the situation will be different in mathematics.[3]

The growing focus on pedagogies that emphasize student voice, student activity, and student agency in the mathematics classroom rather than a charismatic instructor's perfect presentation might be a reflection of the emotional labor of non-dominant groups in this area. Indeed professors from non-dominant groups may find that the "traditional ways of being a mathematics professor" do not work for them. That is, the caricature of a genius mathematician staring at the chalk board or scanning the sea of nameless faces while delivering a flawless lecture may not be the ideal way for all professors to connect with and teach all students. Thus today's mathematics instructors, especially those from non-dominant groups, might tend toward teaching pedagogies that involve more others'-feelings focused emotional work, possibly thus lightening the self-directed emotional work load and the emotional dissonance that might accompany that kind of teaching. In [40], Steurer describes how inquiry-based learning has resonated with her and allowed her to more naturally handle the emotion work of teaching. Conceiving of teaching mathematics as radical care [36] might be another way to reconcile these dissonances.

Academic service of mathematicians is in many ways similar to that of other academics. Though some disciplines may be more open to mentoring and outreach activities than mathematics, and others might more naturally lead to service to the community in other ways (such as an engineering faculty member serving as a pro-bono consultant to a local water treatment facility), I believe that the features of emotion work described in Sect. 3.2 capture many mathematicians' service experiences. Of course some may not have such uniformly negative experiences.

---

[3]I am by no means suggesting that this is a desirable situation; nor am I asserting that this is a choice made by individual instructors. It is often the case that certain faculty find themselves in these situations. It is my belief that nobody should be forced to do more emotional labor than they are willing to do. Unfortunately, faculty from minoritized groups often face the dilemma of either doing the extra emotional work and not being respected for it or rejecting doing the extra emotional work and then suffering the consequences of that decision.

There are many ways to serve, and some of these can be extremely fulfilling. For example, Khadjavi and I worked hard for years to put together a collection of resources for mathematics instructors who want to incorporate social justice issues into their courses [23, 24]. This process was emotionally draining and yet also very satisfying. I also find journal editing work to be completely exhausting but also very fulfilling. I will come back to service in Sect. 5 and there will address this point explicitly.

Thinking of academic research in mathematics, it is clear that mathematicians occasionally engage in research that is personally relevant to them. See, for instance, [26], where Kolba describes research she undertook to better understand twin pregnancies, and [6], where Berger describes her research related to Down's Syndrome. Similarly the work I did with Glass on school districting (see, for instance, [15]) also came out of personal investment in the topic. Nonetheless, I believe I need to move beyond the emotional dimensions already described in Sect. 3.3 to fully engage with the emotional dimensions of mathematical research.

To that end, I will refer to Weidman's list of the four emotional challenges of a mathematical life [47]:

"First of all, the mathematician must be capable of total involvement in a specific problem." That is, mathematics research often demands full focus for extended periods of time, and this is not only mentally exhausting but also emotionally draining. One might feel that one needs to withdraw from other interests, or else one is not doing enough. Each of the self-management dimensions $((a)_{self}–(e)_{self})$ comes into play here.

"Second, the mathematician must risk frustration. Most of the time, in fact, he finds himself, after weeks or months of ceaseless searching, with exactly nothing: no results, no ideas, no energy." A lot of mathematics research work leads to no results of significance. Add to this the challenges of getting published once one does have a significant result, which, for the not-yet-thick-skinned, can get especially disorienting and discouraging. Similarly these are self-management focused $((b)_{self}–(e)_{self})$.

"Next, even the most successful mathematician suffers from lack of appreciation." The mathematics community proudly celebrates its geniuses, but celebrity and genius are fickle [22]. After all, what more can you do once you win a Fields medal? Anything after that will be a let-down. Even those who feel appreciated by their mathematical colleagues may suffer from a dearth of appreciation from family and friends, and the world outside the mathematical one might be totally immune to mathematical glory. This might lead to serious emotional strain $((e)_{self})$.

"Finally, the mathematician must face the fact that he will almost certainly be dissatisfied with himself." Somewhat a corollary of the above, this means that mathematics is huge and the contribution of each individual mathematician is just a small speck. Whatever one does will be small change when compared to some of the giants. Once again this type of dissatisfaction might lead to serious emotional strain $((e)_{self})$.

Weidman's four challenges all require mathematicians to manage their own emotions, corresponding to the dimensions $((a)_{self}–(e)_{self})$. The mathematician who cannot handle all of them at least halfway successfully at least some of the time is bound to be miserable. One might reject some of these challenges as myths and try to disentangle oneself from their hold (see [20] for a call for this kind of a rejection), but that too takes emotional work as these ideas and ideals are quite solidly built into the culture of the discipline.

So far I have explored some specific ways academic mathematicians might engage in emotional labor. Are there any other kinds of emotional labor mathematicians might be called to do? And just who gets to hear that call? What are the consequences of hearing that call? I focus on these questions in the next section.

5 Mathematical Communities, Mentoring Structures, and EDGE

The idea that emotional labor has various human costs is not new; neither is the idea that it is not uniformly a negative for the particular laborer engaged in it, see, for instance, [48]. In this section I explore some of the positive possibilities related to emotional labor in the context of an academic mathematical life. In particular I reflect upon mathematical communities, mentoring structures, and EDGE.

Since 2008, the American Mathematical Society has supported Mathematics Research Communities (MRC). The MRC is "a professional development program offering early-career mathematicians a rich array of opportunities to develop collaboration skills, build a network focused in an active research domain, and receive mentoring from leaders in that area" [2]. In the last few years, the Association for Women in Mathematics has led or supported research networking conferences for women in various fields; see [3] for a list of research networks supported in various ways by the AWM. These programs, and others like them, are all spearheaded by mathematicians who feel called to do the work to create networks, connect people, mentor young mathematicians, and make our community a more welcoming and supportive place for more people.

The Mathematical Association of America has two programs for mentoring junior mathematics faculty. The first one of these, Project NExT (New Experiences in Teaching), "is a professional development program for new or recent Ph.D.s in the mathematical sciences" addressing "all aspects of an academic career: improving the teaching and learning of mathematics, engaging in research and scholarship, finding exciting and interesting service opportunities, and participating in professional activities. It also provides the participants with a network of peers and mentors as they assume these responsibilities" [31]. The MAA Mentoring Network is another mentoring program "aimed at connecting early career mathematicians with experienced mentors working in mathematics" [30].

There are many other mentoring structures built around academic mathematics. One might count among these:

1. The e-Mentoring Network, hosted by AMS blogs, available at https://blogs.ams.org/mathmentoringnetwork/,
2. The Infinite Possibilities Conference, "a national conference designed to promote, educate, encourage and support minority women interested in mathematics and statistics" [9],
3. NSF Mathematics Institutes' Modern Math Workshop at SACNAS, "a pre-conference workshop held at the SACNAS National Conference, intended to encourage undergraduates, graduate students and recent PhDs from underrepresented minority groups to pursue careers in the mathematical sciences and build research and mentoring networks" [43],
4. The Field of Dreams Conference, organized by The National Alliance for Doctoral Studies in the Mathematical Sciences, is a series of annual conferences that "introduces potential graduate students to graduate programs in the mathematical sciences at Alliance schools as well as professional opportunities in these fields. Scholars spend time with faculty mentors from the Alliance schools, get advice on graduate school applications, and attend seminars on graduate school preparation and expectations as well as career seminars" [42].

And of course one cannot forget EDGE. "The EDGE Program is administered by the Sylvia Bozeman and Rhonda Hughes EDGE Foundation with the goal of strengthening the ability of women students to successfully complete PhD programs in the mathematical sciences and place more women in visible leadership roles in the mathematics community. Along with the summer session, EDGE supports an annual conference, travel for research collaborations, travel to present research and other open-ended mentoring activities" [12].

The work involved in each of these programs is varied, but there is a significant emotional labor component. The main job is to connect people to one another, and though a lot of the emotion work is distributed over a large number of people, the main program organizers do a large chunk of it. There is much emotion work that involves the management of the emotions of others; in particular many of the junior mathematicians participating might be feeling insecure and lost or at least mildly confused. The emotional labor involved is mainly about making sure these participants feel a sense of belonging, and a sense of confidence and realistic optimism about their future in academic mathematics.

Digging deeper, one can see the resonances with the types of emotional labor described in Sects. 3.1–3.2. In particular the types of work involve teaching and service. However, people involved do the work willingly. They basically self-select into these roles. This is perhaps one of the main reasons why the emotional labor involved, though still highly burdensome in any objective sense of the word, does also help them feel nourished and fulfilled.

However there are two other reasons I believe.

First the people who put their time and energy into these programs feel called to do this type of work because they believe ideologically and philosophically that it is the right thing to do. Their political and ethical framing of the world puts them in the position to value this kind of work, and this in turn makes the work feel more endurable, more meaningful, and even more joyful.[4] "Meaningful work" is a catchy phrase; see [11] for a working framework for it that revolves around three themes (sense of self, the work itself, and sense of balance), and see [37] for more on the benefits of meaningful work for the laborer. But even leaving related scholarship aside, it is easy to understand how meaningful work can transform strenuous emotional labor into pleasurable and desirable labor. Thinking of the nine dimensions proposed in Sect. 2, one can see that the dimensions of emotional labor activated mainly involved are $(a)_{self}$–$(c)_{self}$ and $(a)_{other}$–$(d)_{other}$. There is little self-deception or misrepresentation of feelings, and there is more or less no emotional dissonance. So perhaps the individuals are exhausted at the end of the day, but they sleep well. This resonates with the work of Morris and Feldman in [32], who found that emotional dissonance was the main component of emotional labor that led to job dissatisfaction; see the relevant quote in Sect. 2.

Secondly and perhaps relatedly, there is often a shared identity component to the decision to dedicate time and energy to a program of this kind. This makes the work meaningful and the emotional dissonance minimal, yes, and in all these ways, this reason may seem similar to the first. But what makes this different is how it interacts with the others'-feelings focused labor dimensions $(a)_{other}$–$(d)_{other}$. The shared identity makes the emotional labor of managing the designated others' emotions a lot easier, as the individual has a better understanding of said emotions of those designated others. This of course does not mean that white women necessarily make the best mentors for white women, black men necessarily make the best mentors for black men, and gay Latinas necessarily make the best mentors for gay Latinas. But it is natural to expect that shared identity makes aspects of the involved emotional labor much easier.

6 Concluding Thoughts

The EDGE Program is an example of projects that demand emotional labor but that also contribute significantly to the well-being of both its participants and its leaders and mentors. In fact several participants become mentors and leaders themselves; see [8]. Though emotionally exhausting, these projects can continue because they fulfill several needs of those that work on them. Instead of expending their energy and emotional well-being on trying to run their departmental and campus committee

---

[4]This certainly applies to other service work, such as my editing work with Khadjavi on [23, 24] and K-12 outreach activities of various mathematicians; see, for instance, [49]. Similarly blogging and hosting other networking sites is a valuable service contribution, where most of the time people who do the needed work engage in it because of political and ethical goals. See in particular the e-Mentoring Network mentioned above as well as the inclusion/exclusion blog of the American Mathematical Society, available at https://blogs.ams.org/inclusionexclusion/.

meetings smoothly despite colleagues who expect them to make coffee or just smile and nod, and these mathematicians can put their efforts into projects that help them connect with each other, find meaning, and thrive in the academy.

On a more pragmatic level, there is perhaps more that can be said. One of the anonymous reviewers of this paper wrote that "the analysis [done here], which has intrinsic value in making visible labor that is generally hidden and unrewarded, has even more value-added as a framework to support these programs." I am sincerely energized by the possibility that this work may help these programs in one way or another. Indeed it may be of some value to the powers that be to comprehend that these programs benefit not only their participants but also their organizers, in ways that enrich their mathematical lives and naturally lighten some of the load of the emotional labor that comes with the territory.

There are several directions where this work could be further enhanced. For example, as one of the reviewers mentioned, there is significant variation across institutions of higher education in the USA in terms of how the three pillars of academic work (teaching, service, research) are to be valued and prioritized. This might add significant complications to the way we can talk about mathematicians' emotional labor. For instance, mathematics faculty in research-focused institutions might be concerned more about their research productivity and might have to perform more emotional work related to this aspect than others (in particular in the categories of $((b)_{self}-(e)_{self})$ as described at the end of Sect. 3.3), while mathematicians in more teaching- and service-oriented institutions might find that their work involves more emotion work along the dimensions of $((b)_{others}, (c)_{others})$ as described in Sects. 3.1–3.2.

As another possible direction, consider that research suggests that "foreign-born women faculty members' patterns of engagement in work activities contradict the gendered division of labor in academia" [29]. Foreign-born faculty make up a significant percentage of the overall American academia (in 2009, nonresident aliens made up 11.5% of the 11,599 new tenure-track faculty members at four-year institutions in the USA, while Asian Americans made 10.5%, African Americans made 0.5%, and Hispanics made 0.4% [25]). Emotional labor in the context of foreign-born faculty might look different; think, for instance, about what kinds of different challenges such faculty might face in the contexts of teaching and advising, and how the benefits described in Sect. 5 may apply to them in different ways. As many US-based mathematicians are also foreign-born (cf. [1, 28]), parts of the discussion in this article on emotional labor in mathematics might need some modifying accordingly.

Acknowledgements I would like to thank the editors of the volume for giving me the opportunity to contribute to this volume. I also very much appreciated their open-mindedness about the uncategorizability of my project and their input and support through the writing process. I should also like to thank the two anonymous reviewers, who offered insightful feedback on the original version of this manuscript, and Brian Katz, who generously read two drafts of the paper and offered his constructive criticism. I believe that their input has much improved this paper. The imperfections that remain are of course all my responsibility; this is the best I could do at this time.

References

1. American Immigration Council, *Foreign-born STEM Workers in the United States*, factsheet available at https://www.americanimmigrationcouncil.org/research/foreign-born-stem-workers-united-states, last accessed on December 20, 2018.
2. American Mathematical Society (AMS), *About the Mathematics Research Communities Program*, available at http://www.ams.org/programs/research-communities/mrc, last accessed on December 18, 2018.
3. Association for Women in Mathematics (AWM), *Research Networks*, available at https://awmadvance.org/research-networks/, last accessed on December 18, 2018.
4. Susan A. Basow and Nancy T. Silberg, "Student Evaluations of College Professors: Are Female and Male Professors Rated Differently?", *Journal of Educational Psychology*, Volume **79** Number 3 (1987), pages 308–314.
5. Marcia L. Bellas, "Emotional Labor in Academia: The Case of Professors", pages 96–110 in [38].
6. Heidi Berger, "The Upside of Down Syndrome: Math is My Superpower!", *Journal of Humanistic Mathematics*, Volume **8** Issue 2 (July 2018), pages 30–37. Available at https://scholarship.claremont.edu/jhm/vol8/iss2/6, last accessed on December 20, 2018.
7. Mark Boylan, "Emotionality and relationship in teaching mathematics: a praxis of embodiment and uncertainty", *For the Learning of Mathematics*, Volume **35** Number 1 (March 2015), pages 41–42.
8. Sarah Bryant and Alejandra Alvarado, "A Mathematician's Journey From Mentee to Mentor: Reflections on the EDGE Program," *Journal of Humanistic Mathematics*, Volume **7** Issue 2 (July 2017), pages 394–400. Available at: http://scholarship.claremont.edu/jhm/vol7/iss2/21, last accessed on December 18, 2018.
9. Building Diversity in Science (BDIS), *Infinite Possibilities Conference*, available at http://www.diversityinscience.org/infinite-possibilities-conference/, last accessed on June 10, 2019. Also see https://mathinstitutes.org/diversity/infinite-possibilities/.
10. Sarah Rose Cavanagh, *The Spark of Learning: Energizing the College Classroom with the Science of Emotion*, West Virginia University Press, Morgantown WV, 2016.
11. Neal Chalofsky, "An emerging construct for meaningful work", *Human Resource Development International*, Volume **6** Issue 1 (2003), pages 69–83.
12. Enhancing Diversity in Graduate Education (EDGE), *The EDGE Program: Enhancing Diversity in Graduate Education*, available at https://www.edgeforwomen.org, last accessed on December 18, 2018.
13. Amani El-Alayli, Ashley A. Hansen-Brown, and Michelle Ceynar, "Dancing Backwards in High Heels: Female Professors Experience More Work Demands and Special Favor Requests, Particularly from Academically Entitled Students", *Sex Roles*, Volume **79** (2018), pages 136–150.
14. Cynthia D. Fisher and Neal M. Ashkanasy, "The emerging role of emotions in work life: an introduction", *Journal of Organizational Behavior*, Volume **21** (2000), pages 123–129.
15. Julie Glass and Gizem Karaali, "Matching Kids to Schools: The School Choice Problem", in *Mathematics for Social Justice: Resources for the College Classroom*, edited by G. Karaali and L.S. Khadjavi (Classroom Resource Materials Volume 60, MAA Press: An Imprint of the American Mathematical Society, 2019), pages 155–170.
16. Alicia A. Grandey, James M. Diefendorff, and Deborah E. Rupp, editors, *Emotional Labor in the 21st Century: Diverse Perspectives on Emotion Regulation at Work*, Routledge, New York, 2013.
17. Cassandra M. Guarino and Victor M. H. Borden, "Faculty Service Loads and Gender: Are Women Taking Care of the Academic Family?", *Research in Higher Education*, Volume **58** (2017), pages 672–694.
18. Sandra Harding, "The Instability of the Analytical Categories of Feminist Theory", *Signs: Journal of Women in Culture and Society*, Volume **11** Number 4 (Summer 1986), pages 645–664.

19. Michael Hardt, "Affective Labor", *boundary 2*, Volume **26** Number 2 (Summer, 1999), pages 89–100.
20. Piper Harron, "On Contradiction", *Journal of Humanistic Mathematics*, Volume **8** Number 2 (July 2018), pages 199–207. Available at: http://scholarship.claremont.edu/jhm/vol8/iss2/22, last accessed on December 20, 2018.
21. Arlie Russell Hochschild, *The Managed Heart: Commercialization of Human Feeling*, University of California Press, Berkeley and Los Angeles, 1983/2003/2012.
22. Gizem Karaali, "On Genius, Prizes, and the Mathematical Celebrity Culture", *The Mathematical Intelligencer*, Volume **37** Issue 3 (2015), pages 61–65.
23. Gizem Karaali and Lily S. Khadjavi, editors, *Mathematics for Social Justice: Resources for the College Classroom*, Classroom Resource Materials Volume 60, MAA Press: An Imprint of the American Mathematical Society, 2019.
24. Gizem Karaali and Lily S. Khadjavi, editors, *Mathematics for Social Justice: Focusing on Quantitative Reasoning and Statistics*, an edited volume of articles and classroom resources, American Mathematical Society Press, forthcoming.
25. Dongbin Kim, Susan Twombly, Lisa Wolf-Wendel, "International faculty in American universities: Experiences of academic life, productivity, and career mobility", in *New Directions for Institutional Research: Refining the focus on faculty diversity in postsecondary institutions* edited by Y.J. Xu (Number 155 (2012), Jossey-Bass, San Francisco, CA) pages 27–46.
26. Tiffany N. Kolba, "A Math Research Project Inspired by Twin Motherhood", *Journal of Humanistic Mathematics*, Volume **8** Issue 2 (July 2018), pages 21–29. Available at https://scholarship.claremont.edu/jhm/vol8/iss2/5, last accessed on December 20, 2018.
27. Vicente M. Lechuga, "Emotional Management and Motivation: A Case Study of Underrepresented Faculty", *New Directions for Institutional Research*, Number **155** (2012), pages 85–98.
28. Zeng Lin, Richard Pearce, and Weirong Wang, "Imported talents: demographic characteristics, achievement and job satisfaction of foreign born full time faculty in four-year American colleges", *Higher Education*, Volume **57** Issue 6 (June 2009), pages 703–721.
29. Ketevan Mamiseishvili, "Foreign-born women faculty work roles and productivity at research universities in the United States", *Higher Education*, Volume **60** Issue 2 (August 2010), pages 139–156.
30. Mathematical Association of America (MAA), *MAA Mentoring Network*, available at https://www.maa.org/news/maa-mentoring-network, last accessed on December 18, 2018.
31. Mathematical Association of America (MAA), *Project NExT*, available at https://www.maa.org/programs-and-communities/professional-development/project-next, last accessed on December 18, 2018.
32. J. Andrew Morris and Daniel C Feldman, "The Dimensions, Antecedents, and Consequences of Emotional Labor", *The Academy of Management Review*, Volume **21** Number 4 (October 1996), pages 986–1010.
33. Gérard Näring, Peter Vlerick, and Bart Van de Ven, "Emotion work and emotional exhaustion in teachers: the job and individual perspective", *Educational Studies*, Volume **38** Number 1 (February 2012), pages 63–72.
34. Karen Pugliesi, "The Consequences of Emotional Labor: Effects on Work Stress, Job Satisfaction, and Well-Being", *Motivation and Emotion*, Volume **23** Issue 2 (June 1999), pages 125–154.
35. Susanne Schultz, translated by Frederick Peters &, "Dissolved Boundaries and "Affective Labor": On the Disappearance of Reproductive Labor and Feminist Critique in *Empire*", *Capitalism Nature Socialism*, Volume **17** Issue 1 (2006), pages 77–82.
36. Ksenja Simic-Muller, "Motherhood and Teaching: Radical Care", *Journal of Humanistic Mathematics*, Volume **8** Issue 2 (July 2018), pages 188–198. Available at https://scholarship.claremont.edu/jhm/vol8/iss2/21, last accessed on December 20, 2018.
37. Michael F. Steger, and Bryan J. Dik, "Work as meaning: Individual and organizational benefits of engaging in meaningful work", in *Oxford Library of Psychology–Oxford Handbook of Positive Psychology and Work* edited by P. A. Linley, S. Harrington, and N. Garcea (Oxford University Press, New York, 2010), pages 131–142.

38. Ronnie J. Steinberg and Deborah M. Figart, editors, *Emotional Labor in the Service Economy*, (Annals of the American Academy of Political and Social Science, Volume **561**), Sage Publications, Thousand Oaks, CA, 1999.

39. Ronnie J. Steinberg and Deborah M. Figart, "Emotional Labor Since The Managed Heart", pages 8–26 in [38].

40. Aliza Steurer, "Inquiry Based Learning: A Teaching and Parenting Opportunity", *Journal of Humanistic Mathematics*, Volume **8** Issue 2 (July 2018), pages 38–59. Available at https://scholarship.claremont.edu/jhm/vol8/iss2/7, last accessed on December 20, 2018.

41. Yvette Taylor and Kinneret Lahad, editors, *Feeling Academic in the Neoliberal University: Feminist Flights, Fights and Failures*, Palgrave Macmillan, Cham, Switzerland, 2018.

42. The National Alliance for Doctoral Studies in the Mathematical Sciences, *Field of Dreams Conferences*, https://mathalliance.org/field-of-dreams-conference/, last accessed on June 10, 2019.

43. The Statistical and Applied Mathematical Sciences Institute, *Modern Math Workshop 2018: October 10–11, 2018*, available at https://www.samsi.info/prog rams-and-activities/2017-2018-education-and-outreach-programs- and-workshops/modern-math-workshop-2018-october-10-11-2018/, last accessed on December 18, 2018.

44. Carolyn J. Thompson and Eric L. Dey, "Pushed to the Margins: Sources of Stress for African American College and University Faculty", *The Journal of Higher Education*, Volume **69** Issue 3 (1998), pages 324–345.

45. Caroline Sotello Viernes Turner, "Women of Color in Academe: Living with Multiple Marginality", *The Journal of Higher Education*, Volume **73** Number 1 (January/February 2002), pages 74–93.

46. United States Bureau of Labor Statistics, *Industries at a Glance: Service-Providing Industries*, available at https://www.bls.gov/iag/tgs/iag07.htm, last accessed on December 18, 2018.

47. Donald R. Weidman, "Emotional Perils of Mathematics", *Science*, New Series, Volume **149** Number 3688 (September 3, 1965), page 1048.

48. Amy S. Wharton, "The Psychosocial Consequences of Emotional Labor", pages 158–176 in [38].

49. Diana White and Lori Ziegelmeier, "Reflections on Math Students' Circles: Two Personal Stories from Colorado", *Journal of Humanistic Mathematics*, Volume **5** Issue 2 (July 2015), pages 110–120. Available at https://scholarship.claremont.edu/jhm/vol5/iss2/10, last accessed on December 20, 2018.

50. Meca Williams, Dionne Cross, Ji Hong, Lori Aultman, Jennifer Osbon, and Paul Schutz, "There Are No Emotions in Math': How Teachers Approach Emotions in the Classroom", *Teachers College Record*, Volume **110** Number 8 (2008), pages 1574–1610.

Part III
Mathematics Teaching

Experiencing Mathematics Abroad

Michelle Craddock Guinn and Bradford Schleben

Abstract Study abroad experiences have been recognized as providing students with valuable opportunities to work with individuals and groups different from themselves, to incorporate diverse viewpoints into their work, and to engage in meaningful experiences outside their culture. This article focuses on one of the mathematics courses in Belmont University's study abroad program that was designed to synthesize course content and authentic learning experiences in order to address the diverse set of student learning outcomes. While improved student engagement in cultural understanding and the promotion of intellectual diversity took on a central role in the course design and assessment, a secondary goal was an improved student perception of mathematics and its application. We examine the course in action by looking at three example assignments, followed by their connections to program experiences, and how these things coordinate to meet the student learning outcomes.

1 Introduction

Global learning opportunities that provide students with international perspectives have become a goal of many university programs, especially those that aim to provide students with opportunities to improve cultural understanding, and promote intellectual diversity. Although these global opportunities are becoming more common, it often remains a challenge to measure competencies and effectiveness in programs that focus on cross-cultural experiences.

Study abroad experiences have, however, been recognized as providing students with valuable opportunities to work with individuals and groups different from themselves, to incorporate diverse viewpoints into their work, and to engage in meaningful experiences outside their culture [4, 5]. In an effort to help educators

M. C. Guinn (✉) · B. Schleben
Belmont University, Nashville, TN, USA
e-mail: Michelle.Guinn@belmont.edu; brad.schleben@belmont.edu

© The Author(s) and the Association for Women in Mathematics 2019
S. D'Agostino et al. (eds.), *A Celebration of the EDGE Program's Impact on the Mathematics Community and Beyond*, Association for Women in Mathematics Series 18, https://doi.org/10.1007/978-3-030-19486-4_13

align study abroad programs with the Essential Learning Outcomes laid out by the Association of American Colleges & Universities [2], the Center for Capacity Building in Study Abroad—a joint project between National Association of Foreign Student Advisers (NAFSA) and the Association of Public and Land-grant Universities (APLU)—argues that meaningfully engaging students in global questions that demand integration of knowledge, skills, and personal and social responsibility is an essential outcome of study abroad programs [6].

The program discussed here is aimed at addressing these concerns by using mathematics as a universal language to better understand and connect people and social issues across different cultures. The program has been implemented in three study abroad trips in Europe and Australia. While improved student engagement in cultural understanding and the promotion of intellectual diversity took on a central role in design and assessment, a secondary goal was an improved student perception of mathematics and its application. This article focuses mainly on one of the mathematics courses in this program abroad developed to benefit student capacity for such engagement while promoting diverse cultural understanding and appreciation. As such, the course was designed to synthesize course content and authentic study abroad learning experiences in order to address the diverse set of student learning outcomes. We will begin with a course overview and the development of student learning outcomes. Next, we examine the course in action by looking at three example assignments, followed by their connections to program experiences, and how these things coordinate to meet the student learning outcomes. Lastly, we will examine qualitative data aimed at examining perceived growth in content and cultural understanding.

2   Course Overview

The authors teach courses at Belmont University, a liberal arts university in Nashville, Tennessee. Students at Belmont must complete a core liberal arts curriculum, including one mathematics course focusing on quantitative literacy and reasoning. Hence, the study abroad course we will examine was designed around the application of basic mathematics skills to the analysis and interpretation of real-world quantitative information in order to tackle problems both relevant to students in their lives and to the communities and cultures in which they were studying. We found that a Math for Social Justice course provided a natural framework for addressing both study abroad and mathematics learning goals.

The course was originally developed by the authors to satisfy the mathematics requirement in the core liberal arts curriculum at Belmont University and was partly inspired by conversations with faculty at other institutions, especially Dr. David T. Kung of St. Mary's College of Maryland. It ran multiple times over the course of four consecutive fall and spring semesters before being adapted to the study abroad format explored here. At the time of this publication, the course will have run three consecutive summer terms in Australia, once in Europe, and will have expanded to include a version in Scandinavia. This course looks to take active learning experiences beyond the institution and home region—highlighting issues of social and economic justice on local, national, and global levels. Further, students would see how mathematics can act as an analytical tool in understanding cultural interdependence, power, and privilege. More succinctly, the goal is to give students means and methods to quantify and interpret complex social issues affecting the people, land, and culture.

As mentioned, this particular Math for Social Justice course has run in the summer as part of an interdisciplinary study abroad program in Australia, during which students spent over three weeks near Brisbane, Townsville, Cairns, and Sydney, as part of a program designed to provide authentic and formative experiences with different communities and cultures. The location selection allows for different experiences in metropolitan cities and smaller coastal towns, with members of multiple Aboriginal groups, and at a Queensland University research station on North Stradbroke Island. The location diversity allows for encounters with a variety of environmental ecosystems, cultures, and peoples—each of which brings different opportunities to quantify issues of change and social justice.

2.1 Student Learning Outcomes and Preprogram Elements

Developing learning outcomes is an essential first step toward effective teaching. Thus, in designing a mathematics course to fit in the study abroad context, it was important to identify learning outcomes that would be effectively served by the course content while also fulfilling the expanded goals of a study abroad course. As appears in [6], NAFSA and the APLU have provided a broad framework in which students should be engaging during a study abroad experience. This framework for Essential Learning Outcomes (ELOs) centers on four areas:

- Personal and Social Responsibility;
- Intellectual and Practical Skills;
- Integrative Learning;
- Knowledge of Human Cultures and the Physical World.

Further, ELOs should be met in an environment which provides perspective and opportunities beyond what is available to students at their home campus. We desired for these changed perspectives to be a result of synthesis between content and cultural learning, as opposed to being inspired solely by non-discipline related experiences abroad. In addition to the above framework, special consideration was given to engage students in deep learning outcomes, as outlined by the National Survey of Student Engagement (NSSE) [10]. These outcomes focus on Higher Order, Integrative, and Reflective experiences which inspired the learning outcome language used in this course. Its inclusion in the design is evidenced in the three activities referenced in this paper: A "Press Release" activity, a "Discussion Lead" assignment, and "Program Journal" component, each of which appears in the Appendix.

In order to lay the groundwork for the above goals, this course would need to establish an effective entry point for a deeper appreciation of mathematical content while also meaningfully engaging students in the outcomes laid out in [6]. We used internal learning objectives, those outlined by other institutions, such as UMASS Lowell [13], and the AAC&U's VALUE rubric for quantitative literacy [1] as guides to help shape the content objectives, particularly focusing on developing curriculum that would be adjustable for different program locations going forward. Considerable attention was given to making the Math for Social Justice course a transferable model for connecting authentic cross-cultural experiences and quantitative reasoning content, while opening the door for a deepened understanding of the complex issues of whichever culture, community, or country students encountered.

Each course operated as a hybrid classroom, containing both online and in-class components. The students had weekly online learning modules (five pre-trip, two during the abroad portion, and one post-trip), accompanied by in-person class meetings multiple times before the trip and interspersed group meetings during the program. This allowed students the opportunity to individually practice and demonstrate skills knowledge while also engaging in progressively more challenging problems, projects, and discussion in a collaborative setting. It also helped ensure sufficient content mastery and exposure to mathematical applications to cultural and social issues prior to being abroad. These elements provided reference points for integrative exposure and encounters with cultures and communities unfamiliar to the students.

Finally, a well-known difficulty study abroad programs face is in combatting the student perception that they are studying but "on vacation" abroad [3]. As many study abroad facilitators and faculty have noted, a disconnect between content and cultural encounters may increase the likeliness of this attitude emerging during the program. In addition to emphasizing the importance of the synthesis of course content and cultural experiences, this was combatted with the inclusion of multiple service-learning components, which have been shown to play an important role in connective learning [9, 11]. The preprogram assignments, discussion, and project preparation sought to help bridge any perceived gap between curricular activities and cultural experiences, priming students to make connections with built-in opportunity for reflection throughout the program.

3   Course in Action

While much of the skills practice was completed individually prior to the program, larger projects and assignments were designed to function as collaborative learning experiences and tasks that could be refined and submitted during the program. Some of these activities utilized small groups in which students submitted a common work, while others had multiple components that were submitted in stages in order to allow for feedback, revision, and discussion. We now look at several of the activities and how they aligned with goals of a quantitative reasoning course, effectively met the underlying ELOs for the study abroad program, and connected meaningfully to site visits and cultural experiences during the program. We will first detail the assignments, each of which is presented in the Appendix. This is by no means an exhaustive list, nor are learning outcome adaptations addressed in only one activity. as they appear in the previous section.

3.1 Activities

The first activity is a "Press Release" assignment that could be modified to address how a variety of social issues—such as poverty rates, incarceration rates, educational performance—are quantified. It could also be scaled up in length and difficulty via an added debate component or a more in-depth analysis. Students were divided into groups, with each group selecting a country comparable to where students would be traveling to abroad (Australia, in this case). The groups were then tasked with collecting data regarding that country or region's energy production coming from renewable sources such as wind or solar over the last 10 years. The groups produced two press releases: one as though they were working for an environmental group that advocates for more funding and support for renewable energy and one in which they were public relations representatives for the country's department of energy. These two statements were to present a case either arguing for increased renewable energy sources or an argument supporting the country's current energy policies. Part of their submission involved a collection of footnotes detailing how statistical measures presented were calculated. Groups were encouraged to be creative with their quantitative arguments, including providing meaningful comparisons to the countries of other groups, while also outlining potential consequences of each side of the argument.

In terms of addressing mathematical content objectives, this activity required students to collect data, form basic statistical measures, and establish evidence to support the positions assigned. In addition, the groups had to take on multiple viewpoints of an important contemporary global issue and attempt to frame the data to serve each side of the debate. Students needed to demonstrate an understanding of the positions presented and the arguments defending each position, as well as be able to decipher the origin of statistics used to support claims. They were assessed on the appropriateness and clarity of data presented, as well as on the accuracy of an accompanied explanation of each statistic presented. This activity also served to assess the group's ability to communicate mathematics effectively. Finally, it served as an opportunity for feedback from both instructor and classmates regarding how effective students were at communicating quantitative information. Polished presentations highlighted the importance that framing plays in communicating quantitative information.

Next was a "Discussion Lead" assignment that required students to research a social, environmental, or ecological issue relevant to the program locations, provide articles and references citing quantitative measures regarding the issue, and then develop the framework for a group discussion which they led near the end of the program. In preparation for this—during each week of the pre-trip portion of the course—students took turns submitting articles or studies containing quantitative data on issues concerning inequality or injustice relevant to Australia and its citizens. Students began developing the inclination to skeptically view how data and mathematics are used in application. They explained concepts like percentage change, correlation, and p-values in the context of social, economic, and environmental issues before and after the mathematical content was rigorously covered in the course. They then engaged in instructor-facilitated discussion about different methods and conclusions of the various articles. For example, in the first week students read articles concerning wealth inequality in Australia, culminating with them learning about Gini coefficients and Lorenz curves as applied to the issue. This practice allowed students to develop their skills of evaluating claims and questioning quantitative evidence prior to the closing "Discussion Lead" assignment. The discussions were required to address the quantitative information in the references, while also challenging the audience to consider how their experiences abroad connected to the topic chosen.

In addition to peer feedback, students were provided a rubric detailing how they would be assessed on four components of their facilitation of discussion: "Content Relevance and Clarity," "Motivation and Connections to Experiences," "Engagement of Classmates," and "Impact Analysis" (including future questions to consider). As "Engagement" was an assessed component, students were encouraged to utilize different active learning strategies they had been exposed to throughout the course (e.g., "Think-Pair-Share"; "One-minute Papers") in order to deepen the discussion. Topics that students selected included: disparities in rates of domestic violence, mental health disorders, and incarceration rates across different demographics in Australia; issues of legal discrimination and land right battles of different indigenous groups; economic and environmental factors involved in bans on plastic bags; and measuring Great Barrier Reef restoration efforts. Each of the discussions presented different challenges and viewpoints in terms of what was being measured, methods used in collecting information, analyses and conclusions drawn from data collected, and the impact on the people and culture. This goal of this assignment was to prime students for connecting the quantifying of social issues with their own experiences abroad, along with highlighting the natural inclination for students to identify and articulate parallels to issues present in their own culture. Further, students were asked to apply content knowledge to determine the degree in which they considered the data presented to be a meaningful representation of reality.

Finally, given that the program aimed to aid students in expanding their understanding of cultures different from their own, the course had a required journaling assignment designed to provide an opportunity for reflection on growth and increased self-awareness. The hope of this activity was to have students identify and describe changes in their own learning, specifically focusing on the contribution of contextual components linked to their cross-cultural experiences. There were structured, common journal prompts assigned as part of the abroad portion of the program, as well as other required entries allowed to be on a topic of the student's choosing. Then, a final prompt, revealed at the end of the program, required students to reflect on the role that quantitative representations played in applying moral reasoning to either social or ethical issues they encountered on the program, while noting any challenges or limitations they encountered in doing so.

Again, the goal was to ensure that the journals went beyond a simple means of recording experiences and instead involve meaningful reflection and assessment of one's own self-awareness, while also providing ample opportunity for autonomy in selecting themes on which connect their own growth. Students completed at least six journal entries throughout the program, each one-to-two pages in length. By this stage of the program, students who had put great effort into quantifying and testing measures of inequality mathematically and critically were tasked with reflecting on their experiences with the people and cultures they had been studying. Further, students were faced with the reality that, although mathematics is essential in measuring and understanding complex issues including cultural differences, it provides only a portion of the insight necessary to understand these differences in full.

3.2 Connecting Activities, Experiences, and ELOs

A challenging aspect of course design in an abroad program is finding and arranging experiences that contribute to student learning and are different than what students may experience on their home campus. A great deal of the program's logistical success in this area is due to Dr. Alison Parker of Belmont University, and the program's organizing partner of Arcadia University. We now look at a selection of the experiences students had in Australia and some ways in which they connected to the Essential Learning Outcomes stated in Sect. 2.1 and the activities outlined in Sect. 3.1.

The "Personal and Social Responsibility" learning outcome encompasses the importance of civic and intercultural knowledge and engagement. As such, it was important for students to complement their quantitative study of complex issues affecting both local and global communities with an ethical reasoning process. For example, in the "Press Release" activity, students are asked to look at different viewpoints of an important issue, the use of renewable energy, and then gather and express quantitative evidence in support of an argument concerning that issue. This emphasis on civic knowledge was present throughout the various "Discussion Lead" topics as well. As a result of differing viewpoints surfacing in discussion, these assignments also required students to interpret and propose hypotheses that indicated a deep understanding of the issues, while engaging in ethical reasoning in determining how to address them.

This outcome was further supported by several abroad experiences. One example was when students met with the CEO and COO of a car-sharing company that utilized statistical analyses to optimize hub locations, reduce emissions, and relieve parking space scarcity in Sydney, Australia. The company not only looked to illuminate the transportation needs of different communities within the city, but it also aimed to help address the real-world challenge of reducing carbon emissions while providing means of affordable transportation. The students also visited an environmental project development firm, Green Collar, that detailed how it applies data analysis in its role as an environmental market investor and sustainable land developer. Unlike the USA, Australia utilizes a carbon credit system and this company works with land owners to make environmentally beneficial farming and agricultural practices more affordable and profitable. While students were exposed to data collection and the development and application of some basic mathematical models, they also witnessed how environmentalism can benefit from the application of some of these mathematical tools. In both of these instances, the experiences not only promoted civic engagement in the issues of sustainability and social responsibility, but they also highlighted the importance of communicating quantitative information effectively, as the companies were often required to convince potential clients, collaborators, and even government officials with compelling quantitative arguments similar to those the students made.

Lastly, within the "Discussion Lead" assignment, several of the topics covered revolved around the challenges facing Australia's unique ecosystems as a result of both humanity's influence and global climate change. With the nonprofit organization Reef Ecologic, students participated in a Citizen Science expedition during which they gathered and analyzed data relevant to an environmental issue of local and national importance—the health of the Great Barrier Reef. They then participated in a case study detailing various proposals aimed at measuring and addressing how pollution and land development affect local ecosystems comprised mainly of wildlife endemic to the area. These assignments and experiences all required students to understand the contexts of critical issues involving sustainability and the environment, a matter of great importance to the land, people, and culture of Australia. Further, students practiced ethical reasoning while witnessing the application of the skills they had developed by the various organizations.

Given the course's focus on quantitative literacy, the activities and experiences almost all had components that addressed the "Intellectual and Practical Skills" learning outcome. Both the "Press Release" and "Discussion Lead" assignments looked to assess the students' quantitative inquiry and analysis skills. Likewise, experiences with the car-sharing company and Green Collar involved the application of these skills in a real-world setting. While these basic quantitative literacy outcomes were important, there were also activities that allowed students to engage in creative and critical thinking processes while practicing oral and written communication, most clearly demonstrated with the "Program Journal" assignment.

In terms of experiences serving these outcomes, the aforementioned Citizen Science expedition provided an authentic opportunity for students to take concepts they had learned examining other people's data and apply them to their own project. Being involved in the entire process, from logistical planning to data collection to analysis, forced students to draw appropriate conclusions based on analysis of data while making judgments about the limitations of data as well. Further, the expedition involved team problem-solving, as groups were tasked with deriving potential solutions and interventions in a post-activity discussion.

Finally, students had the opportunity to engage in meaningful conversation with people of different cultures and communities while abroad. One such opportunity occurred within conversations students had with a member of the Quandamooka Aboriginal people, Matt Burns. As a guest lecturer, Matt delivered an active-learning lecture about his people's culture and the ways in which they had previously and continue to face discrimination and marginalization. He not only provided students with insight and understanding unique to his life experiences, but he also challenged students to describe what they knew about the inequality and injustices facing Aboriginal people. Students who had studied and practiced identifying and quantifying indicators of marginalization of Aboriginal people in the pre-trip component of the "Discussion Lead" activity were now being asked to relay their understanding to someone who had directly encountered and endured them.

Many of the experiences we have described focused on applying skills and knowledge to complex problems. The broad outcomes of "Integrative Learning" and "Knowledge of Human Culture and the Physical and Natural World" were at the heart of each scheduled experience during the program. As these experiences were often pre-empted by relevant pre-trip readings as part of the "Discussion Lead" activity or with faculty-led group discussion, students were repeatedly challenged to take their previous knowledge—regarding both content and cultural—and apply it in new settings. Similarly, the "Program Journal" assignment was meant to be a medium for acknowledgement of the synthesis of current and past learning, as well as an opportunity for students to articulate the connections between content, context, and their own self-awareness. For example, one student responded to the final journal prompt by expressing appreciation for the application of quantitative reasoning, but also frustration and resignation in the limitations of that pursuit because "data will never tell us enough of the story."

Many of the examples in which students not only integrated their knowledge, but also engaged in an activity meant to broaden their cultural understanding, centered around students' experiences connecting with members of different Aboriginal communities. Students had been exposed to a variety of issues affecting Aboriginal people via the "Discussion Lead" activity pre-trip. They had responded to articles examining topics ranging from the prevalence of suicide and depression in certain Aboriginal communities to struggles of these traditional land owners attempting to reclaim their land rights. In these and other contexts, students had experience attempting to quantify the effects of discrimination against these people, but we wanted them to achieve a deeper understanding of these issues of injustice and inequality during their time in Australia.

At different points of the trip, members of both the Quandamooka and Nyawaygi people were willing to discuss with students the challenges their people have faced such as loss of land, liberty, and life. One of the most meaningful aspects of the trip for multiple students, as relayed in the "Program Journal" assignment, was learning from people affected by the inequalities and injustices that students had studied pre-trip. Students were asked to connect the quantitative information on which they have honed their mathematical skills with the experiences and understanding stemming from these conversations. These reflections and subsequent discussions were the least mathematical components of the course, as students grappled with viewpoint of the victims of the injustice and inequality they have been trying to quantify, along with the reality that advantages and power possessed by one culture may often create disadvantages and despair for others.

This was an important component of the student's learning experience, because although they can read and research and attempt to quantify the injustice and inequality affecting people around the world, there is, hopefully, unique under-standing gained in the opportunity to hear from the people affected. Though some students described experiencing a feeling of unease as they grappled with these issues during post-activity discussions, these experiences aimed to serve deeper goals in the addressing of one's own lack of understanding. Whether it was reflection on complex questions about other cultures and diverse groups or cogently connecting life experiences and academic knowledge, the journaling provided another forum for students to articulate their own formation of the synthesis of mathematical content and their experiences.

4   Student Perceptions on Learning

There has been recent research connecting study abroad programs to students' cross-cultural awareness [8] and openness to diversity [7, 12] that provided hope that this program could achieve similar ends. In an effort to measure how successful the program was in achieving its desired outcomes, students were provided with an anonymous, voluntary post-program survey.[1] Given the nature of short-term study abroad programs, the level of student interest in mathematical courses, and the relative recency of the program's development, respondent data is limited. In fact, the data set size ($n < 20$) is still too small to make meaningful conclusions based on quantitative results. However, we feel the nature of the questions and the limited data attained are still of interest in beginning to understanding student perception and also illustrate some potentially useful information in course assessment.

First, students were asked to respond with the degree to which they agreed the program affected them in a number of potential ways. Responses were given on a five-point scale: strongly disagree, disagree, neutral, agree, and strongly agree. The statements below had greater than 90% of students agreeing or strongly agreeing with the statement. Students indicated that they felt the program achieved a variety of goals, including the following:

---

[1]Belmont IRB Exemption, Protocol ID 690.

- Contributed to them developing a more sophisticated way of looking at the world.
- Increased the likelihood they would seek out opportunities to engage with cultures other than their own.
- Was likely to have a lasting impact on their worldview.
- Increased their appreciation of mathematics.
- Helped them to better understand cultural/global issues with increasing complexity.
- Influenced their understanding of cultural values and biases outside of themselves.

The statement that received the highest level of disagreement was "The course caused you to refine your social or political views." This may indicate that, despite wrestling with complex social issues, the problems and solutions examined and discussed were not tied closely to one's own political identity.

Qualitative data were also compiled from short-response questions. Students responded to the following prompts:

- In what ways do you think taking the mathematics class abroad deepened your understanding and appreciation for different cultural and global issues?
- In what ways did the abroad experience affect your interest or appreciation of the mathematical content studied?
- The following is an opportunity to further comment on your experience taking a mathematics course as part of your study abroad. Of particular interest is any effect it had on your appreciation/awareness of cross-cultural issues.

Responses illustrated an appreciation for the intentional connections between course content and cross-cultural experiences and an appreciation for growth and change in students' cultural perspectives. For example, one response focused on achieving deeper understanding in these areas is included here.

"I began to intentionally learn more about other cultures and issues surrounding an area very different from what I was used to. Hearing directly from people in other cultures helped me to gain a deeper understanding of global issues. Another thing that really deepened my understanding and appreciation of global issues was participating in hands on experiences while in Australia."

Along these lines, another student commented, "If it weren't for this class, my perspective on culture, global issues and politics would not have changed for the greater good." These responses demonstrate some degree of growth, while also a level of self-awareness in current and previous cultural perspectives.

Similarly, students noted the importance of connecting cultural activities with course content and the effect doing so had on their growth and learning, with one student writing,

"Being able to explore another culture through math allowed me to view the Australian culture in a way that was interesting to me, and in turn it allowed me to develop a deeper appreciation for the culture there. Had I taken the class in Nashville, I don't think it would have had as much of an impact because I would not have been able to actually see the interactions in that culture."

Other students made similar observations, with one noting,

"Being in a different country and learning about mathematical social justice puts a perspective on a person that no other circumstance could...Learning about the differences and similarities between the two countries helps one realize how far off the stereotypes are. If I wasn't majoring in music, I would be majoring in Math for social justice (if that was a major)!"

Finally, multiple students mentioned a growth in appreciation for mathematics and its role in understanding complex issues, as illustrated by one responder's comments, "Studying abroad with this specific content has helped shape and really educate me on the cross-cultural issues and how to respect and appreciate other cultures." Another student added, "I like math. This trip made me like it more," while yet another indicated she or he would be adding a math minor.

4.1   Future Considerations

Study abroad courses often face some inherent difficulties and limitations, such as limited face-to-face class time or opportunity for active learning experiences. Hence, reflection on our mathematics course abroad has revealed areas of limitation as well as opportunities for improvement. While the course was designed to be adaptable to different programs and locations, it is worth highlighting a few of these issues here.

Looking first at limitations in teaching mathematics abroad, many of the course and program elements that make the synthesis of content and experience effective involve substantial cooperation with groups at the destination location. A clear understanding of course elements and open collaboration with universities and organizations abroad can help optimize the meaningfulness and appropriateness of cultural experiences abroad. However, this can not only be difficult in organizing meaningful site visits or guest speakers for a course, but it may also inflate budgets.

There are also a number of opportunities for improving the mathematics abroad course we have developed moving forward. One opportunity touched on by student comments is providing ample time for connecting discussion and written components of the course. In particular, students felt they could benefit from more scheduled opportunities for discussion after students engaged in reflection and journaling. Previous program design provided students with first an opportunity for discussion and then written reflection. The aim was to give students time to process discussion and experiences before reflecting in their journals, but they may also benefit from a follow-up discussion in which they can compare aspects of their reflection with their classmates. This may prove difficult, however, due to time considerations during abroad programs.

Other potential limitations stem from the class size and issues arising with either small or large enrollments. A larger course or program enrollment can potentially benefit more students, but many activities become less engaging or logistically difficult to implement effectively with a larger number of students. This can significantly impact the effectiveness of activities requiring students to individually communicate mathematical reasoning, as time may be a limiting factor on discussion. On the other side is low-enrollment concern and a matter of driving student interest in mathematics courses abroad. Mathematics courses can often suffer from disinterest or fear, and the added abroad component can be a daunting and determining factor in students avoiding the course.

Overall, the effectiveness of this course, and others like it, hinges on synthesizing content with application to complex problems connected to site visits and encounters while abroad. An important aspect of this is the surrounding discussion and reflection which encourages students to reassess their perspectives and worldview through the dual lenses of both mathematical content and cross-cultural exposure. Quantitative literacy and reasoning are necessary tools for studying and quantifying the elements of power, opportunity, privilege, and interdependence with regard to cross-cultural learning. The perceived value of these mathematics abroad experiences suggests these elements of content exposure and discussion, project application, cultural engagement, and reflective writing provide profound and lasting growth for students in study abroad programs.

Appendix

"Press Release" Activity

Below is the exact version of the assignment that students see, without the example data tables and press release.

Press Release Activity

On the following pages you will find some example data for the total and per capita wind power capacity in megawatts that the select could produce at the end of each year in 2016 and an example press release from the Department of Energy of the United States. Your first task is to collect similar data from the last ten years for both Australia and another country from the provided list. Each group will select a distinct country. Once you have done so, you will be responsible for crafting two press releases.

Side 1

Suppose you worked for the PR department at the Department of Energy for one of the given countries when these data were released (I am requiring that one group will be United States and another will be Australia). Write a press release (≥ 2 paragraphs) that puts these data in the best possible light, supporting your country's energy policy.

Side 2

Suppose you worked for an environmental group that advocates for more support for renewable energy sources like wind. Write a press release that puts these data in the worst possible light, supporting you argument that your country is falling further behind the rest of the world.

Your goal is to establish arguments with evidence drawn from the data. Both of the above releases may involve comparative data as well. Included with your statements should be validation for each statistic you cite. Fore example, if I describe a certain percent change in capacity from 2014–2015, there should be a footnote providing the calculation made to get that percent change value.

Be sure to avoid some of the pratfalls that we have identified in the use of percentages, rates, and ratios (such as averaging percent increase in capacity over a number of years).

"Discussion Lead" Activity

Discussion Lead Assignment

I am asking that you prepare to lead an engaging and informative discussion appropriate for our Math for Social Justice course while on our trip. The scope of the discussion topic is up to you, but I encourage you to use the examples we have done already as a guide. There will be components of this assignment due both before and during our trip, as laid out below:

Pre-trip

Once you have selected your topic, you must provide at least three (3) reputable sources for fellow students to serve as references. These sources should contain quantitative information regarding your topic, giving your fellow students statistics to consider as they prepare to participate in fruitful discussion.

Your source must be uploaded by July 1st to the Blackboard discussion board forum "Discussion Leads". This will allow your fellow students the time to download copies of the articles before the trip, as well as familiarize themselves with the quantitative information on the topic.

While Abroad

I expect this to be approximately a thirty (30) minute discussion (NOT a lecture). The rubric is attached. The main criteria are that this is a quantitative-driven discussion of a social justice issue and the effectiveness at engaging of your fellow classmates in challenging discussion. Good discussions will connect course content with experiences abroad and challenge the participants to consider potential responses and methods for social change, if appropriate.

Expectations for Participation

- Prepare to engage in meaningful discussion centered around quantitative literacy and social justice while in Australia (I may provide further articles).
- Be engaged in group activities, experiences, panels, and lectures to the best of your ability.
- Focus on making connections between course content and cultural experiences abroad.

"Program Journal" Activity

The "Program Journal" activity was completed during the abroad component of the program, allowing students to reflect on their experiences in country. The final journal entry prompt, as mentioned before, was not given to students until the end of the program. It asked students to reflect on the role that quantitative representations played in applying moral reasoning to either social or ethical issues they encountered on the program, while noting any challenges or limitations they encountered in doing so.

Journal Assignment

You will keep a journal chronicling your experiences with mathematics, cultural understanding, and social justice during the travel program. You are required to write six entries over the course of the program, about one full page in length (you may write more), and these may be written or typed. For four of these you will write in response to the experiences of specific days, with the two remaining entries being on topics of your choice but they should relate to mathematics and your experiences in Australia. Your final entry will be revealed while on the program.

Required Experience Journals

Straddie, Swamp vs. Lake Consider procedures for data collection and analysis in the field and compare the two different ecosystems. How are they similar? Different? What did this teach you about the data collection and analysis? About these ecosystems, specifically?

Reef Ecologic As you participate in this larger research project, what is your role? How did your experience and the work you have seen elsewhere on the trip and in the course prepare you to participate in this project? What are your thoughts on the "citizen science" aspect of maintaining reef health? How about the struggle to model and quantify aspects of reef health? Discuss the overall logistics of running an ongoing project of this scale using various volunteer groups and your thoughts/feelings about participating.

Mungalla Station How do the struggles and concerns facing the Nyawaygi people compare to the Quandamooka and Nunukul? How do they connect to issues that we have discussed in class? What issues have you already worked to quantify? How, if at all, did your experience quantifying these issues fall short of illustrating the whole picture? How has this and other similar experiences affected your understanding of issues of injustice and cultural interdependence?

Green Collar What are some of the concepts utilized by Green Collar that you have seen in our course? How does their mission fit in with themes of our course and previous experiences? What aspects of their operation surprised you?

Final Entry Specific prompted response to be done in Sydney.

For your own purposes: Read through your responses to the questions from our pre-departure meeting. Respond to your pre-departure self. How are you the same? How are you different? Describe some of your emotions/thoughts/feelings as you reflect on the last three weeks. What event/activity/encounter/conversation was most impactful for you and how will it change/impact your behavior once you get home?

References

1. Association of American Colleges and Universities, *Quantitative Literacy VALUE Rubric*, (2007), Washington, DC. Retrieved from https://www.aacu.org/sites/default/files/files/VALUE/QuantitativeLiteracy.pdf
2. Association of American Colleges and Universities, *College Learning for the New Global Century*, A report from the National Leadership Council for Liberal Education and American's Promise, (2007), Washington, DC. Retrieved from www.aacu.org/leap/documents/GlobalCentury_final.pdf
3. J. T. Day, *Student Motivation, Academic Viability, and the Summer Language Program Abroad: An Editorial*, Modern Language Journal, **71**, (1987), 261–266
4. M. M. Dwyer, *The Benefits of Study Abroad*, (2017), Retrieved from https://www.iesabroad.org/study-abroad/news/benefits-study-abroad#sthash.ZHeZBNcF.dpbs
5. M. M. Dwyer, *More Is Better: The Impact of Study Abroad Program Duration*, Frontiers: The Interdisciplinary Journal of Study Abroad, **10**, (2004), 151–163
6. K. Hovland, *Global Learning: Aligning Student Learning Outcomes with Study Abroad*, (2010). Retrieved from http://www.nafsa.org/uploadedFiles/NAFSA_Home/Resource_Library_Assets/Networks/CCB/AligningLearningOutcomes.pdf
7. B. Ismail, M. Morgan, and K. Hayes, *Effect of Short Study Abroad Course on Students Openness to Diversity*, Journal of Food Science Education, **1**, (2006), 15–18
8. A. Kitsantas, J. Meyers, *Studying Abroad: Does It Enhance College Student Cross-Cultural Awareness?*, Annual Meeting of Centers for International Business Education and Research, (2001)
9. T. Lewis, R. Niesenbaum, *Extending the Stay: Using Community-Based Research and Service Learning to Enhance Short-term Study Abroad*, Journal of Studies in International Education, (2005), **9**, no. 3, 251–264. https://dx.doi.org/10.1177/1028315305277682

10. National Survey of Student Engagement, *The NSSE 2000 report: National benchmarks of effective educational practice.* (2000).
11. B. Parker, D. Dautoff, *Service-Learning and Study Abroad: Synergistic Learning Opportunities*, Michigan Journal of Community Service Learning, (2007), 40–53
12. J. Wang, A. Peyvandi, and J.M. Moghaddam, *Impact of Short Study Abroad Programs on Students' Diversity Attitude*, International Review of Business Research Papers, **5**, (2009), no.2, 349–357
13. University of Massachusetts Lowell, Essential Learning Outcomes, 2018. https://www.uml.edu/Catalog/Undergraduate/Core-Curriculum/ELO/SRE.aspx

Change Is Hard, But Not Impossible: Building Student Enthusiasm for Inquiry-Based Learning

Jill E. Jordan

Abstract Taking a course that uses an inquiry-based curriculum can be challenging for students who are accustomed to a traditional, lecture-based approach to mathematics instruction. At the end of the course, students who are not fully cognizant of the results of their many hours of hard work may conclude that the teaching approach was ineffective. This article seeks to help instructors who believe in the effectiveness of inquiry-based learning but have trouble getting students on board by giving them specific strategies to help build student confidence and enthusiasm.

1 Introduction

It is the first day of a new semester and a group of students walks into Abstract Algebra I. They're a little nervous about this class, but it's just the first day, so they anticipate getting through it just fine. They know what to expect: the teacher will start with a few definitions and a few examples, maybe a theorem or two, and then give the class an assignment from the chapter "The EDGE Program: 20 Years and Counting" of the text. Most of these students really liked algebra class in high school and are strong math students. This will be no problem for them!

After spending a few minutes going over the usual first-day details, the students have just settled in to take notes when their professor distributes X-shaped pieces of paper. "How many symmetries does this shape have? Think about it for a few minutes on your own, then discuss it with a classmate."[1] Wait, what?! This is not the way a math class is supposed to start. They haven't even learned anything and already they have an assignment!

---

[1] Unit 1, Lesson 1, Task 1 in the Inquiry Oriented Abstract Algebra (IOAA) curriculum [3].

J. E. Jordan (✉)
Houghton College, Houghton, NY, USA
e-mail: Jill.Jordan@Houghton.edu

© The Author(s) and the Association for Women in Mathematics 2019
S. D'Agostino et al. (eds.), *A Celebration of the EDGE Program's Impact on the Mathematics Community and Beyond*, Association for Women in Mathematics Series 18, https://doi.org/10.1007/978-3-030-19486-4_14

167

When class ends, the students have more questions than answers. Their notebooks are filled with scribbles and false starts, and to be honest, they're not sure if anything they wrote is worth keeping. One thing *is* for sure: nothing in there looks like algebra. Students asked some questions but the professor turned the questions right back around to the students instead of giving them the concise answers they were hoping for. Welcome to an inquiry-based learning class.

2   A Primer on Inquiry-Based Learning

Inquiry-based learning (IBL) is a broad term describing an approach to classroom teaching in which students are encouraged to learn through seeking answers to their own or the instructor's questions. Many teaching methods have significant overlap with IBL or refer to a specific system within the category of IBL, including problem-based learning, Moore method, student-centered instruction, discovery-based teaching, problem-solving curriculum, active learning, and cooperative learning.

The question of why I use IBL can be answered quite simply: it works! I know that my students gain a deeper understanding of the material when they confront questions about the mathematics and then struggle individually and together to discover answers to those questions. IBL works particularly well in my abstract algebra course, which draws from a population of motivated mathematics majors. This is a standard junior-level abstract algebra class with multiple pre-requisite courses, and it is typically taken during a student's junior or senior year. Anyone who has decided that mathematics questions are not worth pursuing has abandoned the major before this point. Every time I have taught this course, my students have done whatever I have asked them to do without complaining. And so they spend class time investigating my questions individually, working together, questioning each other, coming to consensus, and finally presenting their results, usually with very little help from me. Through this process, they end up with a better appreciation for each word in a definition, a better understanding of each step in proving a theorem, and better insight into the significance of each example and counterexample. I am often surprised by their deep insights into the material, and I expect that many of these insights would be missed if not for the IBL approach to the class.

As you might expect, the pace of an IBL course is much slower than a traditional lecture-based class. While it takes several class periods for my students to develop the definition of an algebraic group[2], an algebra professor who has chosen a lecture-based approach would have no trouble getting through that definition (along with several others) on the very first day of class. This can pose a problem for courses in which a certain core amount of material must be covered. IBL does not have to be an all-or-nothing proposition, however; while some professors may choose to use

---

[2]Unit 1, Lesson 5, Task 4 in the IOAA curriculum [3].

a comprehensive IBL curriculum, IBL methods can also be sprinkled throughout a traditionally taught course. Professors may choose a level of implementation that fits with their own preferences and the needs of their students, which makes IBL an inherently flexible approach to instruction. I have chosen to teach abstract algebra using a complete IBL curriculum [3], but in other classes, I have opted to mix in elements of IBL when I see opportunities arise.

3   The Problem

I went into my first IBL class with a mix of high hopes and apprehension, but by the end of the course, I was convinced that I had a good thing going. I was thrilled with my students' learning and I could hardly wait to do it all again the next year. Then a few weeks later, I received the student course evaluations and came crashing back to earth. The numerical scores were dismal. Ratings for "Excellent Teacher" and "Excellent Course" were well into the bottom decile. Unfortunately, the student comments were not any better. The students had little positive to say about their experience in the class, and their criticisms were centered around the very IBL methods I had so carefully put into place. I refrained from answering questions immediately, asking students to think them through on their own and then discuss their ideas with each other. When writing proofs, I wanted them to help each other figure out what was wrong and how to fix their mistakes. And yet my students suggested[3]

> Don't let the students sit in silence if they don't understand something. Help them to get to the answer that you want them to have.

and similarly

> ... if the whole class is stuck on a proof, the best course of action may be for you to show us what is wrong, instead of having us just sit and stare at it. Then we just get frustrated and give up.

I requested that students not read ahead in the text so that they could discover the relevant definitions and theorems through inquiry. So, of course, someone pointed out that I should

> [t]ell the class which sections of the book are going to be covered previous to the actual class time. After I spent my break reading the book, I was able to understand more of the concepts that we were going over.

For the weekly problem sets, I intentionally assigned problems with new concepts so that the theme of inquiry would carry over into their work outside of class. Several students shared their thoughts on that decision:

---

[3]Quotes from student course evaluations used with permission from the Houghton College Institutional Review Board.

> Try making the homework on stuff we actually went over in class. There's a thought. Thisis not a 400 level course.

> There seemed to be a rather large disconnect between what was covered in class and what was on the homework.

> It is difficult to do homework on information that we have not learned.

And finally, one student cut right to the chase. His response to being asked how he would change the course:

> More teacher-led classes.

Okay. I get the point, you all hate IBL! Now what?

4 The Solution

I learned early on in my teaching career to not put too much stock in student evaluations. Recent studies (e.g. [2, 4]) have supported this practice, showing that students' evaluations of their teachers are not indicative of how well students learned course material. Despite the poor evaluations, I remained convinced that my students had learned the course material well. And so, if all that mattered to me was how much and how well my students learned mathematics, I would have felt free to ignore the evaluations and continue teaching the way I thought best, making no significant changes to my methods. But the truth is, while student learning is essential, I care about more than that. I want my students to not only learn, but to *know* that they have learned. I want them to be able to appreciate the results of their hard work. I want them to realize that their struggles in the class—yes, even those times when they felt like they were going around in circles without making any progress—made them better mathematicians. And because my goals go beyond student learning, I need to pay close attention to what the student evaluations tell me. This set very clearly told me that students needed to be taught how to appreciate the value of an IBL course. In subsequent semesters of teaching the course, I have incorporated the following into my class with the goal of making my students more aware of how the IBL approach benefits their learning.

Tip #1: Start Off Strong The first time we meet as a class, I spend some time describing IBL to my students. I explain the specific methods I will be using in the class and give the rationale behind the IBL approach. I also distribute a handout going over the same information and strongly recommend that students read the handout before the next class and come back to it as needed throughout the semester.

I also assign my students a self-evaluation, to be turned in at the start of the second class session. I ask my students to write a couple of paragraphs about themselves as mathematicians, including their strengths, weaknesses, and potential areas of improvement. This not only helps to activate their metacognition, but also sets the stage for the end of the semester when they will be asked to write another self-evaluation reflecting on their own growth as a mathematician during the course.

Tip #2: Reiterate Throughout the Course (But Especially at the Beginning)
Students, like all of us, need to hear something many times before it truly sinks in. One of my goals during the first couple of weeks of classes is to remind my students several times per class of the intentionality of the class structure and the philosophy behind an IBL approach to class. For example, when the students are struggling to make progress and starting to feel frustrated, I might say, "I know that this is hard, and you might feel like it is pointless, but this struggle is actually part of the learning process. Don't give up now!" When they have finally figured something out together, they'll hear, "Great job! You figured that out without my help, and you should be very proud! If I had just told you the answer, you wouldn't feel the same sense of accomplishment, and you wouldn't understand it as thoroughly as you do now." Here are some other examples of phrases that you'll hear in my IBL classroom:

- Learning math is a process, and not knowing the answer is part of the process.
- This is what mathematicians do! You're becoming a better mathematician right now![4]
- I could write this on the board, or I could have you guys figure it out. I think you'll learn better if you figure it out for yourselves.
- This is exciting! You're figuring out all of this without me needing to tell you!
- That's a great question! I'll write it on the board so you can all think through it and find the answer together.
- You've really learned the right kind of question to ask!

In keeping with the theme of inquiry, I ask them questions like this:

- Do you see how this is helping you learn?
- Do you feel like you understand this better than if I had told you the answer?
- Do you understand why I wanted you to figure this out for yourself?

These sorts of encouragements and questions, while more frequent at the beginning of the course, never really go away. Students need to be reminded throughout that there is indeed "a method to the madness." Sometimes I start to feel like a broken record, but my students appreciate hearing it again and again.

Tip #3: Homework Counts Too As I mentioned above, I extend the IBL mentality to the students' work outside of class by assigning problem sets that go beyond what we have covered in class, requiring them to learn something new on their own. What I have found is that if students are not told *how* this relates to the IBL approach, they assume it is accidental (or worse, that I am just being a mean professor). I still assign the same types of problems, but now I tell them about my rationale way back at the beginning of the semester and call attention to it again with each new assignment. I indicate which specific problems will require them to go beyond where we have gone in class. I point out that I am not asking them to follow a process that is any

---

[4]I like to use this one when they're stuck, since they don't think of being stuck as *doing* mathematics.

different or more difficult than what they do in class, and as such, they can expect to struggle, but then ultimately to figure it out. Finally, as small groups are frequently utilized in class, I remind them of the benefits of working with classmates, and I encourage them to come to my office hours for additional help.

Tip #4: Ask Them About Their Learning This one sounds a little silly at first. After all, teachers commonly ask students about their learning, and I certainly asked plenty of learning-related questions during that first semester of abstract algebra. But the kind of question I have since added to my repertoire goes beyond the usual "do you understand? Do you need more time? Any questions?" and asks students to delve more deeply into metacognition and reflect on their learning process. I typically start asking these questions about halfway through the course in order to give them enough time to start seeing signs of their own progress. Here are some examples of the types of questions I ask:

- Have you noticed that when I ask a question, you get right to work, whereas at the beginning of the semester you had trouble getting started?
- Can you see how your proofs have improved?
- Do you see any benefits to spending more time and going deeper into the material rather than trying to cover a lot of content?
- Do you think that you learn better by copying information from the board or by working through questions with your classmates?

This "asking" technique culminates at the end of the semester, when I assign a second self-evaluation. Once again, I ask them to think about their strengths, weaknesses, and areas of potential improvement, but this time with a focus on how this course has helped to bring about change. It is good for my students to take some time to put in writing how they have grown throughout the semester, and as their teacher, I love seeing them recognize specific ways that my course has helped them develop as mathematicians.

Tip #5: Don't Stop I mentioned this one before, and yet it seems fitting to say it again. After all, the point here is that even when I feel like I've said it all before, continually encouraging and reminding students that they're making progress, students are ready for more! I have come to believe that part of my job as a teacher is providing my students with a narrative of their experience in my course. Working hard can be either discouraging or invigorating for my students, depending on how I frame it, so I do my best to frame it in the most positive way possible, and I keep doing that right up until the end of the course.

5 Survey Says…

As of this writing, I have taught my IBL abstract algebra class five times since that first disastrous semester. Consider the following results from student evaluations and decide for yourself whether or not my tweaks have made a difference in my students' attitudes toward the class. Since the first semester, the numerical ratings for "Excellent Teacher" and "Excellent Course" on student evaluations have consistently been around the 70th percentile (compared to well below the 10th percentile the first time I taught it), and representative student comments now include:

> I really liked how we worked together on problems and coming up with answers versus it being just a lecture where we take notes.

> I really appreciated the style of this course. I like getting to work individually, in groups, and as the whole class on problems. This way, we learned the definitions and theorems deeply rather than just memorizing.

> I enjoyed trying to first discover things on my own and then working in groups and finally discussing it as a class.

> What I liked best about the course was the requirement/expectation that we prove everything to ourselves rather than simply memorize results.

> My favorite part of the course was that instead of telling us important concepts, [the instructor] pointed us in the right direction and we discovered them for ourselves.

6   In Conclusion

My purpose in sharing about my experience is to help other mathematics teachers who find themselves struggling with student attitudes in their IBL classes. Please use what is helpful and ignore whatever is not. If you are interested in learning more about teaching an IBL abstract algebra course, I enthusiastically recommend the IOAA curriculum [3], which you can read more about on the American Mathematical Society's Math Education blog [1].

References

1. Johnson E, Keene K, Andrews-Larson C (2015) Inquiry-oriented instruction: What it is and how we are trying to help. https://blogs.ams.org/matheducation/2015/04/10/inquiry-oriented-instruction-what-it-is-and-how-we-are-trying-to-help/. Accessed 10 Aug 2018
2. Kornell N, Hausman H (2016) Do the best teachers get the best ratings? Front. Psychol. https://doi.org/10.3389/fpsyg.2016.00570
3. IOAA (2016) Inquiry oriented group theory curriculum. https://taafu.org/ioaa/index.php. Accessed 10 Aug 2018
4. Uttl B, White CA, Gonzalez DW (2017) Meta-analysis of faculty's teaching effectiveness: Student evaluation of teaching ratings and student learning are not related. Studies in Educational Evaluation 54: 22-42. https://doi.org/10.1016/j.stueduc.2016.08.007

Linear Algebra, Secret Agencies, and Zombies: Applications to Enhance Learning and Creativity

Carolyn Otto

Abstract This article will discuss activities and assignments created for a linear algebra course that aim to excite students about learning the course content. The linear algebra course taught at the University of Wisconsin-Eau Claire covers linear algebra, its applications, and serves as the introduction to proofs course. Students often become overwhelmed with all the content and proof techniques in this course. This leads to them not enjoying, nor engaging in, what they are learning and they become discouraged with the material. Throughout the semester, interactive projects are introduced which cover applications of the course content. These activities center around the idea that the students have been recruited to work with the "Zombie Containment Task Force" under the supervision of Agent Frank Larson. Throughout the semester, students must complete several missions to uncover secrets about the workings of the task force, discover knowledge about zombies, and reveal double agents. At the end of the semester, students use clues given throughout the semester to make a final decision which informs them of their future in this fictional world. This article will give the outline of five interconnected projects that are used in the course as well as discuss the implementation of these missions.

1 Project Based Learning

To give a full picture of how I use projects in my course, first I present a background of the linear algebra course at my institution. I will follow the introduction with a discussion of how I incorporate these projects into my teaching approach.

C. Otto (✉)
University of Wisconsin-Eau Claire, Eau Claire, WI, USA
e-mail: ottoa@uwec.edu

© The Author(s) and the Association for Women in Mathematics 2019
S. D'Agostino et al. (eds.), *A Celebration of the EDGE Program's Impact on the Mathematics Community and Beyond*, Association for Women in Mathematics Series 18, https://doi.org/10.1007/978-3-030-19486-4_15

175

1.1   Linear Algebra at the University of Wisconsin-Eau Claire

The linear algebra course taught at the University of Wisconsin-Eau Claire (UWEC) is a four credit-hour course taken by most mathematics/physics majors during their sophomore or junior years. Each semester, there are typically 35–50 enrolled for the course, which could be split between two sections. The course covers the typical linear algebra curriculum, including matrix algebra, systems of linear equations, vector spaces, linear transformations, eigenvalues, and applications of these concepts. The course also covers methods of proof and is treated as the "introduction to proof writing" course for our math majors. New to the curriculum is an overview of student-faculty research and the related opportunities for students.

This course is also used in the department's assessment of the mathematics major with a liberal arts emphasis. Specifically, the course assesses the following two program outcomes:

• Students will be able to write mathematical proofs.
• Students will be able to work independently and collaboratively on mathematical problems.

I have taught this course six times at UWEC. The first time I was in charge of the course, it was only three credit-hours and applications of the material were optional content. The previously stated course description demonstrates how this one semester course is quite dense with content. Within the last several years, our department (with university approval) has increased the number of credits to four in order to add in the applications to the topical outline. The additional credit allows for more time with the application content. Even with the fourth credit, the core content, the proof techniques, the applications, and projects still push the limits on material for a four credit-hour course.

Since the addition of the fourth credit-hour, I have taught our linear algebra course five times. I began by covering a selection of applications from the course textbook [5] by lecturing and assigning problems from those sections. While I did cover all the required material, I found myself not giving the applications the full treatment they deserved in favor of more time with the proof techniques and proof writing.

This is where zombies enter the story. As I taught the course more, I started to incorporate writing projects in my linear algebra sections to get students interested in the material and to engage their creativity. These projects have evolved over time and now focus on covering some of the application sections of the course while creating an ongoing narrative of the students' involvement in a secret government agency, "Zombie Containment Task Force." The specifics of the narrative and projects can be found in Sect. 2.

1.2 Rationales for the Zombie Projects

Sharing my excitement for math is one of the reasons I love being an educator. However, it sometimes takes more than just my excitement to get the students engaged with the material. I create a fun and interesting experience for my linear algebra students by developing creative zombie projects to motivate them to learn and apply the material.

Over the course of my career, I have attended several panels and talks on the incorporation of writing projects into the classroom. Specifically, I attended a Project NExT Session on incorporating writing projects into undergraduate research courses. The speaker mentioned several ways to incorporate projects, how to create effective rubrics, and gave reference to their book [3], which I used to model my rubrics. For more information on Project NExT, please visit the website www.maa.org/programs-and-communities/professional-development/project-next. The idea of incorporating writing projects really appealed to me for this linear algebra course at UWEC, especially considering the two outcomes that this course assesses. While these projects focus on applications and assessment of the second outcome in Sect. 1.1, I have found that more practice writing mathematics helps students to become comfortable writing proofs, which also helps with the first listed outcome in Sect. 1.1. A mathematics focused paper has a different style than a mathematical proof, but it has been my experience that when students are comfortable communicating math in the first form, they have an easier time effectively explaining why a mathematical statement is true. From there we work on transforming their explanation into a proof.

Throughout my course, I give five writing projects to the students (see Sect. 2) with the fourth project, an escape room, as an optional project. I decided to use a zombie theme for these projects. This is for several reasons. First, I love zombies and Halloween, so it is fun for me to write projects involving them. Second, when I first started writing these projects, TV shows, books, and movies such as *Zombieland*, *iZombie*, *World War Z*, and the *Walking Dead* (to name just a few) were popular in the mainstream media. I cannot forget to mention that Colin Adams wrote and published *Zombies & Calculus* [1]. It is worth mentioning that this book and these projects are completely independent and were created without the knowledge of each other[1]. Third, sci-fi topics lead to a little wiggle room when it comes to storytelling and solutions. For example, if I find that 4.58 zombies survive, I don't need to round up. Having 0.58 of a zombie could make sense. Another example my students see is blood flow in the brain changing directions after the injection of a chemical. Again, this scenario could happen in this world.

I also give these projects a zombie theme in order to help motivate the students. When I share my enthusiasm for this fictional world, I demonstrate to my students that I care about the material and am excited to share it with them. The time I

---

[1] How awesome is it to have multiple, creative math and zombie products?

spend developing and testing these projects is evident to the students and seeing my dedication motivates them to learn the material [4]. The idea of telling stories is also a great motivator to retain the information, "organizing a lesson plan like a story is an effective way to help students comprehend and remember" [7]. My projects are unique in regard to the storytelling. It is the aim that students will remember the projects, math, and their experiences with them more than just working problems from the book. For more ideas to incorporate projects with more realistic applications, please see [2].

The primary focus of these projects is the communication of the solutions by the students. I write the problems in a way that requires students to identify what the problems are asking and determine the best method to solve them. I am interested to see if they can explain the solution and what it means in a document written to a person that does not know all the linear algebra details. When I first explain this to the class, I receive comments such as "I thought I was taking a math class not an English class," or "If I wanted to write I would have been an English major." This usually leads to a discussion on why it is important to communicate effectively, especially when working with mathematics. I emphasize the following points:

- an application of linear algebra in the "real world" will not be phrased as "solve this equation,"
- employers need to be able to follow the logic of the solutions,
- many individuals outside of our linear algebra class may not understand math jargon and it is important to know how to communicate ideas effectively, and
- writing down a solution with words helps to communicate the thought process for yourself and the reader.

In addition to the points above, I also discuss collaboration skills at the beginning of the semester, even though students are not in larger groups until the fourth and fifth project. I take the time to explain the importance of working with others. Specifically, I try to foreshadow work scenarios where they will be collaborating with all different types of people and skill levels.

In the next section, I will detail the projects that I give for my course. This article is just an overview of these projects. If you are looking for more specific details, you can contact zombiecontainmenttaskforce@gmail.com and Agent Frank Larson will supply you with more details and materials.

2   The Projects: Missions from the Zombie Containment Task Force

This section provides details about each project/mission: the story, linear algebra component, assignment, the assessment, and other details. In addition, I will provide the general setup of the world of the "Zombie Containment Task Force." It is important to note that in my courses I use wide variety of materials and software in the management and logistics of these projects. If one is interested in creating similar projects, these items are not all necessary and can be altered to fit an instructor's needs.

On the first day of the course, students receive a letter from the mysterious Agent Frank Larson, see Fig. 1. They have been recruited to work for the "Zombie Containment Task Force (ZCTF)." They are informed that throughout the semester they will receive missions that require knowledge of linear algebra to successfully complete.

2.1 Project One: Zombie Classification

Mission The ZCTF computer system has crashed! Information about zombie specimens has been lost, only the paper copies remain. Students are given one of the six medical files for a zombie specimen. In each of these files is general information about a zombie, a photo, an image of "observed blood flow analysis," and information about experimentation that Dr. William had previously performed on the zombie. Using the information supplied, you are to classify your zombie as a crawler, shambler, walker, runner, or thriller by finding the blood flow at the indicator site. The letter written by Agent Frank Larson can be found in Fig. 2.

Linear Algebra Component The blood flow image given in the file gives the fixed and variable blood flow rates of certain veins/arteries in the zombie brain. An example of one of these images can be seen in Fig. 3. Students set up a system of linear equations to solve the system representing this network. The solution they find includes a free variable. The experimentation data given instructs what students should do with that free variable.

Assignment This is an individual assignment. Each student sets up and solves a system of equations based on the information in their file. After they solve the system of equations, they are supplied with one more piece of information: Dr. William was conducting experiments and was able to find one more blood flow value. Using the data, they are instructed to classify their zombie and write a letter to Agent Larson in response to his inquiry that includes the explanation of the solution and the method of solving the problem. This letter is to be typed and submitted to the class's Dropbox folder.

Assessment When the students receive their file folder of information, they are also supplied with a detailed rubric for the assignment. Since this is their first mission/writing assignment, I offer to read a rough draft of their response letter. I specifically require that when they solve the system of equations, they obtain a free variable. Then using Dr. William's experimentation notes, the students can find a value for this variable. I want to demonstrate the convenience of solving the system in general. I make this a specific rubric item, mention it in class, and point it out in the drafts supplied to me as most students will skip this step.

ZOMBIE CONTAINMENT TASK FORCE

Greetings, Linear Algebra Students!

Today you are embarking on a journey filled with matrices, proofs, and so much knowledge using Ron's Elementary Linear Algebra book. I bid you luck during this semester.

There may come a time when I require your assistance. I have spoken with your professor, Dr. Otto, and have given her warning of when I might call upon you all for help. Until that time, she is only allowed to tell you tales about Zombie-Bob, now just Bob, and what the Zombie Containment Task Force (ZCTF) has accomplished in the past.

When required, Dr. Otto and I will give you more information. For now, I will leave you with knowing that the ZCTF is in dire need of your help and is anxiously awaiting your acquisition of some of the vital concepts of linear algebra.

Good luck, students! I will be in contact soon.

Agent Frank Larson

JANUARY 29, 2018

Fig. 1 Introductory letter from Agent Frank Larson which is given to the students on the first day of class. The character of Zombie-Bob, or Z-Bob, is important to the story as he is a character that was turned into a zombie and then took the cure to become human again

ZOMBIE CONTAINMENT TASK FORCE

Greetings, Linear Algebra Student:

My name is Agent Frank Larson and I oversee the Zombie Containment Task Force (ZCTF). Information about the ZCFT is strictly "need to know." I have been in contact with Dr. Otto and she assures me that your linear algebra skills are at the level needed to be of use to the ZCTF. To be frank (no pun intended), I am not as confident. However, we are in dire need of some help and I, for some reason, trust Dr. Otto.

At the beginning of January one of our computer systems was compromised and we lost a large amount of data collected from some of our incapacitated zombie specimens. Luckily, some hard copies of the original examinations of the subjects were discovered. Some of that data is now in your hands.
Unfortunately, the analysis of the data was lost...this is where the ZCTF needs your help.

Data collected about blood flow in the specimen's brain while they were alive can determine the classification of the zombie. I have listed the classifications with information given to me from Dr. William. Before you ask, NO, we cannot just have them run around to classify them. The brains have been harvested and the bodies burned.

Classification	Blood Flow Rates in milliliters per minute at indicator site
Crawler	0-100
Shambler	101-250
Walker	251-450
Runner	451-650
Thriller	650 and up

The indicator site is the vein leaving site B away from site A. Dr. William injected a fluid into the zombie's brain while they were "alive" and observed that some of the other blood flows were dependent on the alteration of one specific site's flow.

We have lost all the classifications of our specimens because of the computer situation. You will find all other relevant information in Dr. William's notes. I do not have time to go into more details. I have left all other instructions with Dr. Otto. I anticipate the proper classification of your zombie with justification by February 19.

Do not fail in this task.

Agent Frank Larson

FEBRUARY 7, 2018

Fig. 2 Agent Larson's instruction letter for the first mission

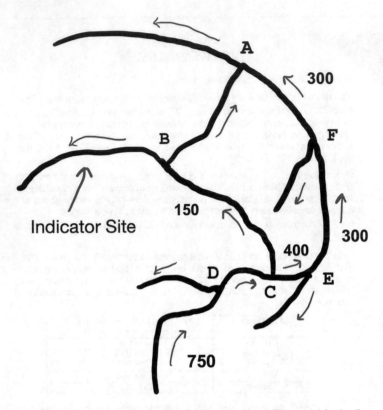

Fig. 3 Example of a brain scan of a zombie showing the blood flows used in the first mission. The letters indicate junction sites, while the numbers indicate the rate of the blood in milliliters per minute

Debriefing I make comments on the letter and supply additional comments on a copy of the rubric the students receive. In addition, students are supplied with the classifications of all six zombies, the capture sites of these zombies, and some identifying feature: a dragon brand or a black umbrella tattoo. This piece of information is a clue for the students which they will use later to make a choice about what happens to their character. These six zombies are used throughout the projects.

Fun Extras To deliver this assignment, I assemble file folders with the information. Each folder has the information sheets stapled in them to resemble a medical file. Students from previous semesters agree to have their photo taken and turned into a zombie and these photos are paper clipped to the file. Lastly, I create a label for the ZCTF that matches their security clearance.

2.2 Project Two: Code Breaking

Mission After completion of the first mission, Agent Frank Larson decides to grant the students a new security clearance! With continuing problems of the main computer systems, employees of the ZCTF have been communicating through coded messages. Two mysterious messages are discovered by Agent Larson's personal assistant Reginald. Students receive one of the two envelopes, each of which contain a set of encrypted messages. Using a set of potential decoding matrices, students must decipher the messages and report back to Agent Larson.

Linear Algebra Component Each message was encoded using one of the several $n \times n$ invertible matrices, known as coding matrices. The messages were partitioned into uncoded row matrices, then multiplied by the coding matrices to obtain coded row matrices. Potential coding matrices are provided to the students from Agent Larson. Students must determine which inverse matrices will used to decode the message and submit the message to Agent Larson.

Assignment This is an individual or partner assignment, students can choose. Each envelope of information includes a coded message broken up into three parts and a collection of five matrices. Students are informed that three of the matrices were used to encrypt the message, one for each part of the message. They need to determine which matrix decodes which part and then decode the message. The description of the mission, detailed instructions on how to decode the message, computations of the decryption, and a conclusion all of which must be typed into a Maple template. Maple is a mathematics algebraic software that serves as a computational environment which students are able to use on any school computer and is the software we use most in class. If trying to adapt this project to one's own needs, the calculations could easily be done on their calculators or in Wolfram Alpha.

Assessment When the students receive their envelopes of information, they are supplied with a detailed rubric for the assignment, see Fig. 4. For this project, I specifically require that they use a Maple template which I supply to them. They are given one class period to work on this project. The classroom in which I teach this course is our department's computer lab so there are an ample number of computers. At UWEC, students are also able to use a virtual lab which allows them to access the Maple software even if they are not on campus. Since the ZCTF lost all their computer files, students need to supply instructions on how to decode the message using Maple. Thus, their Maple worksheet must be able to run and include all relevant computations and instructions.

Debriefing I run all the Maple code to make sure the students' computations work without bugs. I make comments in the Maple code as well as on the rubric. Since there are two different messages, students are given the answers to both when I hand back the assignment. Their Maple code is returned by email so they are able to look at all the comments in the Maple software. The messages set up the main mystery

INSTRUCTIONS FROM DR. OTTO AND AGENT LARSON

The purpose of this project is to apply the method of code-breaker to "real-life" situations. You are to decipher the code given to you using the techniques from Section 2.5. It is expect that you will provide Agent Larson a Maple Worksheet the includes the explanation of the solutions and method of decoding the message.

- You may work by yourself or with a partner. (1-2 people groups!)
- Dr. Otto has provided a Maple Worksheet template on D2L that you should use to write up your assignment. Make sure you use it, but of course, feel free to make it your own.
- This list will be used as a guide to grade your assignment and will be returned to you with comments.
- Put your Maple Worksheet in the Dropbox Folder on D2L labeled "Codes" by Wednesday, February 28 by the beginning of class.

Please feel free to use these checklists as a guide for yourself while writing this assignment.

MATH 324 PROJECT CHECKLIST: MAPLE WORKSHEET

Does this Maple Project:

1. clearly (re)state the problem to be solved?
2. state the answer in a complete sentence which stands on its own?
3. provide a paragraph which explains how the problem was approached?
4. aim its explanations at the appropriate audience?
5. explain how the data is derived, or where it can be found?
6. give acknowledgement where it is due?

In this Project,

1. did the writers use Maple? and the template that was given? follow the instructions in the template?
2. did the writers follow the instructions on the template?
3. are the spelling, grammar, and punctuation correct?
4. are the mathematics correct? include the correct amount of work? too little?
5. did the writers solve the question that was originally asked?

Comments:

Fig. 4 The rubric for the second mission

Decryption of Codes

Code 1:

LAST CONTACT WITH DR. WILLIAM WAS TWO PM ON FRIDAY AT THE UWEC CAMPUS. HIS LAB NOTES AND FORMULAS ARE CURRENTLY SECURE AT THE GAMMA LOCATION.

Z-BOB HAS BEEN NOTIFIED TODAY AND HAS SUCCESSFULLY GONE INTO HIDING.

KEEP COMMUNICATION TO THE OCTOPUS UNIT ONLY. I WILL BE IN TOUCH SOON.

                                    AGENT BELLA LYNN

Code 2:

DR. WILLIAM HAS BEEN SUCCESSFULLY APPREHENDED AND TAKEN TO THE INTERROGATION CENTER. HIS LAB NOTES AND FORMULAS WERE NOT LOCATED.

SUBJECT KNOWN AS Z-BOB HAS ELUDED THE SEARCH AND RECOVER TEAM. THE SEARCH CONTINUES.

AGENT LARSON IS BECOMING SUSPICIOUS. KEEP COMMUNICATION LIMITED.

                                    BU OPERATIVE #TEN

Fig. 5 The deciphered messages from the second mission

for the course: "Who are the Black Umbrella Operatives?" In addition, we find that Dr. William has gone missing. Decoded messages can be found in the appendix section in Fig. 5.

Fun Extras Students meet a new character named Reginald, Agent Larson's personal assistant. I have a member of our math department come in and read a script for my class and hand out the mission to the students.

```
Zombie Internal Body Temperature

Dr. William was studying many aspects of the zombie virus.  In particular, he was
working with Hugo Pecos and Robert Lomax on the "Science of Zombism." While the ZCTF
was able to debunk a lot of that "science," the thermal analysis seems to be correct.
The typical zombie core temp is between 64 and 76 degrees Fahrenheit.  Heat is
released by the various parasites living in the zombie flesh.  This means that the
thermal image camera can detect zombies and is able to distinguish between the undead
and living humans.  The thermal cameras' rays can penetrate man-made structures as
well.  The rays condense into a tetrahedron shape and can detect heat in the whole
figure.


Gamma Location Information

The Gamma Location is located at 333 Gibson Street and 205 S Barstow, Eau Claire, WI
54701.  The height of the building/camera mount is 100ft.  Some of more confidential
documents are housed here.  It is of the utmost importance that the Black Umbrella
spies do not find out about this location!

This location is in plain sight and we work hard to keep the public unaware of its
real purpose. You are able to purchase coffee, enjoy dinner, and book a room in the
spacious hotel.  These purchases directly go to the ZCTF.  Shop and eat local!
```

Fig. 6 Some information provided to the students that will be helpful in this mission. The science behind zombies included can be found at [6]

2.3 Project Three: Location Finding

Mission Agent Frank Larson is quite impressed with the work of the students and increases their security clearance! He informs them that there have been whispers of something known as "Black Umbrella" and that the ZCTF has a secret facility in downtown Eau Claire known as the Gamma Location. Students receive a mini-file of information about sightings and capture locations of zombies as well as the location and specifications of the Gamma Location, see Fig. 6. Using this information and a map of Eau Claire, the main objective is to determine where the Black Umbrella's Interrogation Center is and try to locate Dr. William.

Linear Algebra Component In the information file, students are supplied with a map that includes the observation and capture sites of three zombies as well as the Gamma Location. There is also given a grid printed on a picture of transparency paper that fits over the map. Creating coordinates of all the points of interest, students are able use linear algebra, in particular determinants, to find the area between the zombie capture sites and to find the volume of the tetrahedron that a thermal camera can cover if on top of the Gamma Location. Students are informed that two of the zombies travel in a straight line (but do not know which two). This allows the students to find the points of intersection of the paths that these zombies are following and then can deduce possible locations for the Interrogation Center. In particular there will be three possible locations as there are three distinct combinations of pairs of zombies.

Assignment This is an individual assignment. Each student finds coordinates that represent all the relevant locations and construct matrices in which the determinants will provide the needed values. They are to write a letter to Agent Larson in response to his inquiry that includes the explanation of the solution and the method of solving the problem. This letter is to be typed and submitted to the class's Dropbox folder. For this mission, students are allowed to handwrite the supporting mathematics if they desire with the story justification that the ZCTF is in a rush to find Dr. William.

Assessment When the students receive their mini-file of information, they are also supplied with a detailed rubric for this assignment. At this point in the semester, students are accustomed to reading the rubrics and knowing what information to include. For this specific mission, students need to not only compute the quantities requested, but must construct a map illustrating why their solutions make sense and discuss any error that might occur.

Debriefing Just as with the first mission, I provide feedback on the letter and supply additional comments on a copy of the rubric the students receive. In addition, students are supplied with all the possible locations of the Interrogation Center with a note from Agent Larson. This note congratulates them on a job well done and now a more precise search for Dr. William will be conducted. A list of potential destinations are given to the students with a date on which the search teams will investigate.

Fun Extras My student grader (Agent Bella Lynn) and I had a ton of fun with the delivery of this mission. Before class starts, we tape envelopes to the bottom of their tables. Normally, I close the door to my classroom once class starts but on this day I leave it open. My grader throws in a paper airplane and I read the message inside to the class. The students are instructed to look under their tables to get the mission. They loved this little twist.

2.4 Optional Project Four: The Escape Room

This is an optional project, optional in the sense that I choose whether or not the whole class will participate in this activity. I use factors as time and enthusiasm to decide if the class does this activity. It is meant to be a fun change of pace from the classroom and gives a narrative that advances our story. In the fall this activity usually occurs the day before Thanksgiving. I will give you the general outline of the project and some idea of the problems. Please contact Agent Frank Larson at the agency's email: zombiecontainmenttaskforce@gmail.com for a more thorough debriefing, if you are interested in the specific clues, puzzles, and problems.

Mission With a list of potential locations for the Interrogation Center, Agent Frank Larson is assembling the Octopus Unit (an elite squad in the ZCTF) to search for Dr. William. Students are given temporary clearance to join Agents Terry, Cory, and Rowan in the investigation of Acoustic Cafe. The objective is to recover Dr.

William, find information on Black Umbrella Operatives, and try to discover a mole in the agency. In preparation for the search, Agent Larson tells students to read about the applications of vector spaces in chapter 4 of their textbook [5]. At this point in the semester, students have already worked extensively with vector spaces and this section draws connections between their knowledge of vector spaces and applications to differential equations (which all had seen in the prerequisite course of Calculus II). Students prepare by reading over that chapter, working out problems. It is expected they come to class with a solid understanding of the material. In addition, Agent Larson randomly assigns groups of two or three students which will be the search teams. During the next class, students investigate and try to escape Acoustic Cafe. This cafe is a real location in Eau Claire that students often visit to get snacks, coffee, and work on homework. I supply pictures of the real location if students have never been there so all of the clues make sense.

There are three main tasks/puzzles for this escape room. For each task, groups are given a set of clues, which include edited photos, dials, file folders, and pieces of papers.

- Stopping the Pendulum: Students are locked in the cafe and pendulum blades are descending on them and the other patrons! To stop them students must give Agent Cory the relevant information.

    1. What is the differential equation that models the pendulum? (Can be found in the photographs and from the reading.)
    2. What is a solution set for the above differential equation? (A selection of possible solution sets are in a photo.)
    2b. Verify that the set gives solutions to the differential equation.
    3. Prove that the set is linearly independent using the Wronskian.
  Bonus: Why is this an application of vector spaces?

- Freeing the Barista: The barista is locked behind a glass counter and you see someone lying on the floor. There is a weird dial with an equation in two variables on it and flyers on a bulletin board. Agent Rowan says to perform a rotation of axes to eliminate the xy-term and then classify the conic section the equation represents.

    1. What is the angle of rotation?
    1b. Provide the work to verify the angle.
    2. What conic section is needed?
    2b. Provide the work showing the standard equation of the conic section.
    3. What symbol do students need to give to Agent Rowan to release the barista? (This is found on the dial, the trigonometric watch you're given, and information on the flyers.)
  Bonus: Why is this an application of vector spaces?

- Entering the Code: Dr. William is found lying behind the counter! Agent Terry gets the doctor and says they must leave. There is a keypad by the bathroom door (true fact: Acoustic Cafe has a code to use the bathroom). The barista keeps shouting "PS 3321421" over and over. What could this mean?

1. What is the seven digit code to leave? Students have five attempts. For this question, there have been several clues on photos and paper to help figure it out. This is purely for fun and has no bearing on students' grades.

Assessment It is important to mc that the students know that this project is mostly for fun and for them to get to know their new group members. The largest portion of the grade is participation points. For some group work/homework points, students are to answer Stopping the Pendulum 2b and 3 as well as Freeing the Barista 1b and 2b.

2.5 Project Five: Prediction Models

If students did the escape room, they have all the information to start the final mission and will go to Mission B. If the escape room was not given in class, they go to Mission A.

Mission A With the completion of the previous mission, the Octopus Unit and Agent Bella Lynn create a list of potential locations for the Interrogation Center. It is still believed that Dr. William is being held there. This past Saturday, Agents Cory, Rowan, and Terry led the investigation of Acoustic Cafe, while the rest of the unit checked out the other locations. Dr. William was found at the cafe but remains unconscious. Agent Terry is keeping watch over him. The Octopus Unit found a collection of data sets at the Antique Emporium which belong to the Black Umbrella Organization. Students are to analyze the data to create six prediction models using the methods of least squares. Agent Larson supplies each group a flash drive of the found data, a statement from Z-Bob (see Fig. 7), instructions, peer assessment, and a rubric. Z-Bob is an important character introduced in my Calculus courses and has vital information about the zombie virus. He is highly sought after. Students do not need to have previous knowledge of Z-Bob for these projects.

Mission B After the debacle at Acoustic Cafe, Agent Larson has taken you off the Octopus Unit and back on office work duty. However he does congratulate the students on successfully finding Dr. William! The Octopus Unit searched the other locations and found a collection of data sets at the Antique Emporium which belong to the Black Umbrella Organization. Students are to analyze the data to create six prediction models using the methods of least squares. Agent Larson supplies each group a flash drive of the found data, a statement from Z-Bob, instructions, peer assessment, and a rubric.

Linear Algebra Component Each group is given a flash drive that has six collections of data. Students must use the method of least squares and create six quadratic

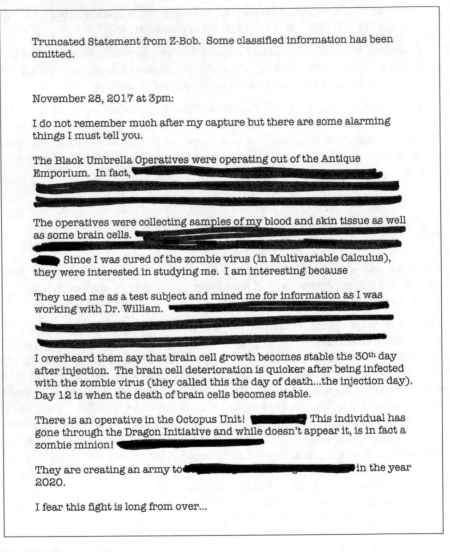

Truncated Statement from Z-Bob. Some classified information has been omitted.

November 28, 2017 at 3pm:

I do not remember much after my capture but there are some alarming things I must tell you.

The Black Umbrella Operatives were operating out of the Antique Emporium. In fact, ▮▮

The operatives were collecting samples of my blood and skin tissue as well as some brain cells. ▮▮▮▮▮▮▮▮▮▮▮▮▮▮▮▮▮▮▮▮▮▮▮▮▮▮▮▮▮

▮▮▮ Since I was cured of the zombie virus (in Multivariable Calculus), they were interested in studying me. I am interesting because

They used me as a test subject and mined me for information as I was working with Dr. William. ▮▮▮▮▮▮▮▮▮▮▮▮▮▮▮▮▮▮▮▮▮▮▮▮▮▮▮▮

I overheard them say that brain cell growth becomes stable the 30th day after injection. The brain cell deterioration is quicker after being infected with the zombie virus (they called this the day of death...the injection day). Day 12 is when the death of brain cells becomes stable.

There is an operative in the Octopus Unit! ▮▮▮▮▮▮▮ This individual has gone through the Dragon Initiative and while doesn't appear it, is in fact a zombie minion! ▮▮▮▮▮▮▮

They are creating an army to ▮▮▮▮▮▮▮▮▮▮▮▮▮▮▮▮▮▮▮▮ in the year 2020.

I fear this fight is long from over...

Fig. 7 Z-Bob's statement

equations to predict values of zombie populations, brain cell regrowth (data can be seen in Fig. 8), and brain cell deterioration.

Assignment This is a group project and the groups are randomly generated. Each group must read the relevant material in the textbook. They also must write a letter in response with the answers as well as a written report on the solutions. Students are given two days in class to work on this letter and report. Also, as with the first mission, I offer to read a draft of these items. The projects must be returned on the flash drive.

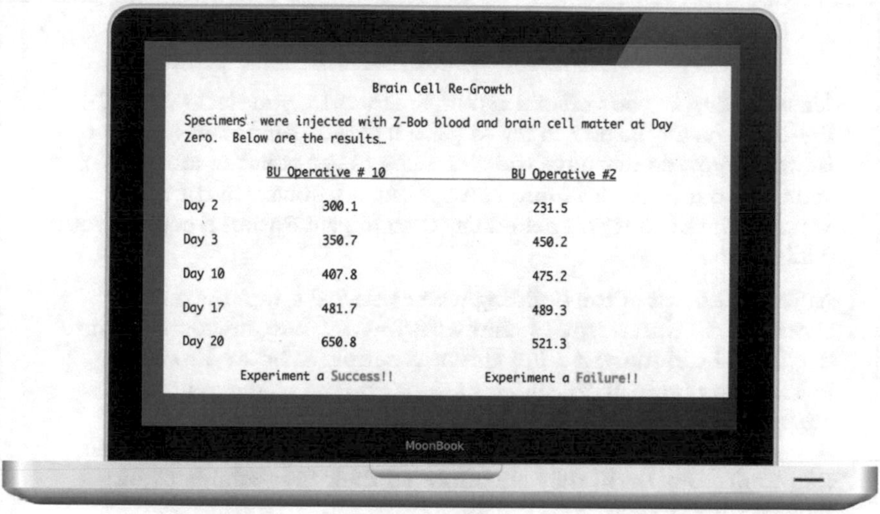

Fig. 8 Data found on a computer at the Antique Emporium

Assessment When the students receive their flash drive of information, they are also supplied with a long, detailed rubric for the assignment and a peer review form. For this project, groups are required to do outside research on the topic of least squares and include references. They are also informed they must use this method, but in the report they must include why this method applies to this specific collection of data (this is the most commonly missed rubric item). Also, there is one model that seems incorrect. The students must explain this as well.

This project helps to assess the second outcome for this course: Students will be able to work independently and collaboratively on mathematical problems. Students must complete a peer evaluation on their group members. My observations and these evaluations will make up 20% of this project grade.

Debriefing As with the previous missions, I make comments on the letter and report and supply additional comments on a copy of the rubric the students received. Finally, students are given information about each of the Agents Terry, Cory, and Rowan. They each individually make a choice about who to give the prediction models and report. On the last day of class, I give them a final letter about what happened to them in this fictional world. See Fig. 9 for an example.

Fun Extras I inform the students of their fates by making them a name badge with a QR code on the back. They can scan the code to get their final outcome and to have a final memento from the course.

What Happened to You?

After turning in your official report to Dr. Otto, you decided to visit the ZCTF headquarters to try to get a look at Agent Frank Larson. However, you do not have the clearance to get past the lobby. On your way out of the building you spot Agent Rowan in the lobby. You remember that you asked Dr. Otto to give Rowan a copy of your final report.

As you head out of the building when you feel a tap on your shoulder. Rowan wants to chat with you and escorts you back into the ZCTF headquarters. For the next couple of hours Rowan informs you more about different departments in the task force. There are also a couple of side conversations about dogs, Agent Rowan loves dogs. It was when Rowan was with their dogs at the Eau Claire Dog Park, they encountered their first zombie (Z_42: Walker with Dragon Brand).

Agent Rowan tells you that the Octopus Unit is impressed with your work, along with Frank Larson. They also tell you that Dr. Otto is the main recruitment officer for ZCTF and she has been using her class for years to secretly recruit new initiates. You are offered a full-time position with the ZCTF and are told you would receive the elusive Opal Clearance and your choice of assignments. The ZCTF can provide excellent benefits, travel opportunities, and four weeks of paid vacation.

You gladly accept. Your first assignment is joining with Agent Bella Lynn and Agent Rowan to identify double agents in the agency. Specifically, you need to find who is BU Operative #10 and #13.

Your new security clearance does not allow you to speak of your mission to anyone. You live out the rest of your Math 324 days quite happy with your new job.

Fig. 9 Example of what may have happened to you at the end of the projects

3 Steps Forward

These projects are living things in my courses; they are constantly evolving to be more effective as teaching tools. This involves changing the problems, adding more to the story, creating "mini" missions, and creating more thorough assessments for the projects themselves. I am currently working with the UWEC's Center of Teaching and Learning (CETL) to create an anonymous assessment survey for each of these projects that will be sent to current and previous students.

My absolute favorite thing about these zombie projects is the students' creativity with writing and storytelling. Students of course have the option of just answering the questions and writing a plain letter back to Agent Larson. However, most of the students really get involved in the world. I have had students create code names for themselves, new departments in the agency, puzzles for Agent Larson, and drawings/comics of the zombies. As such, it is my intention to get a small group of students, from math majors, to art majors, to English majors to write a course booklet about the ZCTF (or a brand new agency) that my calculus and linear algebra students would receive at the beginning of the semester. I have projects made for Calculus I, II, III, and Linear Algebra that would all be included in this volume.

Acknowledgements I would like to thank Cindy Albert from the UWEC's CETL office for all the useful teaching conversations and resources we have shared throughout the years that have helped create these projects. Also, I would like to thank the UWEC's Mathematics Department for giving me the flexibility in my courses to try these projects and the resources (office supplies, flash drives, etc.) that have helped make these projects a little extra special. Finally, I would like to thank and acknowledge Dr. Christopher Davis for all of his time listening to zombie lore and linear algebra ideas.

References

1. Adams, C.: *Zombies and Calculus*. isbn=9781400852017. (2014). Princeton University Press.
2. Chang, Jen-Mei:*A practical approach to inquiry-based learning in linear algebra*. pp. 245–259. International Journal of Mathematical Education in Science and Technology. (2011). https://doi.org/10.1080/0020739X.2010.519795.
3. Crannell, A., LaRose, G., Ratliff, T., & Rykken, E.: *Writing Projects for Mathematics Courses: Crushed Clowns, Cars, and Coffee to Go*. pp. 9–96. Mathematical Association of America. (2004). https://doi.org/10.5948/9781614441137.002
4. Lang, James M: *Small Teaching: Everyday Lessons from the Science of Learning*. pp. 182–193. (2016). Jossey-Bass, Wiley Brand.
5. Larson, R. and Edwards, B.H. and Falvo, D.C: *Elementary Linear Algebra*. isbn=978061800508. (2003). Houghton Mifflin.
6. Lomax, R. and Pecos, H.: *The Science of Zombism*. Zombie Virus. (2018).https://fvza.org/zscience1.html.
7. Willingham, D. *Why don't students like school? A cognitive scientist answers questions about how the mind works and what it means for the classroom*. pp. 67–68. (2014). Jossey-Bass, San Francisco.

Part IV
Mathematics Research

A Multivariate Rank-Based Two-Sample Test Statistic

Jamye Curry, Xin Dang, and Hailin Sang

Abstract The problem of testing whether two samples come from the same or different populations is a classical one in statistics. A new distribution-free test based on standardized ranks for the univariate two-sample problem is studied. Providing a distribution-free (nonparametric) method offers a valuable technique for analyzing data that consists of ranks or relative preferences of data, and of data that are small samples from unknown distributions. The proposed test statistic examines the difference between the average of between-group rank distances and the average of within-group rank distances. This test statistic is closely related to the classical two-sample Cramér–von Mises criterion; however, they are different empirical versions of the same quantity for testing the equality of two population distributions. The advantage of the proposed rank-based test over the classical one is its ease to generalize to the multivariate case. In addition, the motivation of the proposed test is its application to microarray analyses in identifying sets of genes that are differentially expressed in various biological states, such as diseased versus non-diseased. A numerical study is conducted to compare the power performance of the rank formulation test with other commonly used nonparametric tests.

1 Introduction

The objective of this paper is to study and investigate the two-sample problem. A two-sample problem is when two independent random samples are obtained and used to test a hypothesis regarding the populations in which the samples are drawn from. The investigation of such problem, for example, can be used

J. Curry (✉)
Georgia Gwinnett College, Lawrenceville, GA, USA
e-mail: jcurrysavage@ggc.edu

X. Dang · H. Sang
University of Mississippi, University, MS, USA
e-mail: xdang@olemiss.edu; sang@olemiss.edu

© The Author(s) and the Association for Women in Mathematics 2019
S. D'Agostino et al. (eds.), *A Celebration of the EDGE Program's Impact
on the Mathematics Community and Beyond*, Association for Women
in Mathematics Series 18, https://doi.org/10.1007/978-3-030-19486-4_16

to analyze microarray data. Microarray technology allows researchers to examine samples of gene expression levels under diverse circumstances. A common objective in analyzing data from microarray experiments is to identify which genes are differentially expressed which are drawn from samples obtained under various conditions. Researchers test for differentially expressed genes when searching for disease-related genes. For instance, one can compare data of gene expression levels between cancer tissue samples and normal tissue samples to select genes related to the cancer under investigation. The cancer gene will be detected if the average expression levels between the two samples are significantly different.

For many statistical procedures, the goal is to test the null hypothesis of identical distributions versus the location and dispersion (scale) alternatives, which is that the populations are from the same distribution, but have a different measure of central tendency and variability. There are also many statistics used to test for dispersion which tests for the equality of variances of the two populations. However, these procedures are a part of the parametric testing family where there are many assumptions made about the parameters of the underlying distributions. If the assumptions are accurate, then a parametric approach will provide precise estimates of the parameters. On the other hand, if the assumptions are not accurate, then a parametric approach will provide misleading and false results. Thus, we use a nonparametric approach as it is more robust and resistant against outlying data, and has very few assumptions.

Several distribution-free tests such as the Kolmogorov–Smirnov test, the Cramér–von Mises test, and their variations have been proposed and widely used. Suppose $X_1, X_2, \ldots, X_m \overset{iid}{\sim} F$ and $Y_1, Y_2, \ldots, Y_n \overset{iid}{\sim} G$ are two independent random samples with continuous distribution functions F and G, respectively, where the notation iid indicates that the samples are independent and identically distributed. The two-sample problem is to test

$$H_0 : F = G \quad vs \quad H_a : F \neq G. \tag{1}$$

Denote F_m and G_n as the empirical distribution functions of the two samples and H_N as the empirical distribution function of the combined sample, where $N = m + n$. The Kolmogorov–Smirnov (KS) two-sample test statistic is defined as the maximum distance (difference) of the set of distances between the empirical distributions F_m and G_n of the two samples. The classical Cramér–von Mises test statistic has the form

$$T_c = \frac{mn}{N} \int_{-\infty}^{\infty} [F_m(x) - G_n(x)]^2 dH_N(x). \tag{2}$$

This test statistic and its asymptotics have been well studied in the literature, for example, Lehmann [13], Rosenblatt [17], Darling [5], Fisz [8], and Anderson [1].

Both of the KS test statistic and the Cramér–von Mises test statistic are formulated based on the empirical distributions. Baringhaus and Franz [2] studied a test statistic based on the original data. That is,

$$\frac{mn}{N}\left\{\frac{1}{mn}\sum_{i=1}^{m}\sum_{j=1}^{n}|X_i-Y_j|-\frac{1}{2m^2}\sum_{i=1}^{m}\sum_{j=1}^{m}|X_i-X_j|-\frac{1}{2n^2}\sum_{i=1}^{n}\sum_{j=1}^{n}|Y_i-Y_j|\right\}.$$
(3)

This test statistic (3) was motivated by a conjecture considering the distances between points of two types (Morgenstern [14]) and it has a direct generalization to the multivariate case. However, it requires an assumption on the first moment and it is not distribution-free for the univariate case (Baringhaus and Franz [2]). It is worth to note that the test statistic (3) falls in the unified framework on energy statistics studied by Székely and Rizzo [19]. Other similar tests include [7] and [10], although they are derived under different motivations. Fernandez et al. [7] developed a statistic based on the empirical characteristic functions of the observed observations. The statistic uses a weighted integral of the difference between the empirical characteristic function of the two samples. Gretton et al. [10] proposed a test based on a kernel method in which the testing procedure is defined as the maximum difference in expectations over functions evaluated on the two samples. Each of these test statistics are of the form that measures the distance of the samples between the two groups and measures the distance of the samples within the same group, or a combination of the two. All of these test statistics mentioned above are motivations to the proposed test statistic in this paper. The main objective for our approach in investigating the two-sample problem is to test the null hypothesis for the equality of distributions from independent random samples without any parametric assumptions on the underlying populations. We will also observe that the new approach may be applied in the multivariate setting as oppose to only the univariate setting

The layout of the paper is as follows: A rank-based test statistic is proposed for the univariate case followed by a simulation study to investigate the power performance of the test statistic. The section following the univariate simulation study will present the multivariate extension of the proposed rank-based test. The final section includes a brief summary and discussion of the paper.

2   Univariate Case

Here, we propose a new rank-based test of the same form as the test statistic in (3). Nevertheless, it overcomes the limitations of (3). It is formulated based on the ranks of two samples with respect to the combined sample H_N. Denote $R(y, H)$ as the standardized rank of the quantity y with respect to the distribution H, i.e., $R(y, H) = H(y)$. For testing the hypothesis (1), we use the following test statistic:

$$T = \frac{mn}{N}\left\{\frac{1}{mn}\sum_{i=1}^{m}\sum_{j=1}^{n}|R(X_i, H_N) - R(Y_j, H_N)|\right.$$

$$-\frac{1}{2m^2}\sum_{i=1}^{m}\sum_{j=1}^{m}|R(X_i, H_N) - R(X_j, H_N)|$$

$$-\frac{1}{2n^2}\sum_{i=1}^{n}\sum_{j=1}^{n}|R(Y_i, H_N) - R(Y_j, H_N)|\bigg\}. \tag{4}$$

T is interpreted as the difference of the average of between-group rank differences and the average of within-group rank differences. A large value of T indicates the deviation of two groups. The test based on T is distribution-free and does not require any moment condition.

For the balanced samples ($m = n$), one can consider an equivalent but simpler statistic

$$T' = \frac{1}{mn}\sum_{i=1}^{m}\sum_{j=1}^{n}|R(X_i, H_N) - R(Y_j, H_N)|. \tag{5}$$

T' is the average of rank differences between two groups. T and T' are equivalent as $T = nT' - (4n^2 - 1)/(12n)$ when $m = n$.

Later, we show that the test statistic T is closely related to the classical nonparametric Cramér–von Mises criterion T_c. They are different empirical plug-in versions of the same population quantity. The rank-based test statistic and the Cramér–von Mises criterion may not be equal to each other for finite samples, but they are asymptotically equivalent. The advantage of the new rank-based test over the classical one is its ease to generalize to the multivariate case. Multivariate generalizations of Cramér–von Mises tests have been considered by many researchers, but they are either applied on independent data [4], used for testing independence [9], or used in the goodness-of-fit test of the uniform distribution on the transformed data [3]. For the proposed rank-based formulation, generalizations to the multivariate two-sample problem are straightforward by applying notions of multivariate rank functions.

To formulate the rank-based test statistic T in (4), we first establish its population version. We provide a result of the population version, from which we can see the relationship between our test statistic and Cramér–von Mises criterion.

Lemma 1 *Let X and Y be independent continuous random variables from F and G, respectively. Let $H = \tau F + (1-\tau)G$ with $0 \leq \tau \leq 1$ be the mixture distribution, J be the distribution of $R(X, H)$, and K be the distribution function of $R(Y, H)$. Then*

$$\mathbb{E}|R(X, H) - R(Y, H)| = \int_0^1 J(t)(1 - K(t))\,dt + \int_0^1 K(t)(1 - J(t))\,dt. \tag{6}$$

Proof Notice that

$$|R(X, H) - R(Y, H)| = \int_0^1 \Big[I(R(X, H) \le s < R(Y, H))$$

$$+ I(R(Y, H) \le s < R(X, H)) \Big] ds.$$

Since H is continuous and $R(X, H) = H(X)$, $R(Y, H) = H(Y)$, we have $J(x) = F \circ H^{-1}(x)$, $K(x) = G \circ H^{-1}(x)$ for any $x \in [0, 1]$, where $H^{-1}(x) = \inf\{u : H(u) \ge x\}$. Then (6) holds by Fubini's theorem. $\square$

Thus, based on Lemma 1 and the identity

$$\mathbb{E}|R(X, H) - R(Y, H)| - \frac{1}{2}\mathbb{E}|R(X_1, H) - R(X_2, H)| - \frac{1}{2}\mathbb{E}|R(Y_1, H)$$

$$- R(Y_2, H)| = \int_{-\infty}^{\infty} (F(x) - G(x))^2 d(\tau F(x) + (1 - \tau)G(x)), \quad (7)$$

we have the following theorem.

Theorem 1 *Let X, X_1, X_2 and Y, Y_1, Y_2 be independent continuous random variables distributed from F and G, respectively. Let $H = \tau F + (1 - \tau)G$ with $0 \le \tau \le 1$ be the mixture distribution. Then*

$$\mathbb{E}|R(X, H) - R(Y, H)| - \frac{1}{2}\mathbb{E}|R(X_1, H) - R(X_2, H)| - \frac{1}{2}\mathbb{E}|R(Y_1, H)$$

$$- R(Y_2, H)| \ge 0 \quad (8)$$

and the equality holds if and only if $F = G$.

The result of Theorem 1 suggests two possible statistics for testing the hypothesis (1). The first version is the sample plug-in version of the left side of (7). With $\tau = m/N$ and multiplying by mn/N, it is our test statistic defined in (4). H_0 is rejected if the sample version is large, i.e., $T > c_\alpha(m, n)$. The critical value $c_\alpha(m, n)$ is determined by the significance level α and the null distribution of T. The test statistic T is the difference of the average of between-group rank differences and the average of within-group rank differences. A large value of T indicates the deviation of two groups.

The two-sample Cramér–von Mises statistic T_c in (2) is the empirical version of the right side of (7). Hence T and T_c are all plug-in statistics of an equal quantity. Nevertheless, they may take different values. For example, in the case that $m = n = 2$, let the two X realizations be 0 and 2 and the two Y realizations be 1 and 3. It is easy to see that the Cramér–von Mises statistic has value $\frac{1}{4}$ and the test statistic T has value $\frac{1}{8}$.

Next, we will study the properties of T. Let D be $\mathbb{E}|R(X, H) - R(Y, H)| - \frac{1}{2}\mathbb{E}|R(X_1, H) - R(X_2, H)| - \frac{1}{2}\mathbb{E}|R(Y_1, H) - R(Y_2, H)|$, and $\hat{D} = N/(mn)T$. Then we have the following theorem.

Theorem 2 *For $m, n \to \infty$, if $m/(m+n) \to \tau$, then $\hat{D} \to D$ a.s.*

By this theorem and Theorem 1, it is easy to see that our test statistic T is consistent for the alternative $H_a : F \neq G$.

Theorem 3 *Under H_0, T is distribution-free.*

Under H_0, the combined samples $X_1, \ldots, X_m, Y_1, \ldots, Y_n$ constitute a random sample of size N from the distribution $F = G = H$. So any assignment of m numbers to $X_1, \ldots, X_m$ and n numbers to $Y_1, \ldots, Y_n$ from the set of integers $\{1, 2, \ldots, N\}$ is equally likely, i.e., has probability $\binom{N}{m}^{-1}$ which is independent of F. Using the fact that those number assignments have one-to-one linear relationship with the standardized ranks, T is distribution-free.

The exact null distribution of T can be found by enumeration of all possible values of T by considering the $N!/(m!n!)$ orderings of m X's and n Y's. Figure 1 provides the exact null distribution of T for sample sizes $m = n = 7$ and $m = 7, n = 9$ by considering all combinations. However, the exact null distribution is infeasible to obtain for large sample sizes because the number of combinations increases dramatically as m and n increase. For large samples, we can use the Monte Carlo method on all combinations to approximate the null distribution [18]. Also, the limiting distribution of T can be used to determine the critical values of the test.

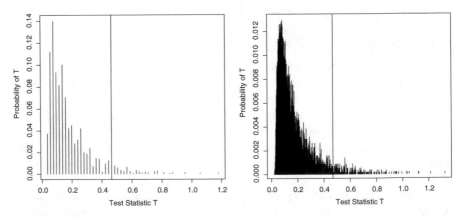

Fig. 1 The exact null distribution of T obtained from all combinations (left: $m = 7, n = 7$; right: $m = 7, n = 9$). The vertical line in each graph indicates the 5% critical value

3 Simulations

There are many nonparametric tests available for the two-sample problem. By the simulation study in this section, we demonstrate the performance of the T test. It is by no means to conduct a comprehensive comparison. Here, we include Kolmogorov–Smirnov test (KS), Wilcoxon rank sum test (W) or Mood's test (M), the empirical likelihood ratio test (ELR) proposed by Gurevich anf Vexler [11], the empirical likelihood test (ELT) proposed by Einmahl and McKeague [6], Baringhaus and Franz's Cramér test (CT) as in (3), and the test studied in Fernándes et al. [7] (DT). It is necessary to note that the CT and DT tests are not distribution-free tests, and their critical values and p-values are based on the Monte Carlo method on permutations in each sample, which is implemented in the R package "Cramer." The R package "dbEmpLikeGOF" is used for the ELR test in which the parameter is set to be 0.1 as suggested in [11]. The critical values of the ELT and our T test are computed through 10^7 random combinations on $\{1, \ldots, N\}$.

Various alternative distributions are considered. For each case, $M = 10,000$ iterations are computed to estimate powers by calculating the fraction of p-values less than or equal to $\alpha = 0.05$, the level of significance. The Monte Carlo errors can be estimated by $\pm 1.96 \sqrt{p(1-p)/M}$. In particular, the size of tests shall maintain in the interval $(0.046, 0.054)$.

Table 1 shows the size and power performance for each test under the normal distributions, where $X_1, \ldots, X_n \sim N(0, 1)$ and $Y_1, \ldots, Y_m \sim N(\Delta, 1)$ with $\Delta = 0$, 0.25, 0.5, 0.75, and 1. When $\Delta = 0$, the KS test is undersized for both the equal and unequal sample size cases; the ELR test is oversized in the equal sample size case and seriously undersized for the unequal sample size case; all other tests keep a desirable size. As expected, the W test is the best among all tests since it is well known to be powerful for the two-sample problem with a constant shift in location, especially when data follow logistic or normal distributions. The CT and ELT tests are comparable to W. The T test is more powerful than the DT, KS, and ELR tests. In the unequal sample size case, the W test is the best followed by the CT test. The ELT and T tests are comparable and significantly better than the DT, KS, and ELR tests.

The experiment is repeated for the t-distribution with 3 degrees of freedom and the result is presented in Table 2. Although the statistical power of the T test is the highest among all tests for all cases, its power differences with the W test or the ELT test are small so that those three tests are comparable.

Table 3 shows the power performance for the Pareto distribution, where $X_1, \ldots, X_n \sim Pa(2, 2)$ and $Y_1, \ldots, Y_m \sim Pa(2+\Delta, 2)$ are generated, with $\Delta = 0$, 0.25, 0.5, 0.75, and 1. The power of the ELR test is much higher than that of all others. For $\Delta = 0.25$, the power of the ELR test is as high as 90%, which is 30% higher than the second best ELT test. A potential reason why the ELR test outperforms the other tests is the idea that the ELR test is based on the empirical likelihood function consisting of components that maximize the likelihood function which satisfies empirical constraints. The T test is the third best one. The power

Table 1 Power performance of each test with significance level $\alpha = 0.05$ for the normal distribution with location alternatives

Δ	KS	W	ELR	ELT	CT	DT	T
0	0.040	0.050	0.057	0.050	0.050	0.051	0.050
	0.041	0.047	0.031	0.047	0.048	0.048	0.049
0.25	0.162	0.228	0.182	0.224	0.226	0.191	0.217
	0.160	0.208	0.119	0.196	0.203	0.171	0.198
0.5	0.534	0.681	0.578	0.670	0.671	0.582	0.652
	0.498	0.621	0.446	0.603	0.615	0.526	0.600
0.75	0.875	0.949	0.902	0.945	0.943	0.901	0.936
	0.851	0.926	0.829	0.919	0.922	0.871	0.912
1	0.988	0.998	0.994	0.997	0.998	0.991	0.996
	0.979	0.995	0.976	0.994	0.994	0.984	0.993

Row 1: $n = m = 50$, Row 2: $n = 50, m = 40$

Table 2 Power performance of each test with significance level $\alpha = 0.05$ for the t_3 with location alternatives

Δ	KS	W	ELR	ELT	CT	DT	T
0	0.036	0.048	0.052	0.048	0.049	0.046	0.047
	0.045	0.054	0.036	0.054	0.055	0.051	0.054
0.25	0.130	0.163	0.124	0.160	0.158	0.142	0.165
	0.135	0.157	0.084	0.152	0.154	0.137	0.162
0.5	0.422	0.488	0.367	0.486	0.481	0.434	0.501
	0.402	0.449	0.268	0.440	0.439	0.394	0.462
0.75	0.776	0.830	0.710	0.825	0.818	0.783	0.836
	0.741	0.786	0.586	0.780	0.776	0.734	0.799
1	0.947	0.966	0.917	0.966	0.965	0.950	0.971
	0.929	0.946	0.842	0.946	0.944	0.925	0.954

Row 1: $n = m = 50$, Row 2: $n = 50, m = 40$

difference between the T test and that of the CT test can be as large as 27% for equal sample sizes and can be as large as 32% for unequal sample sizes.

All considered tests in the experiment for location alternatives are used for scale alternatives except the Wilcoxon test (W), as this is a test for location. Instead, the Mood's test known as a scale test is used and referred to as the M test. Table 4 displays the results when the Y samples of size 50 are generated from $N(0, \Delta)$ or Pareto$(2, 2\Delta)$, where $\Delta = 1, 1.5, 2, 2.5$, and 3. In the normal case, the T test does not compare favorably to all considered tests other than the KS test. It performs significantly better than the KS test, but its power is 2–5 times smaller than that of others. It is interesting to see that the M test outperforms all tests in the normal case, but it is the inferior in the Pareto case. The T test has better performance for Pareto samples than for normal samples due to the heavy tails in Pareto distributions. In the Pareto case, all tests outperform the M test by a great margin and the CT test is the superior. Furthermore, we add one more case in the

Table 3 Power performance of each test with significance level $\alpha = 0.05$ for the Pareto distributions with location alternatives

Δ	KS	W	ELR	ELT	CT	DT	T
0	0.040	0.052	0.058	0.051	0.052	0.050	0.051
	0.044	0.050	0.032	0.050	0.050	0.053	0.052
0.25	0.417	0.443	0.906	0.599	0.205	0.287	0.472
	0.377	0.405	0.815	0.516	0.186	0.265	0.428
0.5	0.960	0.886	0.999	0.978	0.655	0.843	0.968
	0.940	0.859	0.998	0.958	0.598	0.798	0.948
0.75	0.999	0.989	1.000	0.999	0.945	0.993	0.999
	0.998	0.980	1.000	0.998	0.908	0.988	0.997
1	1.000	0.999	1.000	1.000	0.993	1.000	1.000
	1.000	0.998	1.000	1.000	0.988	1.000	1.000

Row 1: $n = m = 50$, Row 2: $n = 50, m = 40$

Table 4 Power performance of each test with significance level $\alpha = 0.05$ for Normal and Pareto scale alternatives, also the case of F = Exp and G = Lognorm

Distribution	Δ	KS	M	ELR	ELT	CT	DT	T
Normal	1	0.039	0.051	0.056	0.047	0.049	0.051	0.047
	1.5	0.118	0.663	0.542	0.238	0.251	0.431	0.138
	2	0.374	0.979	0.965	0.746	0.792	0.915	0.479
	2.5	0.681	0.999	0.999	0.962	0.981	0.994	0.815
	3	0.881	1.000	1.000	0.996	0.999	1.000	0.957
Pareto	1	0.040	0.054	0.055	0.049	0.052	0.049	0.050
	1.5	0.307	0.098	0.378	0.418	0.487	0.356	0.398
	2	0.741	0.165	0.828	0.857	0.909	0.815	0.831
	2.5	0.937	0.214	0.973	0.980	0.992	0.974	0.978
	3	0.988	0.234	0.997	0.998	0.999	0.997	0.998
Exp vs Lgnorm		0.336	0.535	0.654	0.555	0.502	0.315	0.476

simulation in which $X_1, \ldots, X_n \sim exp(1)$ and $Y_1, \ldots, Y_m \sim lognorm(0, 1)$ with sample sizes $m = n = 50$. The Monte Carlo powers of the seven tests are listed in Table 4. In this scenario, the T test performs better than KS and DT, but does not compare as favorably to the CT, W, ELT, and ELR tests.

Based on the simulations above, the T test is not recommended for scale alternatives. The Kolmogorov–Smirnov test is not recommended either. The empirical likelihood ELR test is more suitable for a scale alternative, but is not recommended for a location alternative for symmetric distributions. The T test has a better performance for location alternatives than scale alternatives. It is easy to explain the power performance of the Cramér–von Mises test with the rank-based formulation (4) for the location alternatives. For two samples from the same class distributions (normal distributions, t-distributions or Pareto distributions, and so on), but with different locations, the ranks in the mixture are quite different. Therefore, the corresponding test can easily recognize them and have good power performance. We recommend to apply the T test for location alternatives, especially in the heavy-tailed distributions.

4   Multivariate Extension

The proposed rank-based test statistic is closely related to the two-sample Cramér–von Mises criterion. Both statistics are different sample plug-in forms from a same population quantity. The advantage of our rank test is to allow straightforward generalizations to the multivariate case by using different multivariate rank functions. Among them, the spatial rank is appealing due to its computation ease, efficiency, and other nice properties [15, 16]. The sample version of the spatial rank function with respect to H_N, the empirical distribution of the combined sample $x_1, \ldots, x_m$ and $y_1, \ldots, y_n$ in $\mathbb{R}^d$, is defined as

$$R(x, H_N) = \frac{1}{N} \sum_{i=1}^{N} \frac{x - z_i}{\|x - z_i\|},$$

where $z_i = x_i$ for $i = 1, \ldots, m$, $z_{m+i} = y_i$ for $i = 1, \ldots, n$, and $\| \cdot \|$ is the Euclidian distance. Then the multivariate two-sample spatial rank statistic, denoted by T_M, is defined as

$$T_M = \frac{mn}{N} \{ \frac{1}{mn} \sum_{i=1}^{m} \sum_{j=1}^{n} \|R(x_i, H_N) - R(y_j, H_N)\|$$

$$- \frac{1}{2m^2} \sum_{i=1}^{m} \sum_{j=1}^{m} \|R(x_i, H_N) - R(x_j, H_N)\|$$

$$- \frac{1}{2n^2} \sum_{i=1}^{n} \sum_{j=1}^{n} \|R(y_i, H_N) - R(y_j, H_N)\| \}. \tag{9}$$

The test statistic T_M is the difference of the average of the intra-group rank distances and the average of the inter-group rank distances. A large value of T_M indicates the deviation of the two groups and rejects the null hypothesis. The multivariate counterpart of Theorem 1 states as follows:

Theorem 4 *Let X, X_1, X_2 and Y, Y_1, Y_2 be independent d-variate continuous random vectors distributed from F and G, respectively. Let $H = \tau F + (1 - \tau)G$ with $0 \le \tau \le 1$. Then*

$$\mathbb{E}\|R(X, H) - R(Y, H)\| - \frac{1}{2}\mathbb{E}\|R(X_1, H) - R(X_2, H)\| - \frac{1}{2}\mathbb{E}\|R(Y_1, H)$$

$$- R(Y_2, H)\| \ge 0, \tag{10}$$

where the equality holds if and only if $F = G$.

Proof Let μ be the uniform distribution on the surface of the unit ball $S^{d-1} = \{a \in \mathbb{R}^d : \|a\| = 1\}$. From Theorem 1, we have

$$\mathbb{E}|R(a^T X, H^a) - R(a^T Y, H^a)| - \frac{1}{2}\mathbb{E}|R(a^T X_1, H^a) - R(a^T X_2, H^a)|$$

$$- \frac{1}{2}\mathbb{E}|R(a^T Y_1, H^a) - R(a^T Y_2, H^a)| \geq 0$$

for each $a \in S^{d-1}$, where $H^a = \tau F^a + (1 - \tau)G^a$ with F^a and G^a being the distributions of $a^T X$ and $a^T Y$, respectively. Integration of a with respect to μ obtains (10). Equality holds if and only if for μ-almost all $a \in S^{d-1}$ the distributions of $a^T X$ and $a^T Y$ coincide. For each $t \in \mathbb{R}$ the functions $\mathbb{E}\exp(ita^T X)$ and $\mathbb{E}\exp(ita^T Y)$ with $a \in S^{d-1}$ are continuous. Thus, equality in (10) holds if and only if X and Y have the same characteristic function, hence have the same distribution.   □

The multivariate spatial rank test based on T_M loses the distribution-free property under the null hypothesis. The test relies on the permutation method to determine critical values or to compute p-values. However, the test is robust. For example, it does not require the assumption of finite second moment as the Hotelling's T^2 test. Neither does it require the assumption of finite first moment as the CT test considered by Baringhaus and Franz [2].

A simulation is conducted to compare performance of T_M, CT, and the Hotelling's T^2 under multivariate normal, t_1, and Pareto distributions on $\mathbb{R}^d$ ($d = 2, 5$). Location and scatter alternatives are considered. For location alternatives in normal and t_1 distributions, the parameters of distributions for generating X samples of size $n = 50$ are $\mu = 0$ and $\Sigma_X = I$, while for Y samples with size $m = 50$ are $\mu = (\Delta, \ldots, \Delta)^T$ and $\Sigma_Y = I$, where $\Delta = 0, 0.25, 0.5, 0.75$, and 1. For the Pareto distribution, $X = (X_1, \ldots, X_d)^T$ is generated with each component X_j from Pareto(1,1) and $Y = (Y_1, \ldots, Y_d)^T$ is generated with each component Y_j from Pareto($1 + \Delta$, 1). The R package "Hotelling" is used for the Hotelling's T^2 test. The T_M and CT tests use the permutation method to compute p-values and $M = 10,000$ iterations are computed to estimate powers by calculating the fraction of p-values less than or equal 0.05. The results for the location alternatives are listed in Table 5.

From Table 5, all three tests keep the size 5% well. Powers in $d = 5$ are higher than that in $d = 2$ for each of the three tests under all distributions. In the normal cases, T_M performs slightly worse than the Hotelling's test and CT. The power of T_M is about 2% lower than that of the Hotelling's test and 1% lower than that of CT under H_a when $\Delta = 0.25$ and $\Delta = 0.50$. However, the power gain of T_M over CT and the Hotelling's test is huge in the t_1-distributions. For $\Delta = 0.25$ and $\Delta = 0.5$, T_M is about twice powerful as CT and about triple powerful as the Hotelling's test. The advantage of our proposed T_M over CT and the Hotelling's test is even more significant in the asymmetric Pareto distributions than in the t_1 distributions for the location alternatives.

Table 5 Power performance of T_M, CT, and Hotelling's tests with significance level $\alpha = 0.05$ for multivariate normal, t_1, and Pareto distributions with location alternatives with sample sizes $n = m = 50$

Dist	Dim	Method	$\Delta = 0$	$\Delta = .25$	$\Delta = .50$	$\Delta = .75$	$\Delta = 1$
Norm	$d = 2$	T_M	0.0550	0.3000	0.8688	0.9966	1
		CT	0.0556	0.3090	0.8818	0.9972	1
		Hotelling	0.0518	0.3226	0.8900	0.9976	1
	$d = 5$	T_M	0.0484	0.5178	0.9944	1	1
		CT	0.0500	0.5332	0.9958	1	1
		Hotelling	0.0494	0.5248	0.9942	1	1
t_1	$d = 2$	T_M	0.0538	0.1574	0.4898	0.8212	0.9644
		CT	0.0596	0.0820	0.226	0.4504	0.7134
		Hotelling	0.0546	0.0562	0.0934	0.1360	0.2058
	$d = 5$	T_M	0.0478	0.2382	0.7986	0.9888	0.9996
		CT	0.0546	0.0858	0.288	0.6200	0.8568
		Hotelling	0.0472	0.0742	0.1622	0.2990	0.4608
Pareto	$d = 2$	T_M	0.0492	0.3470	0.8682	0.9886	0.9998
		CT	0.0560	0.1146	0.2850	0.5330	0.7298
		Hotelling	0.0484	0.0986	0.1858	0.3076	0.4188
	$d = 5$	T_M	0.0522	0.2892	0.7942	0.9784	0.9996
		CT	0.0492	0.1142	0.2942	0.5184	0.7128
		Hotelling	0.0528	0.1108	0.2614	0.4462	0.6046

The results for scatter alternatives are listed in Table 6. For multivariate normal and t_1 distributions, we first consider the difference of scatter matrix only on scales. The parameters for X samples are $\mu = 0$ and $\Sigma_X = I$, while for Y samples are $\mu = 0$ and $\Sigma_Y = \Delta I$, where $\Delta = 1, 1.5, 2, 2.5$, and 3. We then consider the alternative with different orientation on the scatter matrices. The scatter matrix is $\begin{pmatrix} 1 & .5 \\ .5 & 1 \end{pmatrix}$ for X samples, while it is $\begin{pmatrix} 1 & -.5 \\ -.5 & 1 \end{pmatrix}$ for Y samples. Hence two components of X are positively correlated and the two components of Y are negatively correlated. The results for orientation difference alternatives are listed in the last column "Orient" in Table 6. In $d = 5$, Σ_X has diagonal elements to be one and off-diagonal elements to be 0.5, and Σ_Y is constructed to have the same eigenvectors as Σ_X and eigenvalues to be the reciprocals of eigenvalues of Σ_X. For Pareto distributions, $X = (X_1, \ldots, X_d)^T$ is generated with each component X_j from Pareto(1,1) and $Y = (Y_1, \ldots, Y_d)^T$ is generated with each component Y_j from Pareto(1, Δ).

From Table 6, all tests maintain the size 5% well. For asymmetric Pareto distributions, CT is slightly better than T_M and T_M is better than the Hotelling's test. For normal and t_1 distributions, the Hotelling's T^2 completely fails in scatter alternatives since it is a test on location difference. CT test is much better than T_M for scale alternatives. Particularly, CT is triple powerful as the T_M in normal case and twice powerful in the t_1 case. This result is not surprising since T_M is based on

Table 6 Power performance of T_M, CT, and Hotelling's tests with significance level $\alpha = 0.05$ for multivariate normal, t_1, and Pareto distributions with scatter alternatives with sample sizes $n = m = 50$

Dist	Dim	Method	$\Delta = 1$	$\Delta = 1.5$	$\Delta = 2$	$\Delta = 2.5$	$\Delta = 3$	Orient
Norm	$d = 2$	T_M	0.0468	0.0640	0.1064	0.1902	0.2992	0.3179
		CT	0.0474	0.0982	0.2716	0.5660	0.8072	0.3016
		Hotelling	0.048	0.0472	0.0598	0.0496	0.0524	0.0493
	$d = 5$	T_M	0.0476	0.0748	0.1272	0.2600	0.4124	0.9678
		CT	0.0470	0.1510	0.5580	0.9192	0.9948	0.8188
		Hotelling	0.045	0.0568	0.0526	0.0576	0.0540	0.0538
t_1	$d = 2$	T_M	0.0486	0.0580	0.0698	0.0948	0.1256	0.2366
		CT	0.0482	0.0680	0.1182	0.1754	0.2370	0.0916
		Hotelling	0.0506	0.0476	0.0488	0.0530	0.0544	0.0495
	$d = 5$	T_M	0.0514	0.0648	0.0900	0.1286	0.1742	0.5896
		CT	0.0512	0.0836	0.1510	0.2320	0.3344	0.1984
		Hotelling	0.0528	0.0494	0.0492	0.0556	0.0550	0.0468
Pareto	$d = 2$	T_M	0.0550	0.6164	0.9802	1	1	–
		CT	0.0540	0.6896	0.9896	1	1	–
		Hotelling	0.0498	0.5148	0.9354	0.9876	0.9976	–
	$d = 5$	T_M	0.0504	0.9158	1	1	1	–
		CT	0.0512	0.9268	0.9996	1	1	–
		Hotelling	0.0566	0.7616	0.9960	0.9998	1	–

the spatial ranks that lose major information on distances or scales. However, when two scatter matrices of distributions are different on orientation, T_M performs better than CT, and especially in the t_1 distribution, the power of T_M is twice or triple as that of CT.

5 Summary and Discussion

In summary, the problem of testing whether two samples come from the same or different populations is a classical one in statistics. We have studied a rank-based test for the univariate two-sample problem, where the test statistic is defined as the difference between the average of between-group rank distances and the average of within-group rank distances. We have shown that, under the null hypothesis, the test statistic is distribution-free, and we recommend to apply the T test for location alternatives, especially in the heavy-tailed distributions.

The proposed rank test statistic is closely related to the two-sample Cramér–von Mises criterion. Both statistics are different sample plug-in forms from the same population quantity. The advantage of our rank test is that it allows for straightforward generalizations to the multivariate case by using different multivariate rank functions.

The test statistic T_M can be applied in the sense of detecting differentially expressed genes. For instance, in a heterogeneous disease, the gene expression level often shows greater variability from one case subject to another, due to the presence of a certain number of distinct disease entities, as compared with the variability seen from one control subject to another. In a single experiment, microarray technology can provide information about hundreds of thousands of gene expression data. Researchers try to test for differentially expressed genes to find genes which are disease-related. That is, compare gene expression levels between cancerous and normal genes, and detect the ones that are differentially expressed. If the mean expression levels are statistically different, then there is a difference between the two groups [12]. This type of data will generally produce multivariate robust data that does not come from a normal distribution. Therefore, the T_M test statistic would be a preferred test in this case to detect such gene expression levels and to analyze such data.

The idea of detecting differentially expressed genes serves as motivation for pursuing the proposed tests. Therefore, a continuation of this work is to study properties of the multivariate T_M test and to utilize the test in application. Also, future work is to further investigate generalizations based on other multivariate rank functions.

References

1. Anderson, T.W. (1962). On the distribution of the two-sample Cramér-von Mises criterion. *Ann. Math. Statist.*, **33**(3), 1148–1159.
2. Baringhaus, L. and Franz, C. (2004). On a new multivariate two-sample test. *J. Multivariate Anal.*, **88**, 190–206.
3. Chiu, S. and Liu, K. (2009). Generalized Cramér-von Mises goodness-of-fit tests for multivariate distributions. *Comput. Stat. Data An.*, **53**, 3817–3834.
4. Cotterill, D. and Csörgő, M. (1982). On the limiting distribution of and critical values for the multivariate Cramér-von Mises Statistic. *Ann. Stat.*, **10**(1), 233–244.
5. Darling, D.A. (1957). The Kolmogorov-Smirnov, Cramér-von Mises tests. *Ann. Math. Stat.*, **28**(4), 823–838.
6. Einmahl, J. and McKeague, I. (2003). Empirical likelihood based hypothesis testing. *Bernoulli*, **9**(2), 267–290.
7. Fernández, V., Jimènez Gamerro, M. and Muñoz Garcìa, J. (2008). A test for the two-sample problem based on empirical characteristic functions. *Comput. Stat. Data An.*, **52**, 3730–3748.
8. Fisz, M. (1960). On a result by M. Rosenblatt concerning the von Mises - Smirnov Test. *Ann. Math. Stat.*, **31**(2), 427–429.
9. Genest, C., Quessy, J.F. and Rémillard, B. (2007). Asymptotic local efficiency of Cramér-von Mises tests for multivariate independence. *Ann. Stat.*, **35**(1), 166–191.
10. Gretton, A., Borgwardt, K.M., Rasch, M.J., Schölkopf, B. and Smola, A. (2008). A kernel method for the two-sample problem, *J. Mach. Learn. Res.*, **1**, 1–10.
11. Gurevich, G. and Vexler, A. (2011). A two-sample empirical likelihood ratio test based on samples entropy. *Stat. Comput.*, **21**, 657–670.
12. Hsu, C.L. and Lee, W.C. (2010). Detecting differentially expressed genes in heterogeneous diseases using half Student's t-test. *International Journal of Epidemiology.*, **39**,1597–1604.

13. Lehmann, E.L. (1951). Consistency and unbiasedness of certain nonparametric tests. *Ann. Math. Stat.*, **22**, 165–179.
14. Morgenstern, D. (2001). Proof of a conjecture by Walter Deuber concerning the distances between points of two types in $\mathbb{R}^d$, *Discrete Math.*, **226**, 347–349.
15. Möttönen J., Oja, H. and Tienari J. (1997). On the efficiency of multivariate spatial sign and rank tests. *Ann. Stat.*, **25**, 542–552.
16. Oja, H. (2010). *Multivariate Nonparametric Methods with R: An Approach Based on Spatial Signs and Ranks*. Springer, New York.
17. Rosenblatt, M. (1952). Limit theorems associated with variants of the von Mises statistic. *Ann. Math. Stat.*, **23**, 617–623.
18. Rubinstein, R. Y. (1981). *Simulation and the Monte Carlo Method*. John Wiley & Sons, New York.
19. Székely, G.J. and Rizzo, M.L. (2013). Energy statistics: A class of statistics based on distances. *J. Stat. Plan. Infer.*, **143**, 1249–1272.

Sufficient Conditions for Composite Frames

Wojciech Czaja and Karamatou Yacoubou Djima

Abstract In many fields of science and technology, multiscale (or multiresolution) analysis is a useful tool to study data via efficient representation at increasingly precise resolution. In this paper, we present a result on composite wavelet frames (or composite frames), a relatively recent development in a long line of multiscale techniques including the celebrated Fourier and wavelet analyses. Composite frames generalize many traditional wavelet-based constructions and can be used to provide new effective schemes, including such successful examples as shearlets, that capture directionality in data, images in particular. Our construction is motivated by wavelets with composite dilations, which were introduced in Guo et al. (Appl Comput Harmon Anal 20:202–236, 2006). We focus on frames, a robust system that generalizes orthogonal bases and can include redundant elements, and show that we can construct composite frames for $L^2(\mathbb{R}^n)$ using two main components: dilation operators from admissible groups and generating functions which are refinable with respect to these groups. We illustrate this theory with the construction of a composite dilation frame based on a Haar wavelet with quincunx dilation and local mollification.

1 Introduction

Composite wavelets, also known as wavelets with composite dilations, are representation systems that generalize classical wavelets, which generated high interest starting the 1980s and continue to prove their usefulness, both in applications and in the theoretical development of new ideas for the analysis of signals, including the

W. Czaja
University of Maryland, College Park, MD, USA
e-mail: wojtek@math.umd.edu

K. Y. Djima (✉)
Amherst College, Amherst, MA, USA
e-mail: kyacouboudjima@amherst.edu

© The Author(s) and the Association for Women in Mathematics 2019
S. D'Agostino et al. (eds.), *A Celebration of the EDGE Program's Impact on the Mathematics Community and Beyond*, Association for Women in Mathematics Series 18, https://doi.org/10.1007/978-3-030-19486-4_17

large datasets with complex structures that are of concern for researchers in many fields today.

The idea of wavelets, and many other systems, is to decompose signals or data (viewed as functions) into basic constituents [6, 8]. In the early twentieth century, the Hungarian mathematician Haar studied the function

$$\psi(x) = \begin{cases} 1 & \text{if } 0 \le x < \frac{1}{2}, \\ -1 & \text{if } \frac{1}{2} \le x < 1, \\ 0 & \text{otherwise.} \end{cases}$$

Haar proved that dilations with dyadic powers combined with integer translates of ψ form an orthonormal basis for $L^2(\mathbb{R})$ [5, 9, 16]. With its simple, orthogonal functions of compact support able to capture local properties of functions in both time and frequency, the Haar wavelet basis opened in new window in the analysis of functions, even though wavelet systems were only thoroughly explored decades later.

In general, if for $\psi \in L^2(\mathbb{R})$ the sequence $\{\psi_{j,k}(x)\}_{j,\,k\in\mathbb{Z}}$, where

$$\psi_{j,k}(x) := 2^{j/2}\psi(2^j x - k), \quad x \in \mathbb{R}, \tag{1}$$

forms an orthonormal basis for $L^2(\mathbb{R})$, the function ψ is called a **wavelet** or **mother wavelet**. The flexibility in the choice of ψ separates wavelets from other representation systems such as the famed, albeit more rigid Fourier basis, and makes wavelets attractive both in theory and application. Mother wavelets with characteristics such as exponential decay or smoothness can compensate for the Haar wavelet poor differentiability properties, which can cause severe errors in certain approximations. In 1989, the introduction of the concept of **multiresolution analysis** (MRA), a useful framework in which functions are approximated via successive, coarse to finer, resolutions, led to a systematic construction of wavelet orthonormal bases by offering conditions for which certain classes of functions generate a wavelet system [22, 23]. All these advantages as well as the rise in computing power at the end of the past century made classical wavelets a potent tool in the approximation of functions or signals, including tasks such as singularities detection, compression, or denoising.

In the early 2000s, further developments occurred in wavelet theory through the addition of directional systems. Directional systems offer significantly more adaptability by incorporating various orientations and elongated shapes with different aspect ratios, as opposed to traditional wavelets that contain only isotropic elements occurring at all scales and locations. This makes directional systems appealing in applications where they can be used for optimal, sparse, representation of functions such as signals or images with certain singularities and geometric features. Well-known examples comprise contourlets [10], curvelets [4], shearlets, and their extensions [7, 11, 13, 14, 18, 19, 21], all of which with proven advantages over existing representations. For example, contourlets efficiently approximate images made of smooth regions separated by smooth boundaries [10] and in very recent work on autism detection using placenta images, shearlets are used in combination with a manifold learning technique, Laplacian eigenmaps, to successfully enhance the appearance of vessels structures [1].

In this paper, we give sufficient conditions to obtain composite frames or frame $\mathbf{G}_A\mathbf{G}$-MRA. Frames are a generalization of orthonormal bases with highly desirable properties such as versatility and robustness to errors in applications. Our frame $\mathbf{G}_A\mathbf{G}$-MRA is generated by consecutively applying dilations formed with elements of subgroups $\mathbf{G}_A$ and $\mathbf{G}$ of the general linear group to a function $\phi \in L^2(\mathbb{R}^n)$ satisfying certain requirements, which we make precise later. Let $GL_n(\mathbb{R})$ denote the **linear group of degree** n, i.e., the set of $n \times n$ invertible matrices over the field $\mathbb{R}$ with the operation of ordinary matrix multiplication and $SL_n(\mathbb{R})$ be the **special linear group of degree** n, i.e., the set of $n \times n$ matrices in $GL_n(\mathbb{R})$ with determinant 1. We denote by $\mathbf{G}$ a finite subgroup of $SL_n(\mathbb{Z})$, and by $\mathbf{G}_A$ the set $\{A^j : j \in \mathbb{Z}\} \subset GL_n(\mathbb{R})$. Our construction is based on Guo et al.'s composite wavelets, defined in [15] as a class of affine systems of the form

$$\mathscr{A}_{\mathbf{G}_A\mathbf{G}} = \{D_A D_B T_k \psi : k \in \mathbb{Z}^n, B \in \mathbf{G}, A \in \mathbf{G}_A\}, \qquad (2)$$

where $\psi \in L^2(\mathbb{R}^n)$, T_k is a translation by $k \in \mathbb{Z}^n$, D_A and D_B are dilation operators, and $\mathbf{G}_A$, $\mathbf{G}$ are countable subgroups of $GL_n(\mathbb{R})$. The dilation operators $\mathbf{G}_A$ and $\mathbf{G}$ lead to wavelets with desired geometric properties such as directionality, elongated shapes, oscillations; in particular, $\mathbf{G}_A$ generally expands or contracts functions in certain directions and $\mathbf{G}$ contains volume-preserving maps in transverse directions.

In [15], the authors give admissibility conditions $\mathbf{G}_A$ and $\mathbf{G}$ such that the system $\mathscr{A}_{\mathbf{G}_A\mathbf{G}}$ is a (multi)wavelet or Parseval frame for $L^2(\mathbb{R}^n)$ using the idea of shift-invariant systems. In our construction, we assume that these conditions hold and focus on constructing a frame multiresolution analysis of scaling functions instead of orthonormal wavelets. In [15], Parseval frames are mentioned but the main result concerns orthonormal wavelets. Our sufficient conditions extend a result that was, so far, only proved for specific orthonormal bases [14, 15]. In particular, our main result, Theorem 16, gives conditions that a function ϕ must satisfy to generate a frame $\mathbf{G}_A\mathbf{G}$. We also use technical arguments that give an insight on how the dilations interact to yield a frame $\mathbf{G}_A\mathbf{G}$-MRA.

This paper is organized as follows: In Sect. 2, we introduce the terminology and notation used throughout the paper and provide some useful results for our subsequent computations. We also give the formal definition of a frame $\mathbf{G}_A\mathbf{G}$-MRA along with admissibility conditions for sets of matrices used as dilation matrices in the frame $\mathbf{G}_A\mathbf{G}$-MRA. In Sect. 3, we prove our main theorem: we assume the existence of a function ϕ whose k-translations, for $k \in \mathbb{Z}^n$, and dilations by admissible matrices form a semi-orthogonal Parseval frame for the space of square integrable functions on $\mathbb{R}^n$ and show that whenever ϕ is refinable, we obtain a frame $\mathbf{G}_A\mathbf{G}$-MRA. This condition of self-similarity for ϕ is related to others in classical wavelets theory [2, 3, 5, 24]. In Sect. 4, we present an example of frame based on the partial mollification of a 2-D Haar wavelet. Our frame differs from a $\mathbf{G}_A\mathbf{G}$-MRA only on a small set and has better smoothness properties. We summarize our results in Sect. 5 and give future directions for our work.

2   Preliminaries

2.1 L^p-Spaces and Fourier Transform

We have already been using $\mathbb{R}^n$ to represent Euclidean spaces; let us denote by $\mathbb{N}$, $\mathbb{C}$, and $\mathbb{Z}^n$, the set of natural numbers, complex numbers, and the n-dimensional integer lattice, respectively. Let x be a column vector representing points in $\mathbb{R}^n$ and ω, a row vector representing points in the frequency domain $\widehat{\mathbb{R}^n}$. Suppose X is an open subset of $\mathbb{R}^n$ or $\mathbb{R}^n$ itself. For $1 \leq p < \infty$, the Banach spaces $L^p(X)$ contain complex-valued functions f for which $|f|^p$ is integrable on X with respect to the Lebesgue measure, i.e.,

$$L^p(X) := \left\{ f : X \longrightarrow \mathbb{C} : f \text{ is measurable and } \int_X |f(x)|^p dx < \infty \right\}.$$

The norm on $L^p(X)$ is

$$\|f\|_p = \|f\|_{L^p(X)} = \left(\int_X |f(x)|^p dx \right)^{1/p}.$$

For $p = \infty$, the **essential supremum** of a function f on X is given by

$$\|f\|_{L^\infty(X)} = \operatorname*{ess\,sup}_{x \in X} f = \inf_{x \in X} \{\lambda \in \mathbb{R} : f(x) \leq \lambda \text{ a.e.}\}.$$

We are particularly interested in the space $L^2(X)$ of complex-valued, square integrable functions on X, with respect to the Lebesgue measure:

$$L^2(X) := \left\{ f : X \longrightarrow \mathbb{C} : f \text{ is measurable and } \int_X |f(x)|^2 dx < \infty \right\},$$

equipped with the norm

$$\|f\|_2 = \|f\|_{L^2(X)} = \left(\int_X |f(x)|^2 dx \right)^{1/2}.$$

In agreement with standard notation, we may sometimes write $\|f\|$ to denote $\|f\|_2$. The space $L^2(X)$ is a Hilbert space and thus has an inner product given by

$$\langle f, g \rangle = \int_X f(x)\overline{g(x)} dx, \quad f, g \in L^2(X).$$

To compute estimates for certain quantities associated with functions in $L^2(X)$, we often use the Cauchy–Schwarz inequality, which states that

$$\int_X |f(x)g(x)|\mathrm{d}x \leq \|f\|_{L^2(X)}\|g\|_{L^2(X)} \text{ for all } f, g \in L^2(X).$$

The discrete analog of $L^2(\mathbb{R}^n)$ is $\ell^2(\mathrm{K})$, the space of square summable scalar sequences with respect to a countable index K:

$$\ell^2(\mathrm{K}) := \left\{ \{x_k\}_{k\in\mathrm{K}}, \ x_k \in \mathbb{C}, \ \mathrm{K} \text{ is countable} : \|f\|^2_{\ell^2(\mathrm{K})} = \sum_{k\in\mathrm{K}} |x_k|^2 < \infty \right\}.$$

The space $\ell^2(\mathrm{K})$ is also a Hilbert space with respect to the inner product

$$\langle \{x_k\}, \{y_k\} \rangle = \sum_{k\in\mathrm{K}} x_k \overline{y_k},$$

where $\{x_k\}_{k\in\mathrm{K}}, \{y_k\}_{k\in\mathrm{K}} \subset \ell^2(\mathrm{K})$. The Cauchy–Schwarz inequality on $\ell^2(\mathrm{K})$ is given by

$$\left| \sum_{k\in\mathrm{K}} x_k \overline{y_k} \right|^2 \leq \sum_{k\in\mathrm{K}} |x_k|^2 \sum_{k\in\mathrm{K}} |y_k|^2, \ \{x_k\}_{k\in\mathrm{K}}, \{y_k\}_{k\in\mathrm{K}} \subset \ell^2(\mathrm{K}).$$

We denote by $\mathbb{T}^n$ the n-dimensional torus $\mathbb{R}^n/\mathbb{Z}^n \simeq [0, 1]^n$. The space of measurable $\mathbb{Z}^n$ periodic function f such that

$$\|f\|^2_{L^2(\mathbb{T}^n)} := \int_{\mathbb{T}^n} |f(x)|^2 \mathrm{d}x < \infty$$

is denoted by $L^2(\mathbb{T}^n)$.

Finally, we define the Fourier transform of functions in $L^2(\mathbb{R}^n)$. Note that the Fourier transform is usually defined for functions in $L^1(\mathbb{R}^n)$, but because we wish to use formulas such as Plancherel's equation without additional assumptions, we adopt the following definition.

Definition 1 The **Fourier transform** of $\mathscr{F} : L^2(\mathbb{R}^n) \longrightarrow L^2(\widehat{\mathbb{R}}^n)$ of a function $f \in L^2(\mathbb{R}^n)$ is given by

$$\widehat{f}(\omega) := \mathscr{F}[f](\omega) = \int_{\mathbb{R}^n} f(x)e^{-2\pi i \omega x}\,\mathrm{d}x, \ \omega \in \widehat{\mathbb{R}}^n. \tag{3}$$

For all $f, g \in L^2(\mathbb{R}^n)$, we have **Plancherel's equation**

$$\langle f, g \rangle = \langle \widehat{f}, \widehat{g} \rangle, \tag{4}$$

and, in particular,

$$\|f\|_{L^2(\mathbb{R}^n)} = \|\widehat{f}\|_{L^2(\widehat{\mathbb{R}^n})}. \tag{5}$$

2.2 Frames

Definition 2 A countable family of elements $\{f_k\}_{k=1}^{\infty}$ in a (separable) Hilbert space $\mathcal{H}$ is a **frame** for $\mathcal{H}$ if for each $f \in \mathcal{H}$ there exist constants $C_L < C_U$, with $0 < C_L < C_U < \infty$, such that

$$C_L \|f\|^2 \leq \sum_{k=1}^{\infty} |\langle f, f_k \rangle|^2 \leq C_U \|f\|^2. \tag{6}$$

The constants C_L and C_U are called the frame bounds: C_L is the **lower frame bound** and C_U is the **upper frame bound**, and they are optimal if C_L is maximal and C_U is minimal. Frames generalize **orthonormal bases** (ONBs) and offer several advantages: (1) without independence and orthogonality restrictions, they can be constructed with varied characteristics that can be custom-made for the application of interest and (2) without uniqueness constraint, they can lead to a more robust representation of vectors or functions in $\mathcal{V}$ during certain processes [17]. Different types of frames can be defined in terms of the value of frame bounds: a frame is **tight** if $A = B$ and a **Parseval frame** is a tight frame with $A = 1$. An orthonormal basis satisfies the frame definition with $C_L = C_U = 1$, so an ONB is also a Parseval frame. In the important area of finite frames (i.e., $\{f_k\}_{k=1,2,\ldots,K}$, where $K \in \mathbb{N}$), a frame is a **finite unit-norm tight frame** (FUNTF) if it is tight and each frame element has norm one.

The following results concern, respectively, the characterization of a frame system via its **synthesis** operator T and the consequence of applying a unitary operator to a frame for $\mathcal{H}$ produces another frame for $\mathcal{H}$. Their proof can be found in [5].

Proposition 1 *A sequence $\{f_k\}_{k=1}^{\infty}$ in $\mathcal{H}$ is a frame for $\mathcal{H}$ if and only if*

$$T : \{c_k\}_{k=1}^{\infty} \longrightarrow \sum_{k=1}^{\infty} c_k f_k$$

is a well-defined mapping from $\ell^2(\mathbb{N})$ onto $\mathcal{H}$.

Proposition 2 *Let* $\{f_k\}_{k=1}^{\infty}$ *be a frame in a Hilbert space* $\mathscr{H}$ *with frame bounds* C_L, $C_U > 0$. *If* $U : \mathscr{H} \longrightarrow \mathscr{H}$ *is a unitary operator, then* $\{Uf_k\}_{k=1}^{\infty}$ *is a frame with frame bounds* C_L, C_U.

2.3 Multiresolution Analysis

Definition 3 A sequence of closed subspaces $\{V_j\}_{j \in \mathbb{Z}}$ of $L^2(\mathbb{R})$ together with a function ϕ is a **MRA** for $L^2(\mathbb{R})$ if the following properties hold:

(i) $\cdots V_{-1} \subset V_0 \subset V_1 \cdots$.
(ii) $\overline{\bigcup_{j \in \mathbb{Z}} V_j} = L^2(\mathbb{R})$ and $\bigcap_{j \in \mathbb{Z}} V_j = \{0\}$.
(iii) $f \in V_j \Longleftrightarrow f(2x) \in V_{j+1}, x \in \mathbb{R}$.
(iv) $f \in V_0 \Longrightarrow f(x - k) \in V_0$, for all $k \in \mathbb{Z}, x \in \mathbb{R}$.
(v) $\{\phi(x - k)\}_{k \in \mathbb{Z}}$ is an orthonormal basis for V_0.

The properties described in Definition 3 are very useful for approximations. For example, if we are looking for the approximation of a function $f \in L^2(\mathbb{R})$ in a certain space V_j and cannot find a satisfying one, we know, by (i), that the V_j's are nested, and this allows us to move to another approximation space $V_{j'}$, $j' \neq j$, via the simple scaling defined in (iii).

Starting with a MRA, one can define, for each $j \in \mathbb{Z}$, the space W_j as the orthogonal complement of V_j in V_{j+1}. It follows that

$$L^2(\mathbb{R}) = \bigoplus_{j \in \mathbb{Z}} W_j.$$

These spaces W_j will satisfy the same dilation property as the V_j's, i.e.,

$$\psi(x) \in W_j \Longleftrightarrow \psi(2x) \in W_{j+1}.$$

To obtain an orthonormal basis $\{\psi_{j,k}(x)\}_{j,k \in \mathbb{Z}}$ for $L^2(\mathbb{R})$, we can use the fact that, via the Fourier transform,

$$\widehat{\phi}(2\omega) = H_0(\omega)\widehat{\phi}(\omega), \text{ a.e. } \omega \in \widehat{\mathbb{R}}^n, \tag{7}$$

where H_0 is a 1-periodic function [5]. A function ϕ that can be written as 7 is said to be **refinable**. In this case, a choice of $\phi \in W_0$ that will generate a wavelet orthonormal basis is

$$\widehat{\psi}(\omega) = \overline{H_0\left(\frac{\omega}{2} + \frac{1}{2}\right)}e^{-\pi i \omega}\widehat{\phi}\left(\frac{\omega}{2}\right).$$

In this work, we are concerned with conditions that guarantee the existence of a frame multiresolution analysis. We will obtain a condition similar to (7) for a frame generated by composite dilations and translations of a scaling function ϕ.

2.4   Dilations and Translations Operators

We formally introduce the operators used to build affine systems with composite dilations. Since we will use both their time and frequency properties, we provide useful commutative relations as well as formulas that describe how the Fourier transform acts on these operators.

Definition 4 Let $f \in L^2(\mathbb{R}^n)$, $x \in \mathbb{R}^n$ and define the following unitary operators on $L^2(\mathbb{R}^n)$:

- The **translation** of f by y, $T_y : L^2(\mathbb{R}^n) \longrightarrow L^2(\mathbb{R}^n)$, where $y \in \mathbb{R}^n$, is given by

$$\left(T_y f\right)(x) = f(x - y).$$

- The **dilation** of f by A, $D_A : L^2(\mathbb{R}^n) \longrightarrow L^2(\mathbb{R}^n)$, where $A \in GL_n(\mathbb{R})$, is given by

$$(D_A f)(x) = |\det A|^{-1/2} f\left(A^{-1}x\right).$$

In particular, when $n = 1$, $A = 2$, $(D_2 f)(x) = \frac{1}{\sqrt{2}} f\left(\frac{x}{2}\right)$ is the standard dyadic dilation, and **G** is the trivial group, i.e., the identity, we have the classical wavelet system:

$$\left\{2^{-j/2}\psi\left(2^{-j/2}x - k\right) : j, k \in \mathbb{Z}\right\}. \tag{8}$$

Translations and dilations commute with one another as follows.

Proposition 5 Let A, $B \in GL_n(\mathbb{R})$, $y \in \mathbb{R}^n$. We have

(i) $D_A^{-1} = D_{A^{-1}}$.
(ii) $D_A D_B = D_{AB}$.
(iii) $D_A T_y = T_{A^{-1}y} D_A$.

We now introduce two unitary operators that will simplify some frequency domain manipulations.

Definition 6 Let $y \in \mathbb{R}^n$, $A \in GL_n(\mathbb{R})$, and $g \in L^2(\widehat{\mathbb{R}}^n)$.

(a) The **Fourier domain dilation** $\widehat{D}_A$ of g is

$$\widehat{D}_A(g)(\omega) = |\det A|^{1/2} g(\omega A), \quad \omega \in \widehat{\mathbb{R}}^n.$$

(b) The operator $M_y : L^2(\widehat{\mathbb{R}}^n) \longrightarrow L^2(\widehat{\mathbb{R}}^n)$ is the **modulation** of g by $y \in \mathbb{R}^n$, given by

$$\left(M_y g\right)(\omega) = e^{-2\pi i \omega y} g(\omega), \ \omega \in \widehat{\mathbb{R}}^n.$$

Remark 7 Observe that the exponent in the modulation operator M_y contains a negative sign. This definition is different from the usual definition of modulation, but it will yield more obvious commutation relationships between operators in the Fourier domain, and will spare us from the task of keeping track of a negative sign in future calculations.

Finally, we have the action of the Fourier transform on dilation and translation operators.

Proposition 8 *Let* $A \in GL_n(\mathbb{R})$, $B \in SL_n(\mathbb{R})$, *and* y, $k \in \mathbb{R}^n$. *Then,*

(i) $\mathscr{F} D_A = \widehat{D}_A \mathscr{F}$.
(ii) $\mathscr{F} T_y = M_y \mathscr{F}$.
(iii) $\mathscr{F} D_A^j D_B = \widehat{D}_{A^j} \widehat{D}_B \mathscr{F}, \ j \in \mathbb{Z}$.

2.5 Shift-Invariant Spaces

Definition 9 A $\mathbb{Z}^n$-**invariant space** (or **shift-invariant space**) of $L^2(\mathbb{R}^n)$ is a closed subspace $V \subset L^2(\mathbb{R}^n)$ such that $T_k V = V$ for each $k \in \mathbb{Z}^n$. Let $\phi \in L^2(\mathbb{R}^n) \setminus \{0\}$. We denote by $\langle \phi \rangle$ the shift-invariant space generated by ϕ:

$$\langle \phi \rangle = \overline{\text{span}} \left\{ T_k \phi : k \in \mathbb{Z}^n \right\}.$$

Definition 10 Let $\mathbf{G}$ be a finite subgroup of $SL_n(\mathbb{Z})$ and let $M \ltimes N$ denote the semi-direct product of two groups M and N. The $\mathbf{G} \ltimes \mathbb{Z}^n$-**invariant spaces** are the closed subspaces $V \subset L^2(\mathbb{R}^n)$ for which $D_B T_k V = V$ for any pair $(B, k) \in \mathbf{G} \ltimes \mathbb{Z}^n$. Let $\phi \in L^2(\mathbb{R}^n)$. The $\mathbf{G} \ltimes \mathbb{Z}^n$-invariant spaces generated by ϕ, denoted $\langle\langle \phi \rangle\rangle$, are defined as

$$\langle\langle \phi \rangle\rangle = \overline{\text{span}} \left\{ D_B T_k \phi : B \in \mathbf{G}, \ k \in \mathbb{Z}^n \right\}.$$

Here, we only consider finite groups $\mathbf{G}$, but one can certainly define shift-invariant spaces for any countable group $\mathbf{G}$, as is done when defining shearlets. For traditional wavelets, we work on the shift-invariant space generated by a function $\psi \in L^2(\mathbb{R}^n)$ under the shifts of $\mathbb{Z}^n$. The observation that $\mathbb{Z}^n$ is the semi-direct product of the trivial group in $SL_n(\mathbb{R})$ and $\mathbb{Z}^n$ is the bridge between shift-invariant spaces for classical wavelet theory and shift-invariant spaces for composite wavelets. For composite wavelets, we consider $L^2(\mathbb{R}^n)$-functions with shifts by the semi-direct product of $\mathbb{Z}^n$ and a discrete, and often non-abelian group, $\mathbf{G}$. In [2, 24], and [3], $\mathbf{G}$ is a crystallographic group, i.e., a group of isometries such as rotations or reflections on $\mathbb{R}^n$. The example that we give in Sect. 5 will also feature such a group, which naturally preserves the MRA structure of the Haar wavelet.

Definition 11 Let $\phi \in L^2(\mathbb{R}^n)$ and let $\mathbf{G}$ be a finite subgroup of $SL_n(\mathbb{Z})$. The set $\Phi_{\mathbf{G}} = \{D_B T_k \phi : B \in \mathbf{G}, k \in \mathbb{Z}^n\}$ is a **semi-orthogonal Parseval frame** for the $\mathbf{G} \ltimes \mathbb{Z}^n$-invariant subspace $\langle\langle \phi \rangle\rangle$ if $\{T_k \phi : k \in \mathbb{Z}^n\}$ is a Parseval frame for $\langle \phi \rangle$ and

$$\langle\langle \phi \rangle\rangle = \bigoplus_{B \in \mathbf{G}} D_B \langle \phi \rangle,$$

i.e., $D_B T_k \phi \perp D_{B'} T_{k'} \phi$ for any $B, B' \in \mathbf{G}$, $B \neq B'$, and $k, k' \in \mathbb{Z}^n$. We also say that ϕ generates the semi-orthogonal Parseval frame $\Phi_{\mathbf{G}}$ in this case.

We can extend Definition 11 to any frame and remove the semi-orthogonality condition for more general shift-invariant spaces. However our example will show that having those conditions simplifies the construction of a frame $\mathbf{G}_A \mathbf{G}$-MRA.

2.6 $\mathbf{G}_A \mathbf{G}$-Multiresolution Analysis

Definition 12 Let $\mathbf{G}$ be a finite subgroup of $SL_n(\mathbb{Z})$. We say that $A \in GL_n(\mathbb{R})$ **normalizes** $\mathbf{G}$ if, for each $B \in \mathbf{G}$, $ABA^{-1} \in \mathbf{G}$.

Definition 13 A matrix A is an **expanding** matrix if all the eigenvalues λ of A satisfy the condition $|\lambda| > 1$.

If A normalizes $\mathbf{G}$ and, in addition, A is an expanding matrix, then the set $\mathbf{G}_A \mathbf{G}$, where $\mathbf{G}_A = \{A^j : j \in \mathbb{Z}\} \subset GL_n(\mathbb{R})$, meets admissibility conditions that guarantee the existence of a Parseval frame of the form (2) for $L^2(\mathbb{R}^n)$. For a more comprehensive discussion of admissibility conditions, refer to [15]. Here, this admissibility condition ensures that we obtain shift-invariance for fundamental domains of $\mathbb{R}^n$.

Definition 14 Let $\mathbf{G}$ be a finite subgroup of $SL_n(\mathbb{Z})$ and $\mathbf{G}_A = \{A^j : j \in \mathbb{Z}\}$, where $A \in GL_n(\mathbb{Z})$ is an expanding matrix. Moreover, assume that A normalizes $\mathbf{G}$. Then, the sequence of closed subspaces $\{V_j\}_{j \in \mathbb{Z}}$ of $L^2(\mathbb{R}^n)$ is a $\mathbf{G}_A \mathbf{G}$-multiresolution analysis ($\mathbf{G}_A \mathbf{G}$-MRA) if the following properties hold:

(i) $D_B T_k V_0 = V_0$, for all $B \in \mathbf{G}, k \in \mathbb{Z}^n$.
(ii) $V_j \subset V_{j+1}$ for each $j \in \mathbb{Z}$, where $V_j = D_A^{-j} V_0$.
(iii) $\bigcap_{j \in \mathbb{Z}} V_j = 0$ and $\overline{\bigcup_{j \in \mathbb{Z}} V_j} = L^2(\mathbb{R}^n)$.
(iv) There exists $\phi \in L^2(\mathbb{R}^n)$ such that $\Phi_{\mathbf{G}} = \{D_B T_k \phi : B \in \mathbf{G}, k \in \mathbb{Z}^n\}$ is a semi-orthogonal Parseval frame for V_0.

3 Sufficient Conditions

We start by giving conditions that guarantee that $V_j \subset V_{j+1}$ for each $j \in \mathbb{Z}$.

Lemma 15 *Assume* $\mathbf{G}$ *is a finite subgroup of* $SL_n(\mathbb{Z})$ *and* $\mathbf{G}_A = \{A^j : j \in \mathbb{Z}\}$, *where* $A \in GL_n(\mathbb{Z})$ *is an expanding matrix. For* $j \in \mathbb{Z}$, *let* V_j *be defined as*

$$V_j = D_A^{-j} \overline{\text{span}} \{\Phi_{\mathbf{G}}\} = \overline{\text{span}} \left\{ D_A^{-j} \Phi_{\mathbf{G}} \right\}. \tag{9}$$

Moreover, let $\phi \in L^2(\mathbb{R}^n)$ *and assume that* $\Phi_{\mathbf{G}} = \{D_B T_k \phi : B \in \mathbf{G}, k \in \mathbb{Z}^n\}$ *is a semi-orthogonal Parseval frame for* V_0. *Then, for* $j \in \mathbb{Z}$,

(i) $D_A^{-j} \Phi_{\mathbf{G}}$ *is a semi-orthogonal Parseval frame for* V_j.
(ii) *A function* $f \in L^2(\mathbb{R}^n)$ *belongs to* V_j *if and only if*

$$f = \sum_{k \in \mathbb{Z}^n} \sum_{B \in \mathbf{G}} c_{k,B} D_{A^{-j}} D_B T_k \phi,$$

for some $\{c_{k,B}\}_{k \in \mathbb{Z}^n, B \in \mathbf{G}} \subset \ell^2(\mathbb{Z}^n)$.
(iii) *A function* $f \in L^2(\mathbb{R}^n)$ *belongs to* V_j *if and only if there exist* $L^2(\mathbb{T}^n)$ *periodic functions* $F_{B,j} \in L^2(\mathbb{T}^n)$, $B \in \mathbf{G}$, *such that*

$$\widehat{f}\left(\omega A^j\right) = \sum_{B \in \mathbf{G}} F_{B,j}(\omega B) \widehat{\phi}(\omega B). \tag{10}$$

Proof

(i) Since $\{T_k \phi : k \in \mathbb{Z}^n\}$ is a Parseval frame and D_A^{-j} is a unitary operator, by Proposition 2, we have that $\left\{ D_A^{-j} T_k \phi : k \in \mathbb{Z}^n \right\}$ is a Parseval frame. For semi-orthogonality, we use again the fact that D_A is unitary, to obtain

$$\left\langle D_A^{-j} D_B T_k \phi, \, D_A^{-j} D_{B'} T_{k'} \phi \right\rangle = \left\langle (D_A^{-j})^* D_A^{-j} D_B T_k \phi, \, D_{B'} T_{k'} \phi \right\rangle$$

$$= \langle D_B T_k \phi, \, D_{B'} T_{k'} \phi \rangle$$

$$= 0,$$

whenever $B \neq B'$, by assumption.
(ii) This is a consequence of (i) combined with Proposition 1.
(iii) Using (ii), for each $f \in V_j$ we have

$$\widehat{f}(\omega) = \mathscr{F}\left[\sum_{k \in \mathbb{Z}^n} \sum_{B \in \mathbf{G}} c_{k,B} D_A^{-j} D_B T_k \phi \right](\omega)$$

$$= \sum_{k\in\mathbb{Z}^n}\sum_{B\in\mathbf{G}} c_{k,B}\mathscr{F}\left[D_A^{-j}D_BT_k\phi\right](\omega).$$

Making use of Proposition 8,

$$\widehat{f}(\omega) = \sum_{k\in\mathbb{Z}^n}\sum_{B\in\mathbf{G}} c_{k,B}\left[\widehat{D}_A^{-j}\widehat{D}_BM_k\widehat{\phi}\right](\omega)$$

$$= \widehat{D}_A^{-j}\sum_{k\in\mathbb{Z}^n}\sum_{B\in\mathbf{G}} c_{k,B}\left[\widehat{D}_BM_k\widehat{\phi}\right](\omega).$$

Then,

$$\widehat{f}\left(\omega A^j\right) = |\det A|^{-j/2}\widehat{D}_A^j\widehat{f}(\omega)$$

$$= |\det A|^{-j/2}\widehat{D}_A^j\left(\widehat{D}_A^{-j}\sum_{k\in\mathbb{Z}^n}\sum_{B\in\mathbf{G}} c_{k,B}\left[\widehat{D}_BM_k\widehat{\phi}\right](\omega)\right)$$

$$= |\det A|^{-j/2}\sum_{k\in\mathbb{Z}^n}\sum_{B\in\mathbf{G}} c_{k,B}\left[\widehat{D}_BM_k\widehat{\phi}\right](\omega)$$

$$= |\det A|^{-j/2}\sum_{B\in\mathbf{G}}\left(\sum_{k\in\mathbb{Z}^n} c_{k,B}e^{-2\pi i\omega Bk}\right)\widehat{\phi}(\omega B).$$

Now, for each $j\in\mathbb{Z}$, $B\in\mathbf{G}$, define

$$F_{B,j}(\omega) = |\det A|^{-j/2}\sum_{k\in\mathbb{Z}^n} c_{k,B}e^{-2\pi i\omega Bk}.$$

To verify that $F_{B,j}$ is in $L^2(\mathbb{T}^n)$, we compute

$$\int_{\mathbb{T}^n}|F_{B,j}(\omega)|^2 d\omega = \int_{\mathbb{T}^n}||\det A|^{-j/2}\sum_{k\in\mathbb{Z}^n} c_{k,B}e^{-2\pi i\omega Bk}|^2 d\omega$$

$$= |\det A|^{-j}\int_{\mathbb{T}^n}\sum_{k\in\mathbb{Z}^n}\sum_{k'\in\mathbb{Z}^n} c_{k,B}\overline{c_{k',B}}e^{-2\pi i\omega Bk}e^{2\pi i\omega Bk'}d\omega.$$

Now $\{c_{k,B}\}\subset \ell^2(\mathbb{Z}^n)$, so

$$\int_{\mathbb{R}^n}|F_{B,j}(\omega)|^2 d\omega = |\det A|^{-j}\sum_{k\in\mathbb{Z}^n}\sum_{k'\in\mathbb{Z}^n} c_{k,B}\overline{c_{k',B}}\int_{\mathbb{T}^n} e^{-2\pi i\omega Bk}e^{2\pi i\omega Bk'}d\omega$$

$$= |\det A|^{-j}\sum_{k\in\mathbb{Z}^n}|c_{k,B}|^2\int_{\mathbb{T}^n}|e^{-2\pi i\omega Bk}|^2(\omega)d\omega,$$

since, for each B, $\left\{e^{-2\pi i \omega Bk}\right\}_{k\in\mathbb{Z}}$ form an orthonormal basis for $L^2\left(\mathbb{T}^n\right)$. Thus

$$\int_{\mathbb{R}^n} |F_{B,j}(\omega)|^2 d\omega = |\det A|^{-j} \sum_{k\in\mathbb{Z}^n} |c_{k,B}|^2 < \infty.$$

Hence, we have that (10) holds with $F_{B,j}(\omega) = |\det A|^{-j/2} \sum_{k\in\mathbb{Z}^n} c_{k,B} e^{-2\pi i \omega Bk}$.

For the other direction, suppose that $f \in L^2(\mathbb{R}^n)$ and there exists a function $F_{B,j}$ such that Eq. (10) is satisfied. Let $d_{B,k}$ be the Fourier coefficients of $F_{B,j}$. If we define $c_{B,k} = |\det A|^{j/2} d_{B,k}$, then we get $f = \sum_{k\in\mathbb{Z}^n} \sum_{B\in G} c_{k,B} D_{A-j} D_B T_k \phi$. Hence, by (ii), $f \in V_j$. □

Next, we obtain sufficient conditions for a $\mathbf{G}_A\mathbf{G}$-MRA by the way of Lemma 15 and a few classical tools from harmonic analysis [5, 9] as well new approaches to deal with the technicalities arising in this more complex setting. In the present paper, we only provide details for (ii), which provide an insight on how the dilations interact to yield a frame $\mathbf{G}_A\mathbf{G}$-MRA.

Theorem 16 *Let* $\mathbf{G}$ *be a finite subgroup of* $SL_n(\mathbb{Z})$ *and* $\mathbf{G}_A = \{A^j : j \in \mathbb{Z}\}$, *where* $A \in GL_n(\mathbb{Z})$ *is an expanding matrix. Assume that* A *normalizes* $\mathbf{G}$. *Assume also that* $\phi \in L^2(\mathbb{R}^n)$ *and that* $\Phi_{\mathbf{G}} = \{D_B T_k \phi : B \in \mathbf{G}, k \in \mathbb{Z}^n\}$ *is a semi-orthogonal Parseval frame for* V_0. *Define* $V_j = D_A^{-j} V_0$. *If* $\widehat{\phi}$ *is continuous, and uniformly bounded on a neighborhood of zero,* $\widehat{\phi} \neq 0$, *and there exist* $\mathbb{T}^n$-*periodic functions* $H_B \in L^\infty(\mathbb{T}^n)$ *such that*

$$\widehat{\phi}(\omega) = \sum_{B\in G} H_B\left(\omega BA^{-1}\right) \widehat{\phi}\left(\omega BA^{-1}\right), \tag{11}$$

then ϕ *generates a semi-orthogonal Parseval frame* $\mathbf{G}_A\mathbf{G}$-*MRA.*

Proof To show that our system satisfies (i) in Definition 14, we simply observe the fact that $\Phi_{\mathbf{G}}$ is a semi-orthogonal Parseval frame for V_0 that implies $V_0 = \overline{\text{span}}\{\Phi_{\mathbf{G}}\}$, which is, by assumption, an $\mathbf{G} \ltimes \mathbb{Z}^n$-invariant subspace of $L^2(\mathbb{R}^n)$, i.e., $D_B T_k V_0 = V_0$. Showing (iii) and (iv) in Definition 14 uses arguments similar in spirit to one found in [9]. The details of these computations will be left out of this paper.

We now focus on Definition 14(ii). Let $f \in V_j$, $j \in \mathbb{Z}$. We want to show that this implies $f \in V_{j+1}$, or equivalently, via Lemma 15, that there exist $\mathbb{T}^n$-periodic functions $F_{B,j+1} \in L^2(\mathbb{T}^n)$ such that Eq. (10) holds. Employing (iii) of Lemma 15 yields

$$\widehat{f}(\omega A^{j+1}) = \widehat{f}((\omega A)A^j) = \sum_{B'\in G} F_{B',j}(\omega AB') \widehat{\phi}(\omega AB'),$$

where $F_{B',j}(\omega)$ is a $\mathbb{T}^n$-periodic, $L^2(\mathbb{T}^n)$ function.

Using Assumption (11),

$$\widehat{f}\left(\omega A^{j+1}\right) = \sum_{B' \in \mathbf{G}} F_{B',j}\left(\omega AB'\right) \cdot$$

$$\left(\sum_{B'' \in \mathbf{G}} H_{B''}\left(\omega AB'B''A^{-1}\right)\widehat{\phi}\left(\omega AB'B''A^{-1}\right)\right).$$

Now, since $\mathbf{G}_A$ normalizes $\mathbf{G}$, we have that $A'B'B''A^{-1} \in \mathbf{G}$. Then, via a simple proof contradiction, we obtain that for distinct B'_1, B'_2 in the first sum, $A'B'_1 B''A^{-1}$ and $A'B'_2 B''A^{-1}$ are also distinct. Then, letting $B = AB'B''A^{-1}$,

$$\widehat{f}\left(\omega A^{j+1}\right) = \sum_{B' \in \mathbf{G}} F_{B',j}\left(\omega AB'\right)\left(\sum_{B \in \mathbf{G}} H_{B''}\left(\omega B\right)\widehat{\phi}\left(\omega B\right)\right)$$

$$= \sum_{B \in \mathbf{G}}\left(\sum_{B' \in \mathbf{G}} F_{B',j}\left(\omega AB'\right) H_{B''}\left(\omega B\right)\right)\widehat{\phi}\left(\omega B\right),$$

where $B'' = A^{-1}(B')^{-1}BA$. Let

$$F_{B,\,j+1}(\omega) = \sum_{B' \in \mathbf{G}} F_{B',j}\left(\omega AB'B^{-1}\right) H_{B''}(\omega).$$

With $F_{B',j}$ and $H_{B''}$ $\mathbb{T}^n$-periodic, $F_{B,\,j+1}$ is $\mathbb{T}^n$-periodic. We obtain $F_{B,\,j+1} \in L^2\left(\mathbb{R}^n\right)$ via the following calculation, in which we denote the norm by η

$$\eta = \int_{\mathbb{T}^n} |\sum_{B' \in \mathbf{G}} F_{B',j}\left(\omega AB'B^{-1}\right) H_{B''}(\omega)|^2 d\omega$$

$$\leq \|H_{B''}\|^2_{L^\infty(\mathbb{T}^n)} \int_{\mathbb{T}^n} |\sum_{B' \in \mathbf{G}} F_{B',j}\left(\omega AB'B^{-1}\right)|^2 d\omega$$

$$\leq \|H_{B''}\|^2_{L^\infty(\mathbb{T}^n)} \#|\mathbf{G}| \int_{\mathbb{T}^n} \sum_{B' \in \mathbf{G}} |F_{B',j}\left(\omega AB'B^{-1}\right)|^2 d\omega,$$

by Minkowski's inequality. Consequently,

$$\int_{\mathbb{T}^n} |F_{B,\,j+1}(\omega)|^2 d\omega < \infty.$$

We omit the proof that the intersection V_j, $j \in \mathbb{Z}$, is empty of that the closure of their union is the entire space $L^2(\mathbb{R}^n)$ as our (standard) argument follows closely that for scalar dilations in [9].

□

4 Example

We present a composite dilation frame based on a Haar-like wavelet associated with the quincunx dilation in $\mathbb{R}^2$. This wavelet appears in [20] and also discussed briefly in [2, 3, 24]. We chose this wavelet because of its simplicity and because it illustrates difficulties that can arise when one tries to obtain more desirable properties for composite dilation systems.

Let A be the quincunx matrix:

$$A = \begin{pmatrix} 1 & -1 \\ 1 & 1 \end{pmatrix} \tag{12}$$

and let $\mathbf{G} = \{B_i : i = 0, \ldots, 7\}$ be the group of symmetries of the unit square. Explicitly,

$$b_0 = \begin{pmatrix} 1 & 0 \\ 0 & 1 \end{pmatrix}, \quad b_1 = \begin{pmatrix} 0 & 1 \\ 1 & 0 \end{pmatrix}, \quad b_2 = \begin{pmatrix} 0 & -1 \\ 1 & 0 \end{pmatrix}, \quad b_3 = \begin{pmatrix} -1 & 0 \\ 0 & 1 \end{pmatrix},$$

$$b_4 = \begin{pmatrix} -1 & 0 \\ 0 & -1 \end{pmatrix}, \quad b_5 = \begin{pmatrix} 0 & -1 \\ -1 & 0 \end{pmatrix}, \quad b_6 = \begin{pmatrix} 0 & 1 \\ -1 & 0 \end{pmatrix}, \quad b_7 = \begin{pmatrix} 1 & 0 \\ 0 & -1 \end{pmatrix},$$

and for $i = 4, 5, 6, 7$, $B_i = -B_{i-4}$.

Consider the region $R_0 = \{(x_1, x_2) : 0 \leq x_1 < 1/2, \ 0 \leq x_2 \leq x_1\}$, i.e., R_0 is the triangular region with the vertices $(0, 0)$, $(\frac{1}{2}, 0)$, and $(\frac{1}{2}, \frac{1}{2})$. For $i = 0, \ldots, 7$, define $R_i = B_i R_0$. The resulting triangles are shown in Fig. 1.

Let ϕ be defined as a scaled on R_0

$$\phi = \sqrt{8} \chi_{R_0}. \tag{13}$$

We will use the following results, proved in [20] and adapted to our setting.

Proposition 17 *Let ϕ be defined as in* (13). *Suppose A is the quincunx matrix in* (12) *and $\mathbf{G} = \{B_i : i = 0, \ldots, 7\}$ is the group of symmetries of the unit square.*

(i) *The system $\Phi_{\mathbf{G}} = \{D_{B_i} T_k \phi : B_i \in \mathbf{G}, k \in \mathbb{Z}^2\}$ is an orthonormal basis for its closed linear span $V_0 \subset L^2(\mathbb{R}^2)$, which comprises all square integrable functions constant on each $\mathbb{Z}^2$-translate of the triangles R_i, $i = 0, \ldots, 7$.*

(ii) *Let $V_j = D_{A^{-j}} V_0$, $j \in \mathbb{Z}$, i.e., V_j contains all $L^2(\mathbb{R}^2)$-functions constant on $A^{-j}\mathbb{Z}^2$ translates of the triangles $A^{-j} R_i$, $i = 0, \ldots, 7$. Then $V_j \subset V_{j+1}$, $j \in \mathbb{Z}$.*

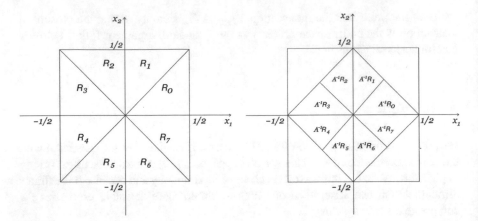

Fig. 1 Fundamental domain R_0 of the scaling function ϕ and its images, on the left, under $B_1, \ldots, B_7$, and on the right, under $B_1, \ldots, B_7$, composed with A^{-1}

(iii) The spaces V_j, $j \in \mathbb{Z}$ form an $\mathbf{G_A G}$-MRA with ϕ, which can be written as

$$\phi = \sqrt{2}\left(D_A^{-1}D_{B_1}\phi(x) + D_{A^{-1}}D_{B_6}T_{(1/2,\,1/2)}\phi(x)\right). \tag{14}$$

Next, we focus on showing that a "smooth" deformation ϕ^ε of ϕ on one side of the triangle R_0 generates a frame for V_0. For $x \in \mathbb{R}^n$, consider a "standard" mollifier

$$\eta(x) = \begin{cases} Ce^{\frac{1}{|x|^2-1}}, & |x| < 1, \\ 0, & \text{otherwise,} \end{cases} \tag{15}$$

where C is chosen so that $\int_{\mathbb{R}^n}\eta(x)\,\mathrm{d}x = 1$.
Given $\varepsilon > 0$, let $\eta_\varepsilon(x) := \frac{1}{\varepsilon^2}\eta\left(\frac{x}{\varepsilon}\right)$. It is easy to see that

$$\int_{\mathbb{R}^2}\eta_\varepsilon(x)\mathrm{d}x = 1,$$

and η_ε is supported in the ball of radius ε.
We define the mollification f^ε of $f : U \longrightarrow \mathbb{R}$, where $U \subset \mathbb{R}^n$ and $f \in L^2(\mathbb{R}^n)$ locally, by the convolution

$$f^\varepsilon(x) := (\eta_\varepsilon * f)(x) = \int_{\mathbb{R}^2}\eta(y)f(x-y)\,\mathrm{d}y, \quad x \in U_\varepsilon,$$

where $U_\varepsilon := \{x \in U : \|x - \partial U\| > \varepsilon\}$. Functions defined in this way are just a particular case of cutoff or bump functions. For given open sets Ω' compactly contained in $\Omega \subset \mathbb{R}^n$, it is well known that there exists a function $\psi \in C_c^\infty(\Omega)$, such that $0 \le \psi \le 1$ on Ω' via mollification.

For our particular case, let $x = (x_1, x_2)$ and consider

$$\rho(x) = \begin{cases} \sqrt{8}, & \frac{1}{2} - \frac{\varepsilon}{2} \le x_1 \le \frac{1}{2} + \frac{\varepsilon}{2}, \ -\frac{\varepsilon}{2} \le x_2 \le x_1, \\ 0, & \text{otherwise.} \end{cases}$$

Define $\rho^\varepsilon(x) = \left(\eta_{\delta/4} * \rho\right)(x)$. Using properties of mollifiers [12], it can be shown that

(1) $\rho^\varepsilon(x) \in C^\infty(\mathbb{R}^2)$.

(2) $\text{supp}(\rho^\varepsilon) = \left\{ (x_1, x_2) : \frac{1}{2} - \varepsilon \le x_1 \le \frac{1}{2} + \varepsilon, \ -\varepsilon \le x_2 \le x_1 \right\}$.

(3) $\text{supp}(\rho^\varepsilon) = \sqrt{8}$ on $\left\{ (x_1, x_2) : -\frac{1}{2} - \frac{\varepsilon}{4} \le x_1 \le \frac{1}{2} + \frac{\varepsilon}{4}, \ -\frac{\varepsilon}{4} \le x_2 \le x_1 + \frac{\varepsilon}{4} \right\}$.

Finally, we define the smooth, one-sided, deformation ϕ^ε of ϕ as $\phi^\varepsilon = \phi + \rho_+^\varepsilon$, where ρ_+^ε is defined as the right side of ρ^ε, that is,

$$\rho_+^\varepsilon(x) = \begin{cases} 1, & \frac{1}{2} \le x_1 \le \frac{1}{2} + \frac{\varepsilon}{4}, \ 0 \le x_2 \le x_1, \\ a(x), & \frac{1}{2} + \frac{\varepsilon}{4} \le x_1 \le \frac{1}{2} + \frac{3\varepsilon}{4}, \ 0 \le x_2 \le x_1, \\ 0, & \text{otherwise.} \end{cases} \tag{16}$$

In this definition, $a(x)$ is a smooth function whose values are in $(0, 1)$ for all x (Fig. 2).

Fig. 2 Domain of Haar scaling function ϕ^ε smoothed on one side and extended via mollification on R_0^ε

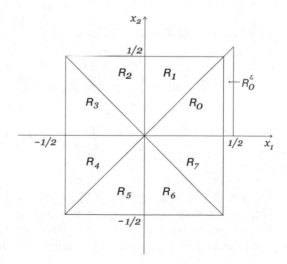

Proposition 3 $\Phi_G^\varepsilon = \{D_{B_i} T_k \phi^\varepsilon : B_i \in \mathbf{G}, k \in \mathbb{Z}^2\}$ *is a frame for* V_0, *i.e., if* $f \in V_0$, *then there exist constant* C_L, $C_U > 0$ *such that*

$$C_L \|f\|_{L^2(\mathbb{R}^2)}^2 \le \sum_{k \in \mathbb{Z}^2} \sum_{i=0}^{7} |\langle f, D_{B_i} T_k \phi^\varepsilon \rangle|^2 \le C_U \|f\|_{L^2(\mathbb{R}^2)}^2. \tag{17}$$

Proof Let $\mu = \sum_{k \in \mathbb{Z}^2} \sum_{i=0}^{7} |\langle f, D_{B_i} T_k \phi^\varepsilon \rangle|^2$. We start with the upper bound

$$\mu = \sum_{k \in \mathbb{Z}^2} \sum_{i=0}^{7} |\langle f, D_{B_i} T_k \phi \rangle + \langle f, D_{B_i} T_k \rho_+^\varepsilon \rangle|^2$$

$$\le 2 \sum_{k \in \mathbb{Z}^2} \sum_{i=0}^{7} \left\{ |\langle f, D_{B_i} T_k \phi \rangle|^2 + |\langle f, D_{B_i} T_k \rho_+^\varepsilon \rangle|^2 \right\}.$$

Since Φ_G is an orthonormal basis for V_0, we have

$$\sum_{k \in \mathbb{Z}^2} \sum_{i=0}^{7} |\langle f, D_{B_i} T_k \phi \rangle|^2 = \|f\|_{L^2(\mathbb{R}^2)}^2,$$

so, we focus on finding a bound for the term

$$\mathcal{R}(f, \varepsilon) = \sum_{k \in \mathbb{Z}^2} \sum_{i=0}^{7} |\langle f, D_{B_i} T_k \rho_+^\varepsilon \rangle|^2$$

$$= \sum_{k \in \mathbb{Z}^2} \sum_{i=0}^{7} |\int_{\mathbb{R}^2} f(x) \rho_+^\varepsilon \left(B_i^{-1} x - k \right) dx|^2.$$

Let $y = B_i^{-1} x - k$. By definition, $\rho_+^\varepsilon(x) = 0$ on $\mathbb{R}^2$ except for $x \in R_0^\varepsilon := \{(x_1, x_2) : 1 \le x_1 \le 1/2 + \varepsilon, 0 \le x_2 \le x_1\}$, as well as $R_i^\varepsilon = B_i R_0^\varepsilon$, for $i = 1, \ldots, 7$. Note that $x = B_i y + B_i k$, which we will write as $x = B_i y - k$ since $B_i k \in \mathbb{Z}^2$, and for any $k_1 \ne k_2$, by invertibility of B_i, $B_i k_1 \ne B_i k_2$. Thus, we have

$$\mathcal{R}(f, \varepsilon) = \sum_{k \in \mathbb{Z}^2} \sum_{i=0}^{7} \left| \int_{R_i^\varepsilon} \rho_+^\varepsilon(y) f\left(B_i^{-1} y - k\right) dy \right|^2$$

$$\leq \sum_{k \in \mathbb{Z}^2} \sum_{i=0}^{7} \|\rho_+^\varepsilon\|_{L^\infty(R_i^\varepsilon)} \left(\int_{R_i^\varepsilon} D_{B_i} T_k |f(y)| \, dy \right)^2$$

$$= \sum_{k \in \mathbb{Z}^2} \sum_{i=0}^{7} \left(\int_{R_i^\varepsilon} D_{B_i} T_k |f(y)| \, dy \right)^2,$$

since, by definition $0 \leq \rho_+^\varepsilon \leq 1$ on R_i^ε. Note that we can write the right-hand side as

$$\sum_{k \in \mathbb{Z}^2} \sum_{i=0}^{7} \left(\int_{\mathbb{R}^2} D_{R_i} T_k |f(y)| \chi_{R_i^\varepsilon}(y) dy \right)^2.$$

Therefore,

$$\mathcal{R}(f, \varepsilon) \leq \sum_{k \in \mathbb{Z}^2} \sum_{i=0}^{7} \left(\int_{\mathbb{R}^2} D_{B_i} T_k |f(y)| \chi_{R_i^\varepsilon}(y) dy \right)^2 \qquad (18)$$

$$= \sum_{k \in \mathbb{Z}^2} \sum_{i=0}^{7} |\langle |f|, D_{B_i} T_k \chi_{R_i^\varepsilon} \rangle|^2.$$

At this stage, it is useful to look at the meaning of (18) geometrically. On R_0, $\mathcal{R}(f, \varepsilon)$ is integrated on translates and images of R_i^ε which overlaps on corners of $[\frac{-1}{2}, \frac{1}{2})^2$ as shown in Fig. 3. On all the other R_i, we have the contribution of at most two other translates of R_i^ε. Therefore, we are integrating most $2|f|$ on each R_i, $i = 0, \ldots, 7$, and their translates. Thus, using Proposition 17,

$$\mathcal{R}(f, \varepsilon) \leq 4\|f\|_{L^2(\mathbb{R}^2)}^2. \qquad (19)$$

Next, we compute the lower bound. We have

$$\mu = \sum_{k \in \mathbb{Z}^2} \sum_{i=0}^{7} |\langle f, D_{B_i} T_k \phi \rangle + \langle f, D_{B_i} T_k \rho_+^\varepsilon \rangle|^2$$

$$\geq \sum_{k \in \mathbb{Z}^2} \sum_{i=0}^{7} |\langle f, D_{B_i} T_k \phi \rangle|^2$$

$$- 2 \sum_{k \in \mathbb{Z}^2} \sum_{i=0}^{7} |\langle f, D_{B_i} T_k \phi \rangle| |\langle f, D_{B_i} T_k \rho_+^\varepsilon \rangle|.$$

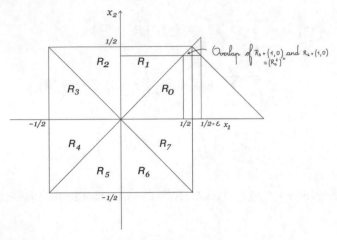

Fig. 3 The region R_0^ε and its images R_i^ε overlap on $\left[\frac{1}{2}, \frac{1}{2}\right]^2$

Using Cauchy–Schwarz, we bound the last double sum as

$$\left(\sum_{k\in\mathbb{Z}^2}\sum_{i=0}^{7}|\langle f,\ D_{B_i}T_k\phi\rangle|^2\right)^{\frac{1}{2}}\left(\sum_{k\in\mathbb{Z}^2}\sum_{i=0}^{7}|\langle f,\ D_{B_i}T_k\rho_+^\varepsilon\rangle|^2\right)^{\frac{1}{2}}.$$

Now, recognizing the second term as $\mathscr{R}\,(f,\ \varepsilon)$ and using (18), we obtain

$$\mathscr{R}\,(f,\ \varepsilon)\leq\sum_{k\in\mathbb{Z}^2}\sum_{i=0}^{7}|\langle|f(w)|,\ D_{B_i}T_k\chi_{R_i^\varepsilon}\rangle|^2.$$

Here, we cannot use the crude upper bound found in (19) because it will make our lower bound negative. Instead, once again, we look at this sum on each translate of $\left[\frac{1}{2}, \frac{1}{2}\right]^2$ and make use of Fig. 3 for more careful estimates. We obtain

$$\mathscr{R}\,(f,\ \varepsilon)\leq\sum_{k\in\mathbb{Z}^2}\sum_{i=0}^{7}\left|\int_{\mathbb{R}^2}|f(x)|D_{B_i}T_k\chi_{(R_i^\varepsilon)'}\,\mathrm{d}x+\int_{\mathbb{R}^2}|f(x)|D_{B_i}T_k\chi_{(R_i^\varepsilon)''}\,\mathrm{d}x\right|^2.$$

Since $f(x)$ is constant on R_i, we can simplify this expression by finding which proportion of $\int_{\mathbb{R}^2}|f(x)|\chi_{R_i}\,\mathrm{d}x$ the terms $\int_{\mathbb{R}^2}|f(x)|\chi_{(R_i^\varepsilon)'}\,\mathrm{d}x$ and $\int_{\mathbb{R}^2}|f(x)|\chi_{(R_i^\varepsilon)''}\,\mathrm{d}x$ represent. These proportions are given by, respectively,

$$p_1:=\frac{\text{Area}((R_i^\varepsilon)')}{\text{Area}(R_1)}=\frac{(1/2)\varepsilon/2}{(1/2)^2/2}=4\varepsilon$$

$$p_2:=\frac{\text{Area}((R_i^\varepsilon)'')}{\text{Area}(R_1)}=\frac{\varepsilon^2/2}{(1/2)^2/2}=4\varepsilon^2.$$

This yields

$$\mathscr{R}(f, \varepsilon) \leq \sum_{k \in \mathbb{Z}^2} \sum_{i=0}^{7} |(p_1 + 2p_2) \int_{\mathbb{R}^2} |f(x)| D_{B_i} T_k \chi_{R_i}(x) dx|^2$$

$$= (8\varepsilon^2 + 4\varepsilon)^2 \|f\|^2_{L^2(\mathbb{R}^2)}.$$

Hence, we have

$$\sum_{k \in \mathbb{Z}^2} \sum_{i=0}^{7} |\langle f, D_{B_i} T_k \phi^{\varepsilon} \rangle|^2 \geq (1 - 2(8\varepsilon^2 + 4\varepsilon)^2) \|f\|^2_{L^2(\mathbb{R}^2)}.$$

We simply pick ε such to get $C_L = (1 - 2(8\varepsilon^2 + 4\varepsilon)^2) > 0$ to obtain a frame. Figure 4 shows that such an ε exists.

$\square$

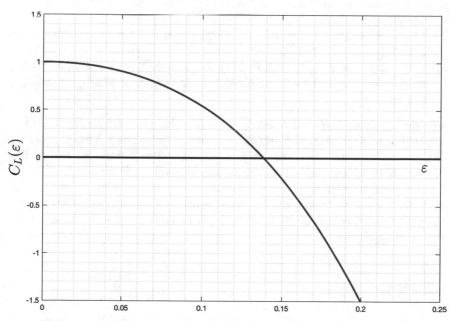

Fig. 4 This is the plot of the term $\theta(\varepsilon) = (1 - 2(8\varepsilon^2 + 4\varepsilon)^2)$ in the lower bound for the approximate $G_A G$-MRA frame, $C_L \geq \theta(\varepsilon)\|f\|$, for any $f \in L^2(\mathbb{R}^n)$. If we pick $\varepsilon \in (0, \frac{1}{4}(\sqrt{1 + \sqrt{2}} - 1))$, with $\frac{1}{4}(\sqrt{1 + \sqrt{2}} - 1) \approx 0.1384$, we obtain a positive C_L

We thus showed that $\Phi_{\mathbf{G}}^{\varepsilon}$ is a frame for the space V_0. Since we have not changed the definition of V_0, the spaces $V_j = A^{-1} V_0$, $j \in \mathbb{Z}$ form a nested sequence for the same reasons as in the orthonormal case. We will say that the scaling function $\phi(x)$ is "approximately refinable." Indeed,

$$\phi^{\varepsilon}(x) = \sqrt{2} \left(D_A^{-1} D_{B-1} \phi^{\varepsilon}(x) + D_A^{-1} D_{B-6} \phi^{\varepsilon}(x) \right),$$

except for two strips S_1^{ε} in $A^{-1} R_1^{\varepsilon} \cup A^{-1} R_6^{\varepsilon} + (1/2 + \varepsilon/2, 1/2 + \varepsilon/2)^T$ and $S_2^{\varepsilon} = \{1/2 + \varepsilon/2 \leq x_1 \leq 1/2 \leq \varepsilon/2, 0 \leq x_2 \leq x_1\}$ which can be made as small as needed by choosing the appropriate ε. This is in part because we did not start with a semi-orthogonal frame, i.e., $\mathbb{Z}^2$-translates of the R_i with the added strips are not disjoint.

5   Conclusion

We showed that, given

(a) a subgroup of dilations $\mathbf{G}_A$, formed by integer powers of an invertible, expanding matrix A,
(b) a finite subgroup of dilations $\mathbf{G}$, formed by invertible matrices with determinant 1, such that A normalizes $\mathbf{G}$, and
(c) a function $\phi \in L^2 (\mathbb{R}^n)$ such that (1) $\Phi_{\mathbf{G}} = \{D_B T_k \phi : B \in \mathbf{G}, k \in \mathbb{Z}^n\}$ is a semi-orthogonal Parseval frame for V_0, a closed subspace of $L^2 (\mathbb{R}^n)$ and (2) ϕ satisfies (11),

we obtain a frame $\mathbf{G}_A\mathbf{G}$-MRA for $L^2 (\mathbb{R}^n)$. We also provided an example of an "approximate" $\mathbf{G}_A\mathbf{G}$-MRA. Although our construction already leads to a scaling function with a significantly improved regularity, in future work, we will mollify ϕ on all sides. We expect the resulting system to also be a frame or a $\mathbf{G}_A\mathbf{G}$-MRA. The extension to other two sides of the domain follows the same pattern and the only remaining issue is around the corners. With the singularity set reduced to the vicinity of three corners, utilizing this new improved function will have a positive impact on applications.

References

1. C. ANGHEL, K. ARCHER, J.-M. CHANG, A. COCHRAN, A. RADULESCU, C. M. SALAFIA, R. TURNER, K. YACOUBOU DJIMA, AND L. ZHONG, *Placental vessel extraction with shearlets, Laplacian eigenmaps, and a conditional generative adversarial network*, in Proceedings of the Conference on Understanding Complex Biological Systems with Mathematics, Springer, 2018.
2. J. BLANCHARD, *Existence and accuracy results for composite dilation wavelets*, Washington University in St. Louis, 2007.

3. J. BLANCHARD AND K. STEFFEN, *Crystallographic Haar-type composite dilation wavelets*, in Wavelets and Multiscale Analysis, J. Cohen and A. I. Zayed, eds., Applied and Numerical Harmonic Analysis, Birkhäuser Boston, 2011, pp. 83–108.
4. E. CANDÈS, L. DEMANET, D. DONOHO, AND L. YING, *Fast discrete curvelet transforms*, Multiscale Modeling & Simulation, 5 (2006), pp. 861–899.
5. O. CHRISTENSEN, *An introduction to frames and Riesz bases*, Applied and Numerical Harmonic Analysis, Birkhäuser Boston, 2003.
6. R. COIFMAN, Y. MEYER, AND V. WICKERHAUSER, *Adapted wave form analysis, wavelet-packets and applications*, in Proceedings of the second international conference on Industrial and applied mathematics, ICIAM 91, Philadelphia, PA, USA, 1992, Society for Industrial and Applied Mathematics, pp. 41–50.
7. W. CZAJA AND E. KING, *Anisotropic shearlet transforms for* $L^2(R^k)$, Mathematische Nachrichten, 287 (2014), pp. 903–916.
8. I. DAUBECHIES, *Orthonormal bases of compactly supported wavelets*, Communications on Pure and Applied Mathematics, 41 (1988), pp. 909–996.
9. ——, *Ten lectures on wavelets*, no. 61 in CBMS/NSF Series in Applied Math., SIAM, 1992.
10. M. DO AND M. VETTERLI, *Contourlets: a directional multiresolution image representation.*, in ICIP (1), 2002, pp. 357–360.
11. G. EASLEY, D. LABATE, AND F. COLONNA, *Shearlet-based total variation diffusion for denoising*, Image Processing, IEEE Transactions on, 18 (2009), pp. 260–268.
12. L. EVANS, *Partial differential equations*, no. 61 in Graduate Studies in Mathematics, American Mathematical Society, Providence, Rhode Island, 2000.
13. P. GROHS, S. KEIPER, G. KUTYNIOK, AND M. SCHÄFER, *Alpha molecules: curvelets, shearlets, ridgelets, and beyond*, vol. 8858, 2013, pp. 885804–885804–12.
14. K. GUO AND D. LABATE, *Optimally sparse multidimensional representation using shearlets*, SIAM Journal on Mathematical Analysis, 39 (2007), pp. 298–318.
15. K. GUO, D. LABATE, W.-Q. LIM, G. WEISS, AND E. WILSON, *Wavelets with composite dilations and their MRA properties*, Applied and Computational Harmonic Analysis, 20 (2006), pp. 202–236.
16. A. HAAR, *Zur theorie der orthogonalen funktionen systeme*, Mathematische Annalen, 69 (1910), pp. 331–371.
17. D. HAN, K. KORNELSON, D. LARSON, AND E. WEBER, *Frames for Undergraduates*, Student mathematical library, American Mathematical Society, 2007.
18. E. KING, *Wavelet and frame theory: frame bounds gaps, generalized shearlets, Grassmannian fusion frames and p-adic wavelets*, University of Maryland, College Park, 2009.
19. E. KING, G. KUTYNIOK, AND X. ZHUANG, *Analysis of inpainting via clustered sparsity and microlocal analysis*, Journal of mathematical imaging and vision, 48 (2014), pp. 205–234.
20. A. KRISHTAL, B. ROBINSON, G. WEISS, AND E. WILSON, *Some simple Haar-type wavelets in higher dimensions*, The Journal of Geometric Analysis, 17 (2007).
21. D. LABATE, W.-Q. LIM, G. KUTYNIOK, AND G. WEISS, *Sparse multidimensional representation using shearlets*, vol. 5914, 2005, pp. 59140U–59140U–9.
22. S. MALLAT, *A theory for multiresolution signal decomposition: the wavelet representation*, IEEE Transactions on Pattern Analysis and Machine Intelligence, 11 (1989), pp. 674–693.
23. ——, *A wavelet tour of signal processing*, AP Professional, London, 1997.
24. B. MANNING, *Composite Multiresolution Analysis wavelets*, Washington University in St. Louis, 2012.

Mathematical Modeling of Immune-Mediated Processes in Coagulation and Anticoagulation Therapy

Erica J. Graham and Ami Radunskaya

Abstract Thousands of people each year succumb to complications from deep vein thrombosis (DVT). DVT occurs when blood flow in the veins is blocked by a clot. Individuals with DVT are at increased risk of experiencing pulmonary embolism (PE), in which small pieces of the clot break off and travel to the lungs. PE can lead to lung damage and even death. Mechanisms implicated in DVT may comprise an imbalance in the body, poor circulation, inflammation, or an immune response. Given the complex interplay of mediating factors in DVT, it is important to understand the role that each plays in thrombus and embolus formation. In this work, we develop and implement a mathematical model of venous clot formation and dissolution by describing interactions between inflammation, blood cells, and various chemical factors. We then use the model to identify factors essential to embolus formation and discuss implications for clinical treatment.

1 Why Do We Need a Mathematical Model of Blood Clotting?

Blood clots in the arteries are typically formed in response to an injury of the blood vessel, where the role of the clot is to stop the bleeding. They are the result of a chain reaction, or *cascade* of chemical events in the blood that is initiated by an injury. In a normal situation, this cascade of reactions results in a layer of activated *platelets*, a type of small blood cell, sticking to the blood vessel wall to stop bleeding. The layer is held together by a mesh made up of a chemical called *fibrin* (see Fig. 1).

E. J. Graham (✉)
Bryn Mawr College, Bryn Mawr, PA, USA
e-mail: ejgraham@brynmawr.edu

A. Radunskaya
Pomona College, Claremont, CA, USA
e-mail: ami.radunskaya@pomona.edu

© The Author(s) and the Association for Women in Mathematics 2019
S. D'Agostino et al. (eds.), *A Celebration of the EDGE Program's Impact on the Mathematics Community and Beyond*, Association for Women in Mathematics Series 18, https://doi.org/10.1007/978-3-030-19486-4_18

237

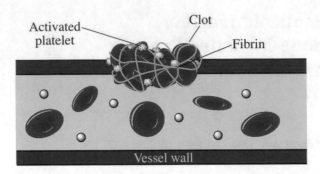

Fig. 1 A typical blood clot

Once the blood vessel wall is healed, the fibrin mesh disintegrates, and the blood clot dissolves.

Problems do arise in this process, however. One type of problem occurs when natural clotting is harmful, such as when a foreign body meant to be helpful triggers the coagulation cascade. This could be the case when a *stent*, or small tube, is placed in an artery to strengthen it and to keep it open. Here, coagulation could be dangerous since it could block the flow of blood to the brain or other organs. In general, such blockage by a clot is referred to as *thrombosis*. *Virchow's Triad*, named after nineteenth century scientist Rudolph Virchow, highlights thrombosis as a collection of pathologies related to altered blood flow rates (stasis), vessel integrity (endothelial injury), and chemical composition of the blood (hypercoagulability) [9, 18].

In order to prevent unwanted blood clots and thrombosis, patients are prescribed *anti-coagulants* such as warfarin, commonly known as "blood thinners." These drugs delay the initiation of the coagulation cascade, or slow its progress, by indirectly affecting the natural chemicals circulating in the blood. A question of concern to the medical community, and to anyone taking blood thinners, is "How much drug should we give, and how often?" Too much anti-coagulant could cause the patient to bleed internally, and too little could result in a life-threatening blockage of the artery. Typically, a patient on anti-coagulants is periodically checked using a prothrombin time (PT) test, from which the *international normalized ratio* or INR is calculated (see Fig. 2). In this test, the patient's blood is first filtered to isolate the plasma, which lacks red blood cells and platelets. Chemical reagents (thromboplastin and calcium) are then added to the filtered sample to initiate the coagulation cascade. A stopwatch is used to determine the time at which a clot forms, i.e., when small strands of fibrin are seen in the liquid. This clotting time is formally referred to as the prothrombin time.

If the prothrombin time is too long relative to normal, the dose of anti-coagulant is reduced; if the time is too short relative to normal, then the dose is increased. The INR is defined as the ratio of the measured clot time to a normal clot time, where "normal" is determined by an international standard developed by the World Health

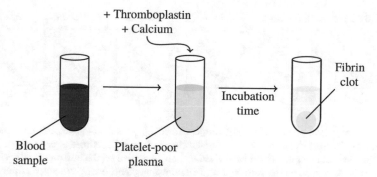

Fig. 2 Schematic of the prothrombin time (PT) test for computing the international normalized ratio (INR)

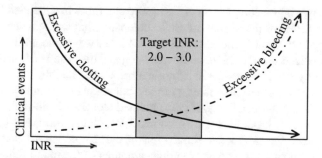

Fig. 3 Representative therapeutic target window for INR in an individual susceptible to abnormal clot formation. Excessive clotting or bleeding may result if the INR is too small or too large, respectively

Organization. Thus, if a patient is at risk for clotting, the INR should be larger than one (we want to slow clot formation to reduce the likelihood of an unwanted clot), but the INR should not be too large, because some clotting is necessary to prevent uncontrolled bleeding. Therefore, the goal of anti-coagulant therapy is to keep the INR larger than, but not too much larger than, one. A value between 2 and 3 is a typical therapeutic target. The PT test is therefore critical in determining the correct dose of anti-coagulant (see Fig. 3).

In our previous work, we addressed the question, "Is the INR an accurate reflection of what is occurring in the body?" We found that the blood flow rate indeed altered coagulation rates, with decreased blood flow producing faster clot formation. [4]. Because the INR is measured in a test-tube setting, there is no effect of blood flow on in vitro clot times. Therefore, if blood flow was decreased, the clot time simulated by the in vitro INR model was longer than the time predicted by the in vivo model. The model predicts that anti-coagulants would be prescribed in doses that are too low, resulting in the possible formation of dangerous blood clots for patients with decreased blood flow, such as those confined to their beds.

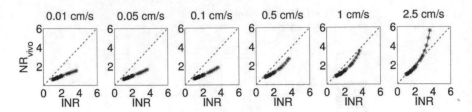

Fig. 4 A comparison of simulated INR results and NR_{vivo} results at increasing flow rates. The graph on the far right shows results at a normal flow rate of 2.5 centimeters per second. In each graph, the INR value, as estimated by our simulation, is plotted on the horizontal axis, while the NR_{vivo} value is plotted on the vertical axis. Different data points on a given graph correspond to different warfarin doses. Since warfarin increases clotting time, higher doses correspond to points with higher INR and NR_{vivo} values. Figure from de Pillis, Graham et al. [4]

In Fig. 4, we show a comparison of the in vivo clot times to simulated INR in vitro values at six different blood flow rates. To compare what we believe is happening inside the body to what is measured in the test tube, we define an in vivo version of the "normalized ratio," which we denote NR_{vivo}. This ratio is defined as the clot time divided by what is considered a "normal" clot time, i.e., the clot time that our model produces at normal flow rates, normal levels of platelets and clotting factors, and with no blood thinner. Thus, if the simulated clot time is normal, then NR_{vivo} would equal 1. If it takes longer than normal for a clot to form, NR_{vivo} would be greater than one. We ran model simulations of clot formation after an injury, for different blood flow rates, and after various amounts of the blood thinner warfarin had been administered. As more warfarin is administered, the clot times increase, as expected. In each case, we also simulated a PT test and recorded the INR value. If the INR test accurately predicted the clotting time in vivo, the resulting data point would lie on the diagonal. If the point lies *above* the diagonal, then the INR test is underestimating the clotting time, which could result in prescribing an excessive amount of blood thinner. If the point lies *below* the diagonal, then the INR test is overestimating the clotting time, and too little blood thinner might be prescribed, potentially resulting in the patient developing a blood clot.

We notice that, for normal flow rates of 2.5 cm per sec, the INR test underestimates the clotting times when these times are relatively long. Since the goal of prescribing warfarin is typically to keep the INR between 2 and 3, we conclude that the test is fairly accurate at normal flow rates, in normal treatment regimes. However, for low flow rates, such as might occur if a patient is inactive for a long period of time, the INR test overestimates the clotting times for most doses of anti-coagulant (see graph on the far left it Fig. 4). This means that the physician will prescribe a dose of warfarin that is smaller than desired, resulting in clots forming. This is, in fact, what is observed: patients who are inactive or who have poor circulation will often develop clots where blood flow is low. This is where the new chapter in our story starts.

As the preceding discussion shows, a mathematical model of blood clotting can give us a more nuanced understanding of clotting times than the PT test typically administered in the clinic. The differences between the clotting times predicted by the in vivo model versus the in vitro model could be critical under clinically relevant conditions such as low blood flow or anti-coagulant treatment. Since we cannot actually measure clotting times in a patient, it is important to develop and validate mathematical models that we can use to predict actual clotting times under a variety of conditions.

2 Deep Vein Thrombosis and Pulmonary Embolisms

Our previous model was successful in capturing the dynamics of a blood clot in an artery that was initiated by an injury to the blood vessel. We were also able to describe the effect of the anti-coagulant drug in different situations. However, the most dangerous types of blood clots are those that occur in the veins, rather than the arteries. This type of blood clot is called *deep vein thrombosis* or DVT. In fact, according to the Center for Disease Control, 60,000 to 100,000 people in the USA die each year of DVT [11]. The veins are different from the arteries: they have different mechanics related to blood flow, and they also carry different concentrations of cells and chemical factors. The goal of the proposed work is to extend our previous work, and to develop a model of coagulation in the veins. We want to describe how clots form in the absence of a specific injury, but rather as the result of some imbalance in the body, from poor circulation, inflammation, and/or an immune response. Figure 5 shows a clot that forms near a valve in a vein in the leg, a common spot for DVT. Sometimes a piece of a large clot in a vein breaks off and travels up to the heart, lungs, or brain, causing serious injury or sudden death. These small pieces are called *embolisms*; when they travel to the lungs they are called *pulmonary embolisms* or PE.

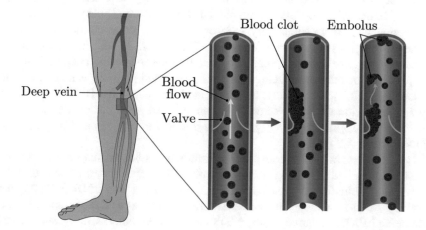

Fig. 5 Deep vein thrombosis and embolus formation

Clinicians would like to be able to determine which patients with DVT are at a high risk for PE, which should be treated with anti-coagulants, which should be left alone, and which need more invasive treatments, such as surgery to remove the clot or to drill a passageway through the clot for a new blood vessel. Understanding the role of the immune system in blood clot formation could help develop strategies for preventing DVTs that would be less likely to cause bleeding complications. The immune system has also been implicated in the dissolution of blood clots [1]. To better understand the risks of PE, we plan to include the dissolution of clots and the formation of embolisms into our model of DVT.

While several different mathematical models of clot formation appear in the literature, very few deal with blood clots that form in the vein, or incorporate the inflammatory response. In this chapter we describe a preliminary model that includes the initiation of clot formation through inflammation, clot dissolution, and the formation of emboli. The model builds on the models in [4, 17] and [6]. In this paper, we focus on the role of the immune cells: their recruitment, their interaction with platelets, and their effect on the initiation of the clotting cascade. Since we are ultimately interested in questions about the treatment of clots, we want the model to also capture the binding of immune cells with platelets, which influence the risk of pulmonary embolisms. For reference, we include a description of all model variables, equations, and parameters in the Appendix.

2.1   The Biological Players in DVT

DVT, and venous thrombosis in general, is a process primarily mediated by the immune system. In contrast to arterial thrombosis, DVT often results from decreased or static blood flow in the absence of injury to the endothelial wall [25]. When blood flow ceases, oxygen levels drop, activating the endothelium, which then expresses various adhesion proteins that can recruit and activate *platelets* and immune cells. When platelets are activated, they become "sticky," and can form aggregates with other platelets. Hypoxia, or low oxygen concentration, also recruits white blood cells, or *leukocytes*, to the site. These immune cells, consisting mostly of *neutrophils*, are activated either by directly adhering to the endothelium or in the presence of other activated neutrophils. Activated platelets can adhere to the endothelial cells, to each other, or to activated immune cells. DVT is established by a combination of tissue factor (TF)-induced (*extrinsic*) coagulation as well as the *intrinsic* coagulation pathway, which is initiated by an activated enzyme called *clotting factor XII* (FXII).

Neutrophil extracellular trap (NET) formation has been discovered as essential to establishing and sustaining venous thrombi. NETs are formed when neutrophils release a combination of proteins and DNA that form a web of DNA [2, 23]. The resulting "traps" can attract and activate more platelets, as well as stimulate both TF release and FXII activation. Thus, NETs can promote both extrinsic and intrinsic coagulation pathways [15]. Once fibrin (recall Fig. 1) is formed, the clot—an aggregate of platelets, NETs, and red blood cells—is stabilized to form a thrombus. Because of the role of NETs in clot formation, venous thrombi do not comprise tightly packed platelet aggregates. As a result, it is more likely that pieces of the clot can break off and travel to other parts of the body, as occurs in PE.

Monocytes are another type of immune cell that is recruited to the wall of the vein when blood flow is drastically slowed or stopped. Monocytes, even more so than neutrophils, are a source of TF, a trigger for the extrinsic coagulation pathway. Together, monocytes and neutrophils make up the population of leukocytes in our model.

3 Model Description

To examine the role of the immune system in venous thrombosis, we develop a mathematical framework of platelet and immune cell dynamics in a cylindrical blood vessel. Specifically, we introduce a preliminary model of immune-mediated platelet aggregation and thrombus formation prior to clot dissolution.

To describe the process that results in DVT, we introduce the following state variables, in addition to those in the model described in [4]. For a complete list of state variables, we refer the reader to the Appendix.

State Variables	
W	Vessel wall (endothelial) activation level, with values between 0 (unactivated) and 1 (fully activated)
L_u^m	Concentration of unactivated and mobile leukocytes
L_a^p	Concentration of activated leukocytes bound to platelets
L_a^e	Concentration of activated leukocytes bound to the endothelium
$[NET]$	Concentration of leukocyte-derived NETs
P_a^b	Concentration of activated platelets bound to other platelets (see [17])
P_a^l	Concentration of activated platelets bound to leukocytes or NETs
P_a^e	Concentration of activated platelets bound to the endothelium
P_a^m	Concentration of activated and mobile platelets
TF	Average concentration of active (decrypted) tissue factor expressed per activated monocyte

The cartoon in Fig. 6 illustrates the meaning of the platelets and white blood cells (*leukocytes*) in their various configurations. These cells as well as smaller molecules (not shown below) flow in and out of the activated zone.

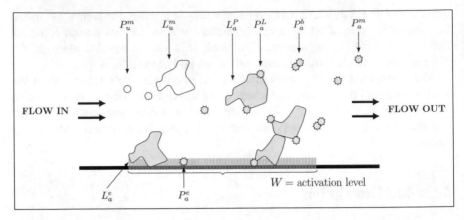

Fig. 6 Behavior of possible cellular interactions in the models. Platelets and leukocytes may flow into and out of the region of interest. They may become activated, bind to an activated endothelium, or bind to each other

3.1 Derivation of the Model Equations

We can represent interactions between platelets and white blood cells using a series of reactions. Reversible reactions are indicated by $\rightleftharpoons$, with the rate parameter for a particular direction indicated above or below the arrow. All other reactions are considered irreversible ($\leftarrow$ or $\rightarrow$). The nine reactions (R1)–(R9) accompanying the species represented in Fig. 6 are listed below, with a slight abuse of notation.

R1: (LEUK FLOW INTO/FROM ZONE) $\quad \xrightarrow{k_{\text{flow}}^{l} L^{\text{up}}} L_u^m \xrightarrow{k_{\text{flow}}^{l}} \varnothing$

R2: (LEUK-TO-ENDO ADHESION) $\quad L_u^m \underset{a_{-1}}{\overset{a_1 \hat{W}}{\rightleftharpoons}} L_a^e \xrightarrow{d_{\text{leuk}}} \varnothing$

R3: (LEUK-TO-PLT BINDING) $\quad L_u^m + P_a^e \underset{a_{-2} \cdot \pi_a^e}{\overset{a_2}{\rightleftharpoons}} L_a^p \underset{a_2}{\overset{a_{-2} \cdot (1 - \pi_a^e)}{\rightleftharpoons}} L_u^m + P_a^l$

R4: (LEUK DEGRADATION) $\quad L_a^p \xrightarrow{d_{\text{leuk}}} \varnothing$

R5: (NET FORMATION) $\quad q L_a^p \xrightarrow{a_3} NET \xleftarrow{a_3^p P_a^m} q L_a^e$

R6: (NET DEGRADATION) $\quad NET \xrightarrow{d_{\text{NET}}} \varnothing$

R7: (PLT-TO-LEUK/NET BINDING) $\quad L_a^e + P_a^m \underset{a_{-5} \cdot \lambda_a^e}{\overset{a_5}{\rightleftharpoons}} P_a^l \underset{a_5}{\overset{a_{-5}(1 - \lambda_a^e)}{\rightleftharpoons}} NET + P_a^m$

R8: (PLT ACTIVATION) $\quad P_u^m \xrightarrow{f(NET, \sum P_a^*, e_2)} P_a^m \xrightarrow{k_{\text{flow}}^p} \varnothing$

R9: (PLT-TO-ENDO ADHESION) $\qquad P_u^m \xrightarrow{k_{adh}^+ \hat{W}} P_a^e \underset{k_{adh}^+ \hat{W}}{\overset{k_{adh}^-}{\rightleftarrows}} P_a^m.$

We use $\hat{W}$ as a surrogate for the more complex description of endothelial activation to be outlined later. The expressions $\pi_a^e = P_a^e/(P_a^e + P_a^l)$ and $\lambda_a^e = L_a^e/(L_a^e + [NET])$ reflect the proportion of unbound platelet–leukocyte complexes assumed to return to an endothelial-bound state. In reaction R8, $f(\cdot)$ is a function of the specified arguments to be defined later, and $\Sigma P_a^* = P_a^b + P_a^e + P_a^l + P_a^m$ is simply the sum of all activated platelets in the system. We use positive subscripts to denote rates of species adhesion and negative subscripts for species dissociation/unbinding.

We can translate the reactions between chemical and cellular species into a system of differential equations. To do this, we make use of *the law of mass action*, which assumes the rate at which a reaction occurs is proportional to the product of the species involved in the reaction. The goal is to identify all reactions containing a particular state variable and use them to define its rate of change. For example, consider L_u^m, which appears in reactions R1–R3. Arrows *approaching* L_u^m indicate an increase in dL_u^m/dt, whereas those *leaving* L_u^m indicate a decrease. The specified parameters are multiplied by the originating species in a particular reaction. For L_u^m, we write

$$\frac{dL_u^m}{dt} = \underbrace{k_{flow}^l L^{up} - k_{flow}^l L_u^m}_{R1} \quad \underbrace{-a_1 \hat{W} L_u^m + a_{-1} L_a^e}_{R2}$$

$$\underbrace{-a_2 L_u^m P_a^e + a_{-2} \frac{P_a^e}{P_a^e + P_a^l} L_a^p}_{R3} \quad \underbrace{-a_2 L_u^m P_a^l + a_{-2} \left(1 - \frac{P_a^e}{P_a^e + P_a^l}\right) L_a^p}_{R4}$$

$$= k_{flow}^l (L^{up} - L_u^m) - a_1 \hat{W} L_u^m + a_{-1} L_a^e - a_2(P_a^e + P_a^l) L_u^m + a_{-2} L_a^p.$$

Applying this methodology to the remaining variables, we can derive the complete set of equations, each of which can be broken down into growth and loss terms. The complete model describes components related to dynamics of the endothelium, leukocytes and NETs, platelets, and coagulation triggers. We explain the equations that are new to this model, and in the next section we discuss some model simulations. For more details on previous model formulations, see [4, 16, 17].

Figure 7 shows interactions between state variables of the model, some of which will flow in and out of the region of interest, where a clot could form, which we will call the *activation region*. Each interaction, as well as flow into and out of the activation region, is represented by a term in the differential equation representing the evolution of the state variable.

Endothelial Activation Let $W(t)$ denote the activation level of the vessel wall endothelium, which is restricted between 0 and 1. Activated white blood cells (L_a^e) activate the endothelium at a maximal rate of a_0. Sustained activation of

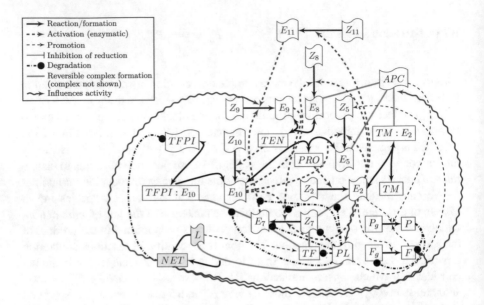

Fig. 7 Diagram representing the biochemical pathways involved in the blood clotting model. Species with "tape" shapes around them, such as L, have flow incorporated in their dynamics as well. Curves with arrows on both ends denote reversible reactions. Figure modified from [4]. The new variables, NET and L, are highlighted in yellow. All of the variables are described in the Appendix

the endothelium can occur through feedback from activated endothelium-bound leukocytes. Endothelium-bound platelets may also provide additional feedback, which we omit here [19]. We assume the magnitude of feedback is concentration-dependent, with Hill coefficient 2 and half-maximal concentration $k_{L_a^e:w}$.

$$\frac{dW}{dt} = a_0 \frac{L_a^{e\,2}}{L_a^{e\,2} + k_{L_a^e:w}^2}(1 - W). \tag{1a}$$

Based on experimental evidence in which flow stasis or reduced flow activates the endothelium [25], we assume initial activation, $W(0)$, of the endothelium is a decreasing function of the average rate of flow through the vessel. That is,

$$W(0) = \mathcal{F}(\bar{v}) = a_w + (1 - a_w) \cdot \frac{1}{1 + (\bar{v}/k_w)^3}, \tag{1b}$$

where $\bar{v} = v/v_{max}$ is the normalized midstream velocity. We assume the endothelium maintains a normally inactive state when $v = v_{max}$, i.e., $\mathcal{F}(1) = 0$, and is fully active under flow stasis, i.e., $\mathcal{F}(0) = 1$. Based on these boundary conditions, we define $a_w = -k_w^3$, where k_w is a shape parameter for the function $\mathcal{F}(\bar{v})$. In other words, k_w determines how sensitive initial endothelium activation is to changes in blood flow.

As activated leukocytes and platelets bind to the endothelium, there is only a fraction remaining that is available for binding by platelets and leukocytes, denoted by W^{avail}, which we define as

$$W^{\text{avail}} = w_{\max} \cdot W - L_a^e - P_a^e. \tag{1c}$$

Leukocyte Dynamics Neutrophils and monocytes are the two white blood cell types implicated in venous thrombosis [3]. For simplicity, we consider a single class of such *leukocytes* that carries out their respective roles in venous thrombogenesis. Leukocytes may be unactivated and mobile (L_u^m), mobile, activated, and bound to platelets (L_a^p), or activated and endothelium-bound (L_a^e). Monocyte-dependent and neutrophil-dependent actions occur according to the observed fractions of these cells in the area of the thrombus. 70–90% of leukocytes in any state are assumed to be neutrophils and the remaining 10–30% are monocytes [3, 25]. Differential activity between the two cell types is distinguished by the fractions q for neutrophils and $1 - q$ for monocytes.

Leukocytes enter the vessel region by flowing in from upstream locations, according to the midstream velocity, reflected in parameter k_{flow}^l. Unactivated leukocytes (L_u^m) can either be activated by binding to endothelial-bound activated platelets (L_u^p) or upon adhesion to the available space on the endothelium (L_a^e). This assumption is based on the idea that the endothelium and platelets play some role in leukocyte activation during thrombogenesis [19]. We also restrict the activating sources based on the notion that leukocytes recruited to the reaction zone typically slow to a roll along the endothelium. The differential equations describing leukocyte dynamics become

$$\frac{dL_u^m}{dt} = \underbrace{k_{\text{flow}}^l (L^{\text{up}} - L_u^m)}_{\substack{\text{flow into/from} \\ \text{zone}}} - \underbrace{[a_1 W^{\text{avail}} + a_2(P_a^e + P_a^l)] \cdot L_u^m}_{\text{ENDO/PLT adhesion}} + \underbrace{a_{-1} L_a^e}_{\substack{\text{ENDO} \\ \text{unbinding}}} + \underbrace{a_{-2} L_a^p}_{\substack{\text{PLT} \\ \text{unbinding}}}, \tag{2}$$

$$\frac{dL_a^e}{dt} = \underbrace{a_1 W^{\text{avail}} L_u^m}_{\text{adhesion}} - \underbrace{a_{-1} L_a^e}_{\text{unbinding}} - \underbrace{q a_3^p P_a^m L_a^e}_{\text{NET formation}} - \underbrace{a_5 P_a^m L_a^e}_{\text{PLTbinding}} \tag{3}$$

$$+ \underbrace{a_{-5} P_a^l \frac{L_a^e}{L_a^e + [NET]}}_{\text{PLTunbinding}} - \underbrace{d_{\text{leuk}} L_a^e [q - \mu(1 - q)]}_{\text{turnover}},$$

$$\frac{dL_a^p}{dt} = \underbrace{a_2(P_a^e + P_a^l) L_u^m}_{\text{PLTbinding}} - \underbrace{q a_3 L_a^p}_{\text{NET formation}} - \underbrace{a_{-2} L_a^p}_{\text{PLTunbinding}} - \underbrace{d_{\text{leuk}} L_a^p [q - \mu(1 - q)]}_{\text{turnover}}, \tag{4}$$

E. J. Graham and A. Radunskaya

where L^{up} denotes the concentration of leukocytes upstream of the activation zone. d_{leuk} is the maximal turnover rate for leukocytes, and we assume monocytes turnover at a fraction μ of the rate of shorter-lived neutrophils [20, 24]. The range for $1/\mu$ in humans can be anywhere between 3 and 12 [20]. Here we arbitrarily set $1/\mu = 6$.

NET formation can occur by interactions between mobile activated platelets and endothelial-bound leukocytes or by platelet-bound activated leukocytes. We omit the possibility that activated neutrophils release NETs in the absence of activated platelets [25]. Leukocytes released from platelet unbinding may either be NETs or endothelial-bound leukocytes, and the release rate is proportional to the fraction present.

$$\frac{\mathrm{d}[NET]}{\mathrm{d}t} = q \underbrace{\left[a_3^p P_a^m L_a^e + a_3 L_a^p \right]}_{\text{formation}} - \underbrace{a_5 P_a^m [NET]}_{\text{PLTbinding}} \tag{5}$$

$$+ \underbrace{a_{-5} P_a^l \frac{[NET]}{L_a^e + [NET]}}_{\text{PLTunbinding}} - \underbrace{d_{\text{NET}}[NET]}_{\text{degradation}}.$$

Platelet Dynamics The pool of mobile inactive platelets is replenished from populations upstream of the vessel. Platelets that bind to each other are now considered as a distinct subset P_a^b. It is assumed that unactivated platelets do not bind to each other. NETs and activated platelets provide feedback to activate inactive platelets [19]. These processes are modeled using mass-action terms with rate constants a_4 and $k_{\text{plt}}^{\text{act}}$, respectively. Thrombin (e_2) may also activate platelets, which is modeled using Michaelis–Menten kinetics with maximal reaction rate $k_{e_2}^{\text{act}}$ and half-maximal concentration $\overline{e_2}$ [16].

$$\frac{\mathrm{d}P_u^m}{\mathrm{d}t} = \underbrace{k_{\text{flow}}^p (P^{\text{up}} - P_u^m)}_{\text{flow into/from zone}} - \underbrace{a_4 [NET] P_u^m}_{\text{activation from NETs}}$$

$$- \underbrace{\left[k_{\text{plt}}^{\text{act}} (P_a^b + P_a^e + P_a^l + P_a^m) + k_{e_2}^{\text{act}} \frac{e_2}{e_2 + \overline{e_2}} \right] \cdot P_u^m}_{\text{PLT/thrombin activation}} \tag{6}$$

$$- \underbrace{k_{\text{adh}}^+ W^{\text{avail}} P_u^m}_{\text{ENDO adhesion}}.$$

As modeled previously, any activated platelets release signals such as ADP that can in turn activate nearby platelets [16, 17]. Mobile activated platelets can adhere at rate k_{coh} to bound platelets according to mass-action dynamics, following [6]. Activated platelets can bind to NETs or other leukocytes (becoming leukocyte-bound platelets, P_a^l). We also assume that platelets bound to the endothelium or to leukocytes, but not to other platelets, can bind a mobile activated platelet. Platelets unbinding from leukocytes return to mobile activated platelets.

$$\frac{dP_a^b}{dt} = \underbrace{k_{\text{coh}} \frac{P_a^e + P_a^l}{P_{\text{max}}} P_a^m}_{\text{PLT cohesion}} \tag{7}$$

$$\frac{dP_a^l}{dt} = \underbrace{a_5 P_a^m (L_a^e + [NET])}_{\text{PLT-LEUK binding}} - \underbrace{a_{-5} P_a^l}_{\text{unbinding}} - \underbrace{a_2 P_a^l L_u^m}_{\text{LEUK binding}} + \underbrace{a_{-2} L_a^p \cdot \frac{P_a^l}{P_a^l + P_a^e}}_{\text{unbinding}} \tag{8}$$

$$\frac{dP_a^m}{dt} = \underbrace{\left[a_4 [NET] + k_{\text{plt}}^{\text{act}} (P_a^b + P_a^e + P_a^l + P_a^m) - k_{e2}^{\text{act}} \frac{e_2}{e_2 + \overline{e_2}} \right] \cdot P_u^m}_{\text{activation}} \tag{9}$$

$$\underbrace{- k_{\text{flow}}^p P_a^m}_{\text{flow out}} \underbrace{- k_{\text{adh}}^+ W^{\text{avail}} P_a^m + k_{\text{adh}}^- P_a^e}_{\text{ENDObinding/unbinding}} \underbrace{- k_{\text{coh}} \frac{P_a^e + P_a^l}{P_{\text{max}}} P_a^m}_{\text{PLTbinding}}$$

$$\underbrace{- a_5 P_a^m (L_a^e + [NET]) + a_{-5} P_a^l}_{\text{LEUKbinding/unbinding}}$$

$$\frac{dP_a^e}{dt} = \underbrace{k_{\text{adh}}^+ W^{\text{avail}} (P_u^m + P_a^m) \quad k_{\text{adh}}^- P_a^e}_{\text{ENDObinding/unbinding}} \underbrace{- a_2 P_a^l L_u^m + a_{-2} L_a^p \cdot \frac{P_a^e}{P_a^l + P_a^e}}_{\text{PLTunbinding}}. \tag{10}$$

Coagulation Triggers Tissue factor (TF) is the primary trigger for the extrinsic coagulation cascade. With respect to venous thrombosis, platelet–leukocyte interactions are believed to initiate tissue factor production in the absence of vessel injury. Activated tissue factor (TF) is expressed by activated monocytes. Specifically, monocytes bound to platelets or the endothelium will express active TF [3]. We assume each individual activated monocyte expresses a constant maximal level of active TF ($[TF]$) with which other coagulation factors may interact. To further simplify our biological assumptions, we omit monocyte-derived microparticle expression of active TF [5].

TF available for initiation of coagulation on an individual monocyte is computed as the difference between the total TF concentration contributed by all monocytes in the region of interest and the different complexes it forms with other factors, as in [16]. Because monocytes, but not neutrophils, express TF, we multiply the activated leukocyte concentration by $1 - q$.

Notation In the following equations, the variables Z_i and E_i denote the amount of unactivated clotting factor i (zymogen) and activated clotting factor i (enzyme), respectively. If the factors are bound to platelets, they are decorated by a superscript m, as in z_7^m. Concentrations are denoted by square brackets, as in $[TF]$ or by lowercase letters for the clotting factors: z_i and e_i. When molecules bind to form a complex, we denote the complex by writing both molecules separated by a colon, as in $Z_{10} : E_7^m$. A more detailed description of all the state variables is given in the

Appendix, with even more detail given in [4].

$$[TF]^{\text{avail}} = (1 - q)\left(L_a^e + L_a^p\right) \cdot \left([TF] - z_7^m - e_7^m - [Z_7^m : E_{10}] - [Z_7^m : E_2]\right) \tag{11}$$

$$- [Z_{10} : E_7^m] - [Z_9 : E_7^m] - [TFPI : E_{10} : E_7^m] - [Z_7^m : E_9]).$$

Clotting factor XII is the primary trigger for the intrinsic (or contact) coagulation cascade. Factor XII can be activated by NETs and activated (mobile or bound) neutrophils [13, 15]. Three primary effects of FXII activation are included. We describe the modifications made to preexisting equations to incorporate these effects.

1. Factor XII activates the intrinsic coagulation pathway upstream of clotting factors XI and IX. To simplify the model, we omit explicit kinetics of factors XII and XI, instead including NET-dependent effects on factor IX activation (the most upstream chemical included in the original model).

$$\frac{dz_9}{dt} = \text{net flow} \pm \text{platelet and other chemical interactions} \tag{12}$$

$$- k_9^{\text{cat}} z_9 [NET]$$

$$\frac{de_9}{dt} = \text{net flow} \pm \text{platelet and other chemical interactions} \tag{13}$$

$$+ k_9^{\text{cat}} z_9 [NET]$$

2. Activated FXII, or FXIIa, can reduce thrombomodulin (TM)-mediated activation of protein C [7] to give APC, activated protein C.

$$\frac{dAPC}{dt} = \text{flow} \pm \text{FVa and FVIIIa interactions} \tag{14}$$

$$+ \frac{k_{PC:TM:e_2}^{\text{cat}}}{1 + \frac{[NET]}{k_{PC:TM:NET}}} \cdot [TM : E_2 : APC]$$

$$\frac{d[TM : E_2]}{dt} = \text{TM–thrombin binding kinetics} \tag{15}$$

$$+ \frac{k_{PC:TM:e_2}^{\text{cat}}}{1 + [NET]/k_{PC:TM:NET}} [TM : E_2 : APC]$$

$$\frac{d[TM : E_2 : APC]}{dt} = \text{TM–thrombin binding kinetics} \tag{16}$$

$$- \frac{k_{PC:TM:e_2}^{\text{cat}}}{1 + [NET]/k_{PC:TM:NET}} [TM : E_2 : APC]$$

3. Factor XIIa may also reduce fibrinolysis, a process which is carried out by the plasmin ($[P]$) enzyme [25].

$$\frac{dF}{dt} = \text{formation from fibrinogen} - \text{cross} - \text{linking}$$

$$- \frac{v_{17}[F][P]}{k_{17a} + [P]} \cdot \frac{k_{e_{12}:F:NET}}{k_{e_{12}:F:NET} + [NET]} \tag{17}$$

In addition, we model the ability of NETs to cleave tissue factor pathway inhibitor (TFPI), which typically works to inhibit intrinsic activation of the coagulation pathway [23].

$$\frac{d[TFPI]}{dt} = \text{flow} \pm \text{FXa interactions} - d_{TFPI} \cdot [NET] \cdot [TFPI]. \tag{18}$$

As modeled previously, we also adjust differential equations that involve activated protein C to omit model dependence on an exposed subendothelium, as a primary assumption is an intact endothelium. Further details may be found in [6].

4 Model Simulations and Discussion

Here we discuss the most salient aspects of the behavior of the model with respect to changes in blood flow rates over a period of 48 h, the composition of resulting clots, and model implications. First we numerically solve the full system of ordinary differential equations using the DLSODA subroutine in Fortran. We assume that veins experience normal shear rates of roughly 200 per second, so that $v_{\max} = 1.0$ microns per second ($\mu\text{m s}^{-1}$) [12]. We initialize the model for a cylindrical blood vessel with radius 0.4 mm and track chemical and cellular interactions in a 40 μm by 40 μm square along the vessel wall. Shear rates ranging from 2 to 200 per second are used to simulate the model. For reference, these rates correspond to midstream blood flow velocities ranging from 0.01 to 1.0 $\mu\text{m s}^{-1}$, respectively. Parameters for the new model components were either based on the literature or selected to achieve reasonable qualitative behavior based on leukocyte and platelet recruitment times [3, 25]. All others were taken from Fogelson et al. [4, 10, 17]. A complete list of parameters is provided in the Appendix (see section "Complete Parameter List").

4.1 Model Behavior Under Reduced Flow

Figure 8 demonstrates the time courses of a subset of our state variables for two different shear rates, 50 and 150 s^{-1} ($v = 0.25$ and $v = 0.75\,\mu\text{m s}^{-1}$, respectively), over an initial 24-h period post-endothelial activation. Recall that the initial level of endothelial activation is determined by the midstream velocity. Variables are shown in normalized levels, scaled by their respective maximal concentrations over the complete 48-h simulation length. In the top row of Fig. 8, we observe an initial rise in inflammation in the clot region, as determined by the concentrations of

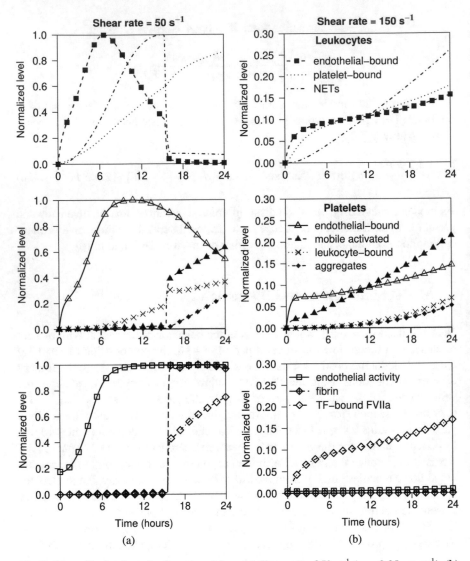

Fig. 8 Normalized trajectories for reduced flow. (**a**) Shear rate of 50 s^{-1} ($v = 0.25\,\mu\text{m s}^{-1}$), (**b**) shear rate of 150 s^{-1} ($v = 0.75\,\mu\text{m s}^{-1}$). *Top*: activated leukocytes; *middle*: activated platelets; *bottom*: endothelial activation level, fibrin, and tissue factor-bound activated factor VII (FVIIa). This figure indicates that platelets and white blood cells are recruited to the activated region substantially more quickly and in greater concentrations at lower shear, i.e., slower blood flow, rates

bound leukocytes and NETs. Leukocytes bind to the endothelium and activated platelets immediately, whereas NETs are released shortly after leukocyte activation. The relative, though not absolute, rate of increase in endothelium-bound cells is higher than that of platelet-bound cells. In the middle row, platelets rapidly bind to the endothelium and activated leukocytes. In addition, platelets and leukocytes are initially recruited to the endothelium fairly early on, i.e., within the first hour of the simulation. In the bottom row, we note that changes in endothelial activation occur much more quickly under dramatically reduced flow conditions. Under low flow, a sharp rise in TF-bound activated factor VII (FVIIa) initializes the clot cascade, ultimately leading to pronounced fibrin production. Finally, low flow coagulation dynamics are more dramatic within a shorter time frame relative to the delayed dynamics experienced as flow rates approach the normal end of the spectrum.

In a more detailed comparison of flow-specific model behavior, we note that the timeline illustrated under shear rate 50 s^{-1} is more consistent with descriptions of inflammation-mediated coagulation processes than the 150 s^{-1} case. Specifically, rapid and large-scale binding of leukocytes and platelets to the endothelium occurs within the first hour of the simulation. This is followed by several hours' worth of additional recruitment and binding, with leukocytes achieving maximal numbers earlier in the process. Once leukocytes and platelets are sufficiently activated, they serve as feedback mediators for additional cellular recruitment and endothelial activation, setting the stage for the TF-mediated intrinsic coagulation cascade [3].

4.2 Clot Composition

To determine the effect of inflammation and blood flow on the structure of the resulting clot, we observe the composition of platelets and leukocytes over the duration of the simulation. These results are summarized in Fig. 9a, in which we determine which cell types dominate the volume in the reaction zone hours 1, 3, 6, 9, and 12 into the simulation and every 12 h thereafter. For frame of reference, we also note that an average leukocyte has a diameter that is roughly 4–5 times that of a platelet. In addition, because fibrin is the primary determinant of clot stability, we show the levels of fibrin present at these time points with green shading.

Clot volumes in the simulation primarily comprise platelet-bound leukocytes, followed by leukocyte-bound platelets. Other activated cells contribute minimally to the overall clot structure. This makes sense, as we consider a reaction zone that is significantly smaller than the blood vessel. As such, the surface area that leukocytes and platelets have to bind in a single layer is limited, whereas cellular-bound leukocytes and platelets may bind in several layers through successive interactions.

Both cell types increase upon endothelial activation, but the degree to which this occurs varies with blood flow rate. For example, severe flow reduction, as with a 2 s^{-1} shear rate, produces a smaller clot with very low levels of fibrin. Slightly higher flow rates (e.g., 10 and 50 s^{-1} shear rates) result in a bigger clot, with greater recruitment of activated leukocytes earlier in the coagulation process and progressively higher fibrin levels. As flow rates approach normal (e.g., 100 and 150 s^{-1} shear rates), the clot volume is composed of mostly leukocytes, whose numbers decline dramatically the closer the flow rate is to normal. Leukocyte-bound platelet volumes, on the other hand, appear to change less dramatically with changes in flow.

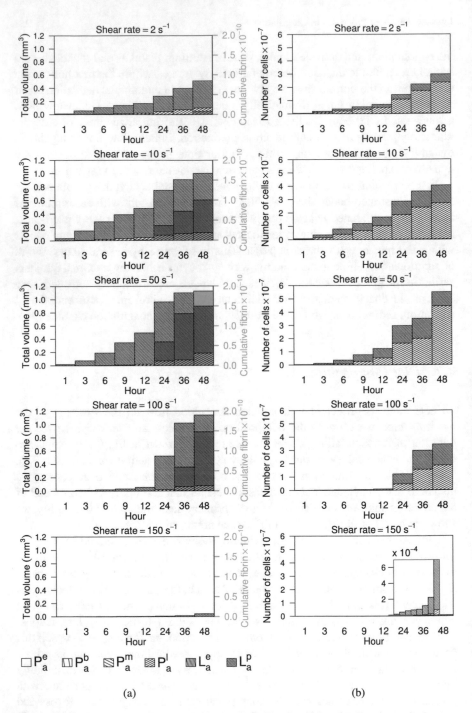

Fig. 9 (a) Composition of cellular volumes in clot region for activated leukocytes and platelets. Both cell types are assumed spherical with diameters of 8 and 2 μm for leukocytes and platelets, respectively. (b) Cell number distribution for activated leukocytes and platelets. At very low blood flow, the clot contains fewer platelets and leukocytes, along with low levels of fibrin; thus, it is fairly unstable. With low, but not extremely low flow, larger clots can result in very high levels of fibrin, resulting in increased stability. With near-normal (higher) flow rates, clotting is negligible

Figure 9b illustrates the absolute cell number composition of the various clots produced. No considerable differences in absolute number of platelets bound to leukocytes are observed for very low flow (1%–5% of normal), but less extreme reductions (e.g., 25% of normal) result in larger cell-bound platelet-to-leukocyte ratios, which then decline as flow rates are increased further. These results suggest that variable mechanisms are responsible for establishing blood clots depending on the magnitude of reduction. Specifically, extremely low flow results in smaller but less stable clots, whereas low flow can produce larger clots with very high levels of fibrin. Minimal flow reductions cannot produce an appreciable clot during the length of the simulation. We hypothesize that similar behavior to other flow rates would be eventually exhibited in the higher flow cases, but to a smaller magnitude and over a much longer time span.

4.3 Implications for Embolization

To understand how inflammatory processes might influence embolus formation mathematically, we examine trajectories in the two-dimensional phase plane for multiple shear rates (Fig. 10). In particular, we focus on initial and terminal processes in venous thrombus formation. That is, we examine behavior of endothelial activation of leukocytes and platelets and of fibrin formation, respectively. The largest number of endothelial-bound leukocytes occurs with a shear rate of 50 s^{-1}, whereas the largest number of platelets bound to the endothelium occurs for the slowest shear rate examined. As flow rates normalize, leukocyte and platelet activity on the endothelium is drastically reduced. The maximal fibrin concentration is achieved for a shear rate of 100 s^{-1}, and the largest concentration of mobile activated platelets occurs with shear rate 2 s^{-1}. Both fibrin and mobile activated platelets are minimal for near-normal blood flow.

Beyond 40 h, fibrin levels exhibit a rebound late in the simulation for moderately reduced flow rates (not shown), accompanied by a decline in mobile activated platelets. This is due to the limit placed on the height of cellular layers necessary to accommodate the simplified flow structure in the vessel.

These results illustrate an important aspect of inflammation and coagulation. Consistent with experimental observations [3, 25], we find that nearly complete flow stasis results in a non-standard clot characterized by very little fibrin, but a large number of activated mobile platelets. In terms of embolization, the stickiness of such platelets increases the possibility that downstream processes may be triggered by platelet activation. In addition, the mobility of these platelets indicates that they may travel to other parts of the body, such as in the case of venous thromboembolism.

Based on these results, we hypothesize that the stickiness of mobile activated platelets are likely triggers of downstream processes that would exacerbate the hypercoagulable state created by endothelial activation. Further, the mobility of these platelets would allow them to travel to other parts of the body, as in the case of venous thromboembolism. Reduced clot stability resulting from very low fibrin concentrations also enhances the potential for heterotypic platelet–leukocyte

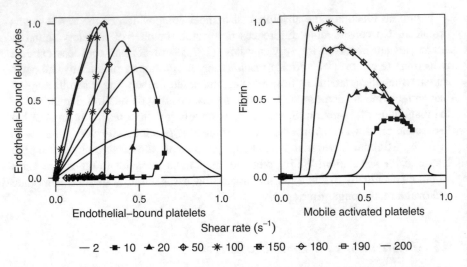

Fig. 10 Phase plane trajectories over 40 h post-endothelial activation. Variables normalized to maximum value achieved over all simulations. *Left*: Endothelial-bound leukocytes as a function of endothelial-bound platelets. *Right*: Fibrin as a function of mobile activated platelets. Moderately, but not extremely, low shear rates generate clots with high fibrin concentrations appearing at relatively late times. Extremely low shear rates produce less fibrin-rich clots, but generate maximal fibrin concentrations at earlier times

aggregates to break off from the primary clot structure. The current framework does not accommodate the possibility for this latter process, but rather provides a foundation for future extensions.

5   Next Steps

We have formulated a model of clot formation in the vein in which leukocytes play an important role. Preliminary numerical solutions of the model show that an intact yet activated endothelium requires leukocyte and platelet recruitment to stimulate coagulation. Results indicate that leukocytes are important in initiating coagulation under low, but not extremely low, flow conditions. Based on the shear rates considered, we find that reductions in blood flow of 50–95% support venous thrombus formation in a primarily immune-mediated manner. In contrast, nearly complete flow stasis supports platelet-mediated coagulation, where leukocytes seem to play a limited role in thrombogenesis. Our preliminary results qualitatively agree with experimental observations that highlight major differences between thrombus formation under complete flow stasis and flow reduction in the veins.

Due to the negative feedback loop in the clotting network, where the formation of plasmin, a clot inhibitor, is promoted by the same factors that initiate the clot, the process of clot resolution begins once a clot is formed (see Fig. 7, lower right corner). If a thrombus has persisted, there must be a continued trigger, i.e., via leukocytes and NETs, as we have described in our model. Anti-coagulants are helpful for preventing clot formation, but they have very little effect on existing thrombi. Traditional treatments for large blood clots involve surgical removal, or using chemicals or lasers to destroy the clot. These treatments, which destroy the clots quickly, can be dangerous, and are associated with morbidity and other complications [22]. Thus, treatments that encourage a natural resolution of the clot are needed. These treatments are likely to involve the processes that we explore in this model.

As we mentioned in the previous section, our preliminary results do have implications for the formation of embolisms, where small pieces of the main thrombus can break off and travel through the bloodstream, perhaps initiating new clots or causing life-threatening blockages in the lungs or brain. However, in order to understand the challenges of treating an existing thrombus in the vein, we need to model a large clot that completely obstructs the vessel, as well as the resolution of the clot.

In light of these concerns, our first step will be to expand the model to allow for the occlusion of the vein. Next, we will compare simulated results to experimental and clinical results in order to validate the model. For example, the results shown in Fig. 9 where the composition of the clot is pictured over time can be compared to the cell composition of actual thrombi. Emerging new techniques for the real-time imaging of thrombi will allow us to validate and calibrate our models of the interactions between leukocytes, NETs, and various clotting factors [14, 21]. Finally, we can use the validated mathematical model to explore the effects of varying conditions, testing immuno-modulation scenarios that could inform new treatments [3, 22].

Acknowledgements This project builds on previous work that was initiated at the WhAM! workshop hosted at the Institute for Mathematics and Its Applications in September 2013. We thank Trachette Jackson for co-organizing the workshop, our teammates for being generally fabulous, the IMA for hosting the event, the NSF for funding, and the AWM for facilitating the workshop itself.

Appendix

Model Variables

The model consists of interactions between 86 state variables, many of them taken from [4]. A schematic of these interactions is given in Fig. 7. We gather here descriptions of all of the state variables for ease of reference; for more details, refer to [4].

The largest group of state variables consists of clotting factors (molecules) and complexes that are involved in the coagulation cascade. These factors are often denoted by roman numerals, e.g., "factor IX," or "factor II." Clotting factors can be either *inactive*, or *active*, typically denoted by an "a," as in factor IX (inactive) and factor IXa (active). Here, we use Z_i to denote the inactive clotting factor i

(zymogen) and E_i to denote the active factor (enzyme). Concentrations are shown with square brackets, or, to simplify the notation, concentrations of factors are denoted by lowercase letters. So $[Z_9] = z_9$ and $[E_9] = e_9$ represent the concentrations of factor IX and factor IXa, respectively. Factors that are bound to platelets are denoted with a superscript m, as in e_9^m or z_9^m. Certain enzymes can bind to specific sites on platelets. These specifically bound enzymes are denoted with superscripts $m*$ or h, as in e_{11}^{m*}, e_{11}^h or, when both binding sites are used, $e_{11}^{h,m*}$.

In addition to these clotting factors, we have the following variables, some of which have already been listed in Sect. 3.

F	Fibrin
F_g	Fibrinogen (fibrin precursor)
P	Plasmin
P_g	Plasminogen (plasmin precursor)
$TFPI$	Tissue factor pathway inhibitor
APC	Activated protein C
TM	Thrombomodulin
W	Vessel wall (endothelial) activation level, with values between 0 (unactivated) and 1 (fully activated)
$[NET]$	Concentration of leukocyte-derived NETs
TF	Average concentration of active (decrypted) tissue factor expressed per activated monocyte
L_u^m, L_a^p, L_a^e	Concentration of unactivated and mobile, activated and bound to platelets, and activated and endothelium-bound leukocytes, respectively
$P_a^b, P_a^l, P_a^e, P_a^m$	Concentration of activated platelets bound to other platelets, leukocytes, endothelium, or unbound (mobile), respectively

Complexes are shown as molecules joined by a colon, as in $TM : E_2$ or $TM : E_2 : APC$. Some complexes have special names:

$$TEN = \text{VIII:IXa} = \text{platelet-bound tenase}$$
$$PRO = \text{Va:Xa} = \text{prothrombinase.}$$

Model Equations

Here we give the complete list of model equations incorporating immune-mediated mechanisms to coagulation in the veins. Following [6], we eliminate the endothelium-dependent kinetic equations from the model in [4], as the current model does not distinguish between sub- and intact endothelium. As such, the following variables are omitted, with the necessary reactions appearing in the modified equations: e_2^{ec}, APC^{ec}, e_9^{ec}, e_{10}^{ec}. Specific changes from the previous model [4] which models arterial coagulation (in vivo framework) are highlighted in purple.

$$\frac{dW}{dt} = a_0 \frac{L_a^{e\,2}}{L_a^{e\,2} + k_{L_a^e:w}^2}(1 - W), \quad W(0) = f(\bar{v}) = a_w + (1 - a_w) \cdot \frac{1}{1 + (\bar{v}/k_w)^3} \tag{1}$$

$$W^{\text{avail}} = w_{\max} \cdot W - L_a^e - P_a^e \tag{2}$$

$$\frac{dL_u^m}{dt} = k_{\text{flow}}^l(L^{\text{up}} - L_u^m) - \left(k_{\text{leuk}}^+ W^{\text{avail}} + a_2 P_a^e\right) \cdot L_u^m + a_{-1}L_a^e + a_{-2}L_a^p \tag{3}$$

$$\frac{dL_a^e}{dt} = k_{\text{leuk}}^+ W^{\text{avail}} L_u^m - a_{-1}L_a^e - qa_3^p P_a^m L_a^e - a_5 P_a^m L_a^e + a_{-5}P_a^l \frac{L_a^e}{L_a^e + [NET]} \tag{4}$$

$$\quad - d_{\text{leuk}}L_a^e[q - (1 - q)/6]$$

$$\frac{dL_a^p}{dt} = a_2 P_a^e L_u^m - qa_3 L_a^p + a_2 P_a^l * L_u^m - a_{-2}L_a^p - d_{\text{leuk}}L_a^e[q - (1 - q)/6] \tag{5}$$

$$\frac{d[NET]}{dt} = q\left[a_3^p P_a^m L_a^e + a_3 L_a^p\right] - a_5 P_u^m * [NET] + a_{-5}P_a^l \frac{[NET]}{L_a^e + [NET]} \tag{6}$$

$$\frac{dP_u^m}{dt} = k_{\text{flow}}^p(P^{\text{up}} - P_u^m) + k_{\text{adh}}^- P_a^e \tag{7}$$

$$\quad - \left[a_4[NET] + k_{\text{plt}}^{\text{act}}(P_a^b + P_a^e + P_a^l + P_a^m) + k_{e2}^{\text{act}}\frac{e_2}{e_2 + \overline{e_2}} + k_{\text{adh}}^+ W^{\text{avail}}\right] \cdot P_u^m$$

$$\frac{dP_a^b}{dt} = k_{\text{coh}}(P_a^e + P_a^l)P_a^m \tag{8}$$

$$\frac{dP_a^l}{dt} = a_5 P_a^m(L_a^e + [NET]) - a_{-5}P_a^l - a_2 P_a^l L_u^m + a_{-2}L_a^p \cdot \frac{P_a^l}{P_a^l + P_a^e} \tag{9}$$

$$\quad + d_{\text{leuk}}L_a^e[q - (1 - q)/6]$$

$$\frac{dP_a^m}{dt} = -k_{\text{flow}}^p P_a^m + \left[a_4[NET] + k_{\text{plt}}^{\text{act}}(P_a^b + P_a^e + P_a^l + P_a^m)k_{e2}^{\text{act}}\frac{e_2}{e_2 + \overline{e_2}}\right] \cdot P_u^m \tag{10}$$

$$\quad - k_{\text{adh}}^+ W^{\text{avail}} P_a^m + k_{\text{adh}}^- P_a^e - k_{\text{coh}}(P_a^e + P_a^l)P_a^m - a_5 P_a^m(L_a^e + [NET]) + a_{-5}P_a^l$$

$$\frac{dP_a^e}{dt} = k_{\text{adh}}^+ W^{\text{avail}}(P_u^m + P_a^m) - k_{\text{adh}}^- P_a^e - a_2 P_a^l L_u^m + a_{-2}L_a^p \cdot \frac{P_a^e}{P_a^l + P_a^e} \tag{11}$$

$$\frac{dz_9}{dt} = -k_9^{\text{on}}z_9 p_9^{\text{avail}} + k_9^{\text{off}}z_9^m + k_{\text{flow}}(z_9^{\text{up}} - z_9) - k_{z9:e_7^m}^+ z_9 e_7^m + k_{z9:e_7^m}^-[Z_9 : E_7^m] \tag{12}$$

$$\quad - k_{z9:e_{11}}^+ z_9 e_{11}^h + k_{z9:e_{11}}^-[Z_9 : E_{11}^h] - k_{z9:e_{11}}^+ z_9 e_{11} + k_{z9:e_{11}}^-[Z_9 : E_{11}] - k_9^{\text{cat}} \cdot [NET]z_9$$

$$\frac{de_9}{dt} = -k_9^{\text{on}}e_9 p_9^{\text{avail}} + k_9^{\text{off}}e_9^m - k_9^{\text{on}}e_9 p_{91}^{\text{avail}} + k_9^{\text{off}}e_9^{m*} + k_{\text{flow}}(e_9^{\text{up}} - e_9) \tag{13}$$

$$\quad - k_{AT:e_9}^{\text{in}}e_9 + k_{z9:e_7^m}^{\text{cat}}[Z_9 : E_7^m]$$

$$\quad + (k_{z7:e_9}^{\text{cat}} + k_{z7:e_9}^-)[Z_7 : E_9] - k_{z7:e_9}^+ z_7 e_9 + (k_{z_7^m:e_9}^{\text{cat}} + k_{z_7^m:e_9}^-)[Z_7^m : E_9]$$

$$\quad - k_{z_7^m:e_9}^+ z_7^m e_9 + k_{z9:e_{11}}^{\text{cat}}[Z_9 : E_{11}^h] + k_{z9:e_{11}}^{\text{cat}}[Z_9 : E_{11}] - k_9^{\text{cat}} \cdot [NET]e_9$$

$$\frac{de_2}{dt} = -k_2^{on} p_{2s}^{avail} e_2 + k_2^{off} e_2^m + k_{flow}(e_2^{up} - e_2) + k_{z_2^m:PRO}^{cat}[Z_2^m : PRO] \tag{14}$$

$$- k_{AT:e_2}^{in} e_2 + (k_{z_5:e_2}^{cat} + k_{z_5:e_2}^-)[Z_5 : E_2]$$

$$- k_{z_5:e_2}^+ z_5 e_2 + (k_{z_8:e_2}^{cat} + k_{z_8:e_2}^-)[Z_8 : E_2] - k_{z_8:e_2}^+ z_8 e_2$$

$$+ (k_{z_7:e_2}^{cat} + k_{z_7:e_2}^-)[Z_7 : E_2] - k_{z_7:e_2}^+ z_7 e_2$$

$$+ (k_{z_7^m:e_2}^{cat} + k_{z_7^m:e_2}^-)[Z_7^m : E_2]$$

$$- k_{z_7^m:e_2}^+ z_7^m e_2 + (k_{z_{11}:e_2}^{cat} + k_{z_{11}:e_2}^-)[Z_{11} : E_2] - k_{z_{11}:e_2}^+ z_{11} e_2$$

$$+ (k_{z_{11}:e_2}^{cat} + k_{z_{11}:e_2}^-)[E_{11}^h : E_2]$$

$$- k_{z_{11}:e_2}^+ e_{11}^h e_2 - k_{TM}^+ e_2[TM]^{avail} + k_{TM}^-[TM : E_2]$$

$$\frac{d[TFPI]}{dt} = -k_{TFPI:e_{10}}^+ e_{10}[TFPI] + k_{TFPI:e_{10}}^-[TFPI : E_{10}] \tag{15}$$

$$+ k_{flow}([TFPI]^{up} - [TFPI]) - d_{TFPI} \cdot [NET] \cdot [TFPI]$$

$$\frac{dAPC}{dt} = (k_{e_5^m:APC}^{cat} + k_{e_5^m:APC}^-)[APC : E_5^m] - k_{e_5^m:APC}^+ e_5^m[APC] \tag{16}$$

$$(k_{e_5:APC}^{cat} + k_{e_5:APC}^-)[APC : E_5] - k_{e_5:APC}^+ e_5[APC]$$

$$(k_{e_8^m:APC}^{cat} + k_{e_8^m:APC}^-)[APC : E_8^m] k_{e_8^m:APC}^+ e_8^m[APC]$$

$$(k_{e_8:APC}^{cat} + k_{e_8:APC}^-)[APC : E_8] k_{e_8:APC}^+ e_8[APC] - k_{flow}[APC]$$

$$+ \frac{k_{PC:TM:e_2}^{cat}}{1 + \frac{[NET]}{k_{PC:TM:NET}}} \cdot [TM : E_2 : APC]$$

$$\frac{d[TM : E_2]}{dt} = k_{TM}^+ e_2[TM]^{avail} - k_{TM}^-[TM : E_2] \tag{17}$$

$$+ \frac{k_{PC:TM:e_2}^{cat}}{1 + [NET]/k_{PC:TM:NET}}[TM : E_2 : APC]$$

$$\frac{d[TM : E_2 : APC]}{dt} = -\frac{k_{PC:TM:e_2}^{cat}}{1 + [NET]/k_{PC:TM:NET}}[TM : E_2 : APC] \tag{18}$$

$$\frac{d[F]}{dt} = \frac{v_{14} e_2[Fg]}{k_{14a} + [Fg]} - \frac{v_{16}[F]e_2}{k_{16a} + e_2} - \frac{v_{17}[F][P]}{k_{17a} + [P]} \cdot \frac{k_{e_{12}:F:NET}}{k_{e_{12}:F:NET} + [NET]} \tag{19}$$

$$\frac{d}{dt} z_7 = -k_7^{on} z_7[TF]^{avail} + k_7^{off} z_7^m - k_{z_7:e_2}^+ z_7 e_2 + k_{z_7:e_2}^-[Z_7 : E_2] - k_{z_7:e_{10}}^+ z_7 e_{10} \tag{20}$$

$$+ k_{z_7:e_{10}}^-[Z_7 : E_{10}] - k_{z_7:e_9}^+ z_7 e_9 + k_{z_7:e_9}^-[Z_7 : E_9] + k_{flow}(z_7^{up} - z_7)$$

$$\frac{d}{dt} e_7 = -k_7^{on} e_7[TF]^{avail} + k_7^{off} e_7^m + k_{z_7:e_2}^{cat}[Z_7 : E_2] + k_{z_7:e_{10}}^{cat}[Z_7 : E_{10}] \tag{21}$$

$$+ k_{z_7:e_9}^{cat}[Z_7 : E_9] + k_{flow}(e_7^{up} - e_7)$$

$$\frac{d}{dt} z_7^m = k_7^{on} z_7[TF]^{avail} - k_7^{off} z_7^m - k_{z_7^m:e_{10}}^+ z_7^m e_{10} + k_{z_7^m:e_{10}}^-[Z_7^m : E_{10}] \tag{22}$$

$$- k_{z7:e_2}^+ z_7^m e_2 + k_{z7:e_2}^- [Z_7^m : E_2] - k_{z_7^m:e_9}^+ z_7^m e_9 + k_{z_7^m:e_9}^- [Z_7^m : E_9]$$

$$\frac{\mathrm{d}}{\mathrm{d}t} e_7^m = k_7^{\mathrm{on}} e_7 [TF]^{\mathrm{avail}} - k_7^{\mathrm{off}} e_7^m \tag{23}$$

$$- k_{TFPI:e_{10}:e_7^m}^+ e_7^m [TFPI : E_{10}] + k_{TFPI:e_{10}:e_7^m}^- [TFPI : E_{10} : E_7^m]$$

$$+ k_{z_7^m:e_{10}}^{\mathrm{cat}} [Z_7^m : E_{10}] + k_{z_7^m:e_2}^{\mathrm{cat}} [Z_7^m : E_2]$$

$$+ (k_{z_{10}:e_7^m}^{\mathrm{cat}} + k_{z_{10}:e_7^m}^-)[Z_{10} : E_7^m] - k_{z_{10}:e_7^m}^+ z_{10} e_7^m$$

$$+ (k_{z_9:e_7^m}^{\mathrm{cat}} + k_{z_9:e_7^m}^-)[Z_9 : E_7^m] - k_{z_9:e_7^m}^+ z_9 e_7^m$$

$$+ k_{z_7^m:e_9}^{\mathrm{cat}} [Z_7^m : E_9]$$

$$\frac{\mathrm{d}}{\mathrm{d}t} z_{10} = -k_{10}^{\mathrm{on}} z_{10} p_{10}^{\mathrm{avail}} + k_{10}^{\mathrm{off}} z_{10}^m - k_{z_{10}:e_7^m}^+ z_{10} e_7^m + k_{z_{10}:e_7^m}^- [Z_{10} : E_7^m] + k_{\mathrm{flow}}(z_{10}^{\mathrm{up}} - z_{10}) \tag{24}$$

$$\frac{\mathrm{d}}{\mathrm{d}t} e_{10} = -k_{10}^{\mathrm{on}} e_{10} p_{10}^{\mathrm{avail}} + k_{10}^{\mathrm{off}} e_{10}^m - k_{z_{10}:e_7^m}^{\mathrm{cat}} [Z_{10} : E_7^m] \tag{25}$$

$$+ (k_{z_7:e_{10}}^{\mathrm{cat}} + k_{z_7:e_{10}}^-)[Z_7 : E_{10}] - k_{z_7:e_{10}}^+ z_7 e_{10}$$

$$+ (k_{z_7^m:e_{10}}^{\mathrm{cat}} + k_{z_7^m:e_{10}}^-)[Z_7^m : E_{10}] - k_{z_7^m:e_{10}}^+ z_7^m e_{10}$$

$$- k_{TFPI:e_{10}}^+ e_{10}[TFPI] + k_{TFPI:e_{10}}^- [TFPI : E_{10}]$$

$$- k_{AT:e_{10}}^{in} e_{10} - k_{\mathrm{flow}} e_{10}$$

$$\frac{\mathrm{d}}{\mathrm{d}t} z_{10}^m = k_{10}^{\mathrm{on}} z_{10} p_{10}^{\mathrm{avail}} - k_{10}^{\mathrm{off}} z_{10}^m \tag{26}$$

$$- k_{z_{10}^m:TEN}^+ [TEN] z_{10}^m + k_{z_{10}^m:TEN}^- [Z_{10}^m : TEN]$$

$$- k_{z_{10}^m:TEN}^+ [TENS] z_{10}^m + k_{z_{10}^m:TEN}^- [Z_{10}^m : TENS]$$

$$\frac{\mathrm{d}}{\mathrm{d}t} e_{10}^m = k_{10}^{\mathrm{on}} e_{10} p_{10}^{\mathrm{avail}} - k_{10}^{\mathrm{off}} e_{10}^m \tag{27}$$

$$+ k_{z_{10}^m:TEN}^{cat} [Z_{10}^m : TEN] + k_{z_{10}^m:TEN}^{cat} [Z_{10}^m : TENS]$$

$$- k_{e_5^m:e_{10}^m}^+ e_{10}^m e_5^m + k_{e_5^m:e_{10}^m}^- [PRO] - k_{z_5^m:e_{10}^m}^+ z_5^m e_{10}^m + (k_{z_5^m:e_{10}^m}^{\mathrm{cat}} + k_{z_5^m:e_{10}^m}^-)[Z_5^m : E_{10}^m]$$

$$- k_{z_8^m:e_{10}^m}^+ z_8^m e_{10}^m + (k_{z_8^m:e_{10}^m}^{\mathrm{cat}} + k_{z_8^m:e_{10}^m}^-)[Z_8^m : E_{10}^m]$$

$$\frac{\mathrm{d}}{\mathrm{d}t} z_5 = -k_5^{\mathrm{on}} z_5 p_5^{\mathrm{avail}} + k_5^{\mathrm{off}} z_5^m - k_{z_5:e_2}^+ z_5 e_2 + k_{z_5:e_2}^- [Z_5 : E_2] + k_{\mathrm{flow}}(z_5^{\mathrm{up}} - z_5) \tag{28}$$

$$\frac{\mathrm{d}}{\mathrm{d}t} e_5 = -k_5^{\mathrm{on}} e_5 p_5^{\mathrm{avail}} + k_5^{\mathrm{off}} e_5^m + k_{z_5:e_2}^{\mathrm{cat}} [Z_5 : E_2] + k_{e_5:APC}^- [APC : E_5] \tag{29}$$

$$- k_{e_5:APC}^+ [APC] e_5 - k_{\mathrm{flow}}(e_5^{\mathrm{up}} - e_5)$$

$$\frac{\mathrm{d}}{\mathrm{d}t} z_5^m = k_5^{\mathrm{on}} z_5 p_5^{\mathrm{avail}} - k_5^{\mathrm{off}} z_5^m - k_{z_5^m:e_{10}^m}^+ z_5^m e_{10}^m + k_{z_5^m:e_{10}^m}^- [Z_5^m : E_{10}^m] \tag{30}$$

$$- k_{z_5:e_2^m}^+ z_5^m e_2^m + k_{z_5:e_2^m}^- [Z_5^m : E_2^m]$$

$$\frac{d}{dt}e_5^m = k_5^{on}e_5 p_5^{avail} - k_5^{off}e_5^m - k_{e_5^m:e_{10}^m}^+ e_{10}^m e_5^m \tag{31}$$

$$+ k_{e_5^m:e_{10}^m}^- [PRO] + k_{z_5^m:e_{10}^m}^{cat}[Z_5^m : E_{10}^m] + k_{z5:e_2^m}^{cat}[Z_5^m : E_2^m]$$

$$+ k_{e_5^m:APC}^- [APC : E_5^m] - k_{e_5^m:APC}^+ [APC]e_5^m$$

$$\frac{d}{dt}z_8 = -k_8^{on}z_8 p_8^{avail} + k_8^{off}z_8^m + k_{flow}(z_8^{up} - z_8) - k_{z8:e_2}^+ z_8 e_2 + k_{z8:e_2}^-[Z_8 : E_2] \tag{32}$$

$$\frac{d}{dt}e_8 = -k_8^{on}e_8 p_8^{avail} + k_8^{off}e_8^m + k_{flow}(e_8^{up} - e_8) + k_{z8:e_2}^{cat}[Z_8 : E_2] - k_8^{deg}e_8 \tag{33}$$

$$+ k_{e_8:APC}^-[APC : E_8] - k_{e_8:APC}^+[APC]e_8$$

$$\frac{d}{dt}z_8^m = k_8^{on}z_8 p_8^{avail} - k_8^{off}z_8^m - k_{z_8^m:e_2^m}^+ z_8^m e_2^m + k_{z_8^m:e_2^m}^-[Z_8^m : E_2^m] \tag{34}$$

$$- k_{z_8^m:e_{10}^m}^+ z_8^m e_{10}^m + k_{z_8^m:e_{10}^m}^-[Z_8^m : E_{10}^m]$$

$$\frac{d}{dt}e_8^m = k_8^{on}e_8 p_8^{avail} - k_8^{off}e_8^m + k_{z_8^m:e_{10}^m}^{cat}[Z_8^m : E_{10}^m] + k_{z_8^m:e_2^m}^{cat}[Z_8^m : E_2^m] \tag{35}$$

$$- k_8^{deg}e_8^m + k_{e_8^m:APC}^-[APC : E_8^m] - k_{e_8^m:APC}^+[APC]e_8^m$$

$$- k_{e_8^m e_9^m}^+ e_8^m e_9^m + k_{e_8^m e_9^m}^-[TEN] - k_{e_8^m e_9^m,*}^+ e_8^m e_9^m + k_{e_8^m e_9^m}^-[TEN^*]$$

$$\frac{d}{dt}z_9^m = k_9^{on}z_9 p_9^{avail} - k_9^{off}z_9^m - k_{z_9^m:e_{11}^{h,m}}^+ z_9^m e_{11}^{h,m} + k_{z9:e_{11}}^-[Z_9^m : E_{11}^{h,m}] \tag{36}$$

$$- k_{z9:e_{11}}^+ z_9^m e_{11}^{m*} + k_{z9:e_{11}}^-[Z_9^m : E_{11}^{m*}]$$

$$\frac{d}{dt}e_9^m = k_9^{on}e_9 p_9^{avail} - k_9^{off}e_9^m + k_{z9:e_{11}}^{cat}[Z_9^m : E_{11}^{h,m}] + k_{z9:e_{11}}^{cat}[Z_9^m : E_{11}^{m*}] \tag{37}$$

$$- k_{e_8^m e_9^m}^+ e_8^m e_9^m + k_{e_8^m e_9^m}^-[TEN]$$

$$\frac{d}{dt}z_2 = -k_2^{on}p_2^{avail}z_2 + k_2^{off}z_2^m + k_{flow}(z_2^{up} - z_2) \tag{38}$$

$$\frac{d}{dt}z_2^m = k_2^{on}p_2^{avail}z_2 - k_2^{off}z_2^m - k_{z_2^m:PRO}^+[PRO]z_2^m + k_{z_2^m:PRO}^-[Z_2^m : PRO] \tag{39}$$

$$\frac{d}{dt}e_2^m = k_2^{on}p_{2s}^{avail}e_2 - k_2^{off}e_2^m \tag{40}$$

$$+ (k_{z5:e_2^m}^{cat} + k_{z5:e_2^m}^-)[Z_5^m : E_2^m] - k_{z5:e_2^m}^+ z_5^m e_2^m$$

$$+ (k_{z_8^m:e_2^m}^{cat} + k_{z_8^m:e_2^m}^-)[Z_8^m : E_2^m] - k_{z_8^m:e_2^m}^+ z_8^m e_2^m$$

$$+ (k_{z11:e_2}^{cat} + k_{z11:e_2}^-)[Z_{11}^m : E_2^m] - k_{z11:e_2}^+ z_{11}^m e_2^m$$

$$+ (k_{z11:e_2}^{cat} + k_{z11:e_2}^-)[E_{11}^{h,m*} : E_2^m] - k_{z11:e_2}^+ e_{11}^{h,m*} e_2^m$$

$$\frac{d}{dt}[TEN] = k_{e_8^m e_9^m}^+ e_8^m e_9^m - k_{e_8^m e_9^m}^-[TEN] \tag{41}$$

$$+ (k_{z_{10}^m:TEN}^{cat} + k_{z_{10}^m:TEN}^-)[Z_{10}^m : TEN] - k_{z_{10}^m:TEN}^+ z_{10}^m[TEN]$$

$$\frac{d}{dt}[PRO] = k^+_{e^m_5:e^m_{10}} e^m_{10} e^m_5 - k^-_{e^m_5:e^m_{10}}[PRO] \tag{42}$$

$$- k^+_{z^m_2:PRO} z^m_2 [PRO] + (k^-_{z^m_2:PRO} + k^{cat}_{z^m_2:PRO})[Z^m_2 : PRO]$$

$$\frac{d}{dt}[TFPI : E_{10}] = -k_{\text{flow}}[TFPI : E_{10}] + k^+_{TFPI:e_{10}} e_{10}[TFPI] \tag{43}$$

$$- k^-_{TFPI:e_{10}}[TFPI : E_{10}] \tag{44}$$

$$+ k^-_{TFPI:e_{10:e^m_7}}[TFPI : E_{10} : E^m_7] - k^+_{TFPI:e_{10}:e^m_7} e^m_7 [TFPI : E_{10}]$$

$$\frac{d}{dt}[TFPI : E_{10} : E^m_7] = -k^-_{TFPI:e_{10:e^m_7}}[TFPI : E_{10} : E^m_7] \tag{45}$$

$$+ k^+_{TFPI:e_{10}:e^m_7} e^m_7 [TFPI : E_{10}] \tag{46}$$

$$\frac{d}{dt}[Z_7 : E_2] = k^+_{z7:e_2} z_7 e_2 - (k^{cat}_{z7:e_2} + k^-_{z7:e_2})[Z_7 : E_2] - k_{\text{flow}}[Z_7 : E_2] \tag{47}$$

$$\frac{d}{dt}[Z_7 : E_{10}] = k^+_{z7:e_{10}} z_7 e_{10} - (k^{cat}_{z7:e_{10}} + k^-_{z7:e_{10}})[Z_7 : E_{10}] - k_{\text{flow}}[Z_7 : E_{10}] \tag{48}$$

$$\frac{d}{dt}[Z^m_7 : E_{10}] = k^+_{z^m_7:e_{10}} z^m_7 e_{10} - (k^{cat}_{z^m_7:e_{10}} + k^-_{z^m_7:e_{10}})[Z^m_7 : E_{10}] \tag{49}$$

$$\frac{d}{dt}[Z^m_7 : E_2] = k^+_{z7:e_2} z^m_7 e_2 - (k^{cat}_{z7:e_2} + k^-_{z7:e_2})[Z^m_7 : E_2] \tag{50}$$

$$\frac{d}{dt}[Z_{10} : E^m_7] = k^+_{z10:e^m_7} z_{10} e^m_7 - (k^{cat}_{z10:e^m_7} + k^-_{z10:e^m_7})[Z_{10} : E^m_7] \tag{51}$$

$$\frac{d}{dt}[Z^m_{10} : TEN] = k^+_{z^m_{10}:TEN}[TEN] z^m_{10} - (k^{cat}_{z^m_{10}:TEN} + k^-_{z^m_{10}:TEN})[Z^m_{10} : TEN] \tag{52}$$

$$\frac{d}{dt}[Z_5 : E_2] = k^+_{z5:e_2} z_5 e_2 - (k^{cat}_{z5:e_2} + k^-_{z5:e_2})[Z_5 : E_2] - k_{\text{flow}}[Z_5 : E_2] \tag{53}$$

$$\frac{d}{dt}[Z^m_5 : E^m_{10}] = k^+_{z^m_5:e^m_{10}} z^m_5 e^m_{10} - (k^{cat}_{z^m_5:e^m_{10}} + k^-_{z^m_5:e^m_{10}})[Z^m_5 : E^m_{10}] \tag{54}$$

$$\frac{d}{dt}[Z^m_5 : E^m_2] = k^+_{z5:e^m_2} z^m_5 e^m_2 - (k^{cat}_{z5:e^m_2} + k^-_{z5:e^m_2})[Z^m_5 : E^m_2] \tag{55}$$

$$\frac{d}{dt}[Z^m_8 : E^m_{10}] = k^+_{z^m_8:e^m_{10}} z^m_8 e^m_{10} - (k^{cat}_{z^m_8:e^m_{10}} + k^-_{z^m_8:e^m_{10}})[Z^m_8 : E^m_{10}] \tag{56}$$

$$\frac{d}{dt}[Z^m_8 : E^m_2] = k^+_{z^m_8:e^m_2} z^m_8 e^m_2 - (k^{cat}_{z^m_8:e^m_2} + k^-_{z^m_8:e^m_2})[Z^m_8 : E^m_2] \tag{57}$$

$$\frac{d}{dt}[Z_8 : E_2] = k^+_{z8:e_2} z_8 e_2 - (k^{cat}_{z8:e_2} + k^-_{z8:e_2})[Z_8 : E_2] - k_{\text{flow}}[Z_8 : E_2] \tag{58}$$

$$\frac{d}{dt}[APC : E^m_8] = k^+_{e^m_8:APC}[APC] e^m_8 - (k^{cat}_{e^m_8:APC} + k^-_{e^m_8:APC})[APC : E^m_8] \tag{59}$$

$$\frac{d}{dt}[Z_9 : E^m_7] = k^+_{z9:e^m_7} z_9 e^m_7 - (k^{cat}_{z9:e^m_7} + k^-_{z9:e^m_7})[Z_9 : E^m_7] \tag{60}$$

$$\frac{d}{dt}[Z^m_2 : PRO] = k^+_{z^m_2:PRO} z^m_2 [PRO] - (k^{cat}_{z^m_2:PRO} + k^-_{z^m_2:PRO})[Z^m_2 : PRO] \tag{61}$$

$$\frac{d}{dt}[APC:E_5^m] = k_{e_5^m:APC}^+[APC]e_5^m - (k_{e_5^m:APC}^{cat} + k_{e_5^m:APC}^-)[APC:E_5^m] \tag{62}$$

$$\frac{d}{dt}[Z_7:E_9] = k_{z_7:e_9}^+ z_7 e_9 - (k_{z_7:e_9}^{cat} + k_{z_7:e_9}^-)[Z_7:E_9] \tag{63}$$

$$\frac{d}{dt}e_9^{m*} = k_9^{on} p_{91}^{avail} e_9 - k_9^{off} e_9^{m*} + k_{e_8^m:e_9^m}^-[TEN^*] - k_{e_8^m:e_9^m}^+ e_8^m e_9^{m*} \tag{64}$$

$$\frac{d}{dt}[TEN^*] = -k_{e_8^m:e_9^m}^-[TEN^*] + k_{e_8^m:e_9^m}^+ e_8^m e_9^{m*} \tag{65}$$
$$+ (k_{z_{10}^m:TEN}^{cat} + k_{z_{10}^m:TEN}^-)[Z_{10}^m:TEN^*] - k_{z_{10}^m:TEN}^+[TEN^*]z_{10}^m$$

$$\frac{d}{dt}[Z_{10}^m:TEN^*] = k_{z_{10}^m:TEN}^+[TEN^*]z_{10}^m - (k_{z_{10}^m:TEN}^{cat} + k_{z_{10}^m:TEN}^-)[Z_{10}^m:TEN^*] \tag{66}$$

$$\frac{d}{dt}[APC:E_5] = k_{e_5:APC}^+ e_5[APC] - (k_{e_5:APC}^{cat} + k_{e_5:APC}^-)[APC:E_5] \tag{67}$$

$$\frac{d}{dt}[APC:E_8] = k_{e_8:APC}^+ e_8[APC] - (k_{e_8:APC}^{cat} + k_{e_8:APC}^-)[APC:E_8] \tag{68}$$

$$\frac{d}{dt}z_{11} = k_{flow}(z_{11}^{up} - z_{11}) - k_{z_{11}}^{on} z_{11} p_{11}^{avail} + k_{z_{11}}^{off} z_{11}^m - k_{z_{11}:e_2}^+ z_{11} e_2 + k_{z_{11}:e_2}^-[Z_{11}:E_2] \tag{69}$$

$$\frac{d}{dt}e_{11} = k_{flow}(e_{11}^{up} - e_{11}) - k_{e_{11}}^{on} e_{11} p_{111}^{avail} + k_{e_{11}}^{off} e_{11}^{m*} \tag{70}$$
$$- k_{z_9:e_{11}}^+ z_9 e_{11} + (k_{z_9:e_{11}}^- + k_{z_9:e_{11}}^{cat})[Z_9:E_{11}] + k_{z_{11}:e_2}^{cat}[E_{11}^h:E_2]$$

$$\frac{d}{dt}z_{11}^m = k_{z_{11}}^{on} z_{11} p_{11}^{avail} - k_{z_{11}}^{off} z_{11}^m - k_{z_{11}:e_2}^+ z_{11}^m e_2^m + k_{z_{11}:e_2}^-[Z_{11}^m:E_2^m] \tag{71}$$

$$\frac{d}{dt}e_{11}^{m*} = k_{e_{11}}^{on} e_{11} p_{111}^{avail} - k_{e_{11}}^{off} e_{11}^{m*} \tag{72}$$
$$- k_{z_9:e_{11}}^+ e_{11}^{m*} z_9^m + (k_{z_9:e_{11}}^- + k_{z_9:e_{11}}^{cat})[Z_9^m:E_{11}^{m*}] + k_{z_{11}:e_2}^{cat}[E_{11}^{h,m*}:E_2^m]$$

$$\frac{d}{dt}[Z_{11}^m:E_2^m] = k_{z_{11}:e_2}^+ z_{11}^m e_2^m - (k_{z_{11}:e_2}^- + k_{z_{11}:e_2}^{cat})[Z_{11}^m:E_2^m] \tag{73}$$

$$\frac{d}{dt}[Z_9^m:E_{11}^{m*}] = k_{z_9:e_{11}}^+ z_9^m e_{11}^{m*} - (k_{z_9:e_{11}}^- + k_{z_9:e_{11}}^{cat})[Z_9^m:E_{11}^{m*}] \tag{74}$$

$$\frac{d}{dt}[Z_{11}:E_2] = k_{flow}([Z_{11}:E_2]^{up} - [Z_{11}:E_2]) + k_{z_{11}:e_2}^+ z_{11} e_2 - (k_{z_{11}:e_2}^- + k_{z_{11}:e_2}^{cat})[Z_{11}:E_2] \tag{75}$$

$$\frac{d}{dt}[Z_9:E_{11}] = k_{flow}([Z_9:E_{11}]^{up} - [Z_9:E_{11}]) + k_{z_9:e_{11}}^+ z_9 e_{11} - (k_{z_9:e_{11}}^- + k_{z_9:e_{11}}^{cat})[Z_9:E_{11}] \tag{76}$$

$$\frac{d}{dt}e_{11}^h = k_{flow}(e_{11}^{h,up} - e_{11}^h) - k_{e_{11}}^{on} e_{11}^h p_{111}^{avail} + k_{e_{11}}^{off} e_{11}^{h,m*} - k_{z_{11}}^{on} e_{11}^h p_{11}^{avail} + k_{z_{11}}^{off} e_{11}^{h,m} \tag{77}$$
$$- k_{z_9:e_{11}}^+ z_9 e_{11}^h + (k_{z_9:e_{11}}^- + k_{z_9:e_{11}}^{cat})[Z_9:E_{11}^h] + k_{z_{11}:e_2}^{cat}[Z_{11}:E_2]$$
$$- k_{z_{11}:e_2} e_{11}^h e_2 + k_{z_{11}:e_2}^-[E_{11}^h:E_2]$$

$$\frac{d}{dt}e_{11}^{h,m} = k_{z_{11}}^{on} e_{11}^h p_{11}^{avail} - k_{z_{11}}^{off} e_{11}^{h,m} \tag{78}$$

$$-k^+_{z_9:e_{11}}z_9^m e_{11}^{h,m} + (k^-_{z_9:e_{11}} + k^{cat}_{z_9:e_{11}})[Z_9^m : E_{11}^{h,m}] + k^{cat}_{z_{11}:e_2}[Z_{11}^m : E_2^m]$$

$$\frac{d}{dt}e_{11}^{h,m*} = k^{on}_{e_{11}}e_{11}^h p_{111}^{avail} - k^{off}_{e_{11}}e_{11}^{h,m*} - k^+_{z_{11}:e_2}e_{11}^{h,m*}e_2^m + k^-_{z_{11}:e_2}[E_{11}^{h,m*} : E_2^m] \tag{79}$$

$$\frac{d}{dt}[Z_9 : E_{11}^h] = k_{flow}([Z_9 : E_{11}^h]^{up} - [Z_9 : E_{11}^h]) + k^+_{z_9:e_{11}}z_9 e_{11}^h - (k^-_{z_9:e_{11}} + k^{cat}_{z_9:e_{11}})[Z_9 : E_{11}^h] \tag{80}$$

$$\frac{d}{dt}[Z_9^m : E_{11}^{h,m}] = k^+_{z_9:e_{11}}z_9^m e_{11}^{h,m} - (k^-_{z_9:e_{11}} + k^{cat}_{z_9:e_{11}})[Z_9^m : E_{11}^{h,m}] \tag{81}$$

$$\frac{d}{dt}[E_{11}^h : E_2] = k_{flow}([E_{11}^h : E_2]^{up} - [E_{11}^h : E_2]) + k^+_{z_{11}:e_2}e_{11}^h e_2 - (k^-_{z_{11}:e_2} + k^{cat}_{z_{11}:e_2})[E_{11}^h : E_2] \tag{82}$$

$$\frac{d}{dt}[E_{11}^{h,m*} : E_2^m] = k^+_{e_{11}^{h,m*}e_2^m}e_{11}^{h,m*}e_2^m - (k^-_{e_{11}^{h,m*}e_2^m} + k^{cat}_{e_{11}^{h,m*}e_2^m})[E_{11}^{h,m*} : E_2^m] \tag{83}$$

$$\frac{d}{dt}[Fg] = -\frac{v_{14}e_2[Fg]}{k_{14a} + [Fg]} - \frac{v_{15}[P][Fg]}{k_{15a} + [P]} + d_{Fg}(1 - [Fg]) \tag{84}$$

$$\frac{d}{dt}[F] = \frac{v_{14}e_2[Fg]}{k_{14a} + [Fg]} - \frac{v_{16}[F]e_2}{k_{16a} + e_2} - \frac{v_{17}[F][P]}{k_{17a} + [P]} \cdot \frac{k_{e_{12}:F:NET}}{k_{e_{12}:F:NET} + [NET]} \tag{85}$$

$$\frac{d}{dt}[Pg] = -\frac{v_{21}e_2[Pg]}{k_{21a} + e_2} - \frac{v_{23}[APC][VKH_2][Pg]}{(k_{23a}([APC] + c_{37a}) + [VKH_2][APC])}$$
$$- \frac{v_{22}[F][Pg]}{k_{22a} + [F]} + d_{Pg}(1 - [Pg]) \tag{86}$$

$$\frac{d}{dt}[P] = \frac{v_{21}e_2[Pg]}{k_{21a} + e_2} + \frac{v_{23}[APC][VKH_2][Pg]}{(k_{23a}([APC] + c_{37a}) + [VKH_2][APC])}$$
$$+ \frac{v_{22}[F][Pg]}{k_{22a} + [F]} - d_P[P]. \tag{87}$$

Platelet Variables

Activated platelet dynamics are calculated as in Fogelson et al. [10]. At the beginning of each step, we store the total active platelet concentration $PL_a^i = [P_a^e]ps/p + [P_a^m] + P_a^b + P_a^l$, and the total active leukocyte population: $L_a^i = L_a^p + L_a^e \cdot ls/leuk$. Then the change in platelet concentration during a given time step is calculated as

$$dP_L = [P_a^e]ps/p + [P_a^m] + P_a^b + P_a^l - PL_a^i.$$

The current height of the platelet layer, nhc, is updated using

$$nhc = nhc + (nhc + nh)(dP_L \cdot p + 4 * (L_a^p + L_a^e \cdot ls/leuk - L_a^i)leuk) \cdot 6 \times 10^9$$

and, from this, the volume of the chemical boundary layer, v_{ch}, and the platelet volume, v_{pl}, are calculated:

$$v_{ch} = (nh + nhc * n_{nhc})/(nh + nhc^i n_{nhc}) \qquad v_{pl} = (nh + nhc)/(nh + nhc^i).$$

These volumes are then used to adjust the platelet and chemical concentrations. Then platelet binding site availabilities are updated as follows:

$$p_2 = p_2 + d_{PL}(p \cdot np_2/s_2) \tag{88}$$

$$p_5 = p_5 + d_{PL}(p \cdot np_5/s_5) \tag{89}$$

$$p_8 = p_8 + d_{PL}(p \cdot np_8/s_8) \tag{90}$$

$$p_9 = p_9 + d_{PL}(p \cdot np_9/s_9) \tag{91}$$

$$p_{10} = p_{10} + d_{PL}(p \cdot np_{10}/s_{10}) \tag{92}$$

$$p_{11} = p_{11} + d_{PL}(p \cdot np_{11}/s_{11}) \tag{93}$$

$$p_{111} = p_{111} + d_{PL}(p \cdot np_{111}/s_{111}). \tag{94}$$

Additionally, the binding site availabilities for surface binding interactions are defined as follows (each step):

$$p_{PLAS}^{\text{avail}} = p_{PLAS} - [P_a^e] \tag{95}$$

$$p_{10}^{\text{avail}} = p_{10} - z_{10}^m - e_{10}^m - [Z_{10}^m : TEN] - [Z_5^m : E_{10}^m] \tag{96}$$
$$- [Z_8^m : E_{10}^m] - [PRO] - [Z_2^m : PRO] - [Z_{10}^m : TEN^*]$$

$$p_5^{\text{avail}} = p_5 - z_5^m - e_5^m - [Z_5^m : E_{10}^m] - [Z_5^m : E_2^m] - [APC : E_5^m] \tag{97}$$
$$- [PRO] - [Z_2^m : PRO]$$

$$p_8^{\text{avail}} = p_8 - z_8^m - e_8^m - [TEN] - [Z_8^m : e_{10}^m - [Z_8^m : E_2^m] - [Z_{10}^m : TEN] \tag{98}$$
$$- [APC : E_8^m] - [TEN^*] - [Z_{10}^m : TEN^*]$$

$$p_9^{\text{avail}} = p_9 - z_9^m - e_9^m - [TEN] - [Z_{10}^m : TEN] - [Z_9^m : E_{11}^{h,m*}] - [Z_9^m : E_{11}^{m*}] \tag{99}$$

$$p_2^{\text{avail}} = p_2 - z_2^m - [Z_2^m : PRO] - [Z_{11}^m : E_2^m] - [E_{11}^{h,m*} : E_2^m] \tag{100}$$

$$p_{2s}^{\text{avail}} = p_{2s} - e_2^m - [Z_5^m : E_2^m] - [Z_8^m : E_2^m] - [Z_{11}^m : E_2^m] - [E_{11}^{h,m*} : E_2^m] \tag{101}$$

$$p_{91}^{\text{avail}} = p_{91} - e_9^{m*} - ([TEN^*] + [Z_{10}^m : TEN^*]) \tag{102}$$

$$p_{11}^{\text{avail}} = p_{11} - z_{11}^m - e_{11}^{h,m*} - [Z_9^m : E_{11}^{h,m*}] - [Z_{11}^m : E_2^m] \tag{103}$$

$$p_{111}^{\text{avail}} = p_{111} - e_{11}^{h,m*} - e_{11}^{m*} - [Z_9^m : E_{11}^{m*}] - [E_{11}^{h,m*} : E_2^m] \tag{104}$$

$$[TM]^{\text{avail}} = [TM] - [TM : E_2] - [TM : E_2 : APC]. \tag{105}$$

Complete Parameter List

New Model Parameters

See Table 1.

Table 1 List of parameters used in the inflammation-based modification in the present work

Parameter	Units	Description	Value
v	cm s^{-1}	Midstream velocity	Varied
q	#	Neutrophil fraction of leukocytes	$0.7 - 0.9$ [3]
w_{max}		Maximal vessel wall surface area available for adhesion	v-dependent
k_{adh}^{+}	M^{-1}s^{-1}	Platelet adhesion to endothelium	$5 \times 10^8 = k^+$ (adjusted from [4, 17])
k_{adh}^{-}	s^{-1}	Platelet unbinding from endothelium	0 [17]
a_0	s^{-1}	Endothelial activation rate	5×10^7
a_1	M^{-1}s^{-1}	leukocyte-to-endothelium adhesion rate	2×10^{10}
a_{-1}	s^{-1}	Leukocyte–endothelium separation rate	0
a_2	M^{-1}s^{-1}	Leukocyte-to-platelet adhesion rate	2×10^{10}
a_{-2}	s^{-1}	Leukocyte–platelet separation rate	0
a_3	s^{-1}	NETosis from platelet-bound leukocytes	9.26×10^{-5}
a_3^p	M^{-1}s^{-1}	Platelet-dependent NETosis by endothelial-bound leukocytes	2.315×10^8
a_4	M^{-1}s^{-1}	NET-mediated platelet activation	3×10^8
a_5	M^{-1}s^{-1}	Platelet-to-leukocyte/NET adhesion rate	2×10^{10}
a_{-5}	s^{-1}	Platelet-to-leukocyte/NET separation rate	0
$k_{PC:TM:NET}$	nM	Half-maximal NET inhibition of protein C activation	1000
k_9^{cat}	M^{-1}s^{-1}	NET-mediated (via FXIIa) FIX activation	1×10^7
$k_{e12:F:NET}$	nM	Half-maximal NET inhibition of fibrinolysis	1000
d_{TFPI}	M^{-1}s^{-1}	NET-dependent TFPI degradation	1×10^7
d_{NET}	s^{-1}	NET degradation rate	1.07×10^{-5} [8]
d_{leuk}	s^{-1}	Neutrophil degradation rate	4×10^{-5}
k_{coh}		Platelet cohesion rate	2×10^5 [6, 17]

Where available, references are provided. In all other cases, parameter values are chosen to provide representative qualitative behavior

Original Model Parameters

See Tables 2, 3, and 4.

Table 2 Modified list of reaction parameters for the clotting pathway except for vitamin K components

Parameter	Value	Units
$k^+_{TFPI:e_{10}}$	1.6×10^7	$M^{-1} s^{-1}$
$k^-_{TFPI:e_{10}}$	0.00033	s^{-1}
$k^+_{TFPI:e^m_7}$	10^7	s^{-1}
$k^-_{TFPI:e^m_7}$	0.0011	$M^{-1} s^{-1}$
$k^m_{z10:e^m_7}$	4.5×10^{-7}	M
$k^{cat}_{z10:e^m_7}$	1.15	s^{-1}
$k^-_{z10:e^m_7}$	1	s^{-1}
$k^+_{z10:e^m_7}$	$(k^-_{z10:e^m_7} + k^{cat}_{z10:e^m_7})/k^m_{z10:e^m_7}$	M
$k^m_{z^m_7:e_{10}}$	1.2×10^{-6}	M
$k^{cat}_{z^m_7:e_{10}}$	5	s^{-1}
$k^-_{z^m_7:e_{10}}$	1	s^{-1}
$k^+_{z^m_7:e_{10}}$	$(k^-_{z^m_7:e_{10}} + k^{cat}_{z^m_7:e_{10}})/k^m_{z^m_7:e_{10}}$	M
$k^m_{z7:e_{10}}$	10^{-6}	M
$k^{cat}_{z7:e_{10}}$	5	s^{-1}
$k^-_{z7:e_{10}}$	1	s^{-1}
$k^+_{z7:e_{10}}$	$(k^-_{z7:e_{10}} + k^{cat}_{z7:e_{10}})/k^m_{z7:e_{10}}$	$M^{-1}s^{-1}$
$k^{cat}_{z7:e_2}$	0.061	s^{-1}
$k^-_{z7:e_2}$	1	s^{-1}
$k^m_{z7:e_2}$	2.7×10^{-6}	M
$k^+_{z7:e_2}$	$(k^-_{z7:e_2} + k^{cat}_{z7:e_2})/k^m_{z7:e_2}$	$M^{-1}s^{-1}$
$k^m_{z9:e^m_7}$	2.4×10^{-7}	M
$k^{cat}_{z9:e^m_7}$	1.15	s^{-1}
$k^-_{z9:e^m_7}$	1	s^{-1}
$k^+_{z9:e^m_7}$	$(k^-_{z9:e^m_7} + k^{cat}_{z9:e^m_7})/k^m_{z9:e^m_7}$	$M^{-1}s^{-1}$
$k^m_{z^m_5:e^m_{10}}$	1.04×10^{-8}	M
$k^{cat}_{z^m_5:e^m_{10}}$	0.046	s^{-1}
$k^-_{z^m_5:e^m_{10}}$	1	s^{-1}
$k^+_{z^m_5:e^m_{10}}$	$(k^-_{z^m_5:e^m_{10}} + k^{cat}_{z^m_5:e^m_{10}})/k^m_{z^m_5:e^m_{10}}$	$M^{-1}s^{-1}$
$k^m_{z5:e^m_2}$	7.1×10^{-8}	M
$k^{cat}_{z5:e^m_2}$	0.23	s^{-1}
$k^-_{z5:e^m_2}$	1	s^{-1}

(continued)

Table 2 (continued)

Parameter	Value	Units
$k^+_{z5:e^m_2}$	$(k^-_{z5:e^m_2} + k^{\text{cat}}_{z5:e^m_2})/k^m_{z5:e^m_2}$	$M^{-1}s^{-1}$
$k^m_{z8:e_2}$	2×10^{-7}	M
$k^{\text{cat}}_{z8:e_2}$	0.9	s^{-1}
$k^-_{z8:e_2}$	1	s^{-1}
$k^+_{z8:e_2}$	$(k^-_{z8:e_2} + k^{\text{cat}}_{z8:e_2})/k^m_{z8:e_2}$	$M^{-1}s^{-1}$
$k^+_{e^m_5:e^m_{10}}$	10^8	$M^{-1}s^{-1}$
$k^-_{e^m_5:e^m_{10}}$	0.01	s^{-1}
$k^+_{e^m_8:e^m_9}$	10^8	$M^{-1}s^{-1}$
$k^-_{e^m_8:e^m_9}$	0.01	s^{-1}
$k^m_{z_{10}:TEN}$	1.6×10^{-7}	M
$k^{\text{cat}}_{z_{10}:TEN}$	20	s^{-1}
$k^-_{z_{10}:TEN}$	1	s^{-1}
$k^+_{z_{10}:TEN}$	$(k^-_{z_{10}:TEN} + k^{\text{cat}}_{z_{10}:TEN})/k^m_{z_{10}:TEN}$	$M^{-1}s^{-1}$
$k^m_{z^m_2:PRO}$	3×10^{-7}	M
$k^{\text{cat}}_{z^m_2:PRO}$	30	s^{-1}
$k^-_{z^m_2:PRO}$	1	s^{-1}
$k^+_{z^m_2:PRO}$	$(k^-_{z^m_2:PRO} + k^{\text{cat}}_{z^m_2:PRO})/k^m_{z^m_2:PRO}$	$M^{-1}s^{-1}$
$k^m_{z^m_8:e^m_2}$	2×10^{-7}	M
$k^{\text{cat}}_{z^m_8:e^m_2}$	0.9	s^{-1}
$k^-_{z^m_8:e^m_2}$	1	s^{-1}
$k^+_{z^m_8:e^m_2}$	$(k^-_{z^m_8:e^m_2} + k^{\text{cat}}_{z^m_8:e^m_2})/k^m_{z^m_8:e^m_2}$	$M^{-1}s^{-1}$
$k^m_{z^m_8:e^m_{10}}$	2×10^{-8}	M
$k^{\text{cat}}_{z^m_8:e^m_{10}}$	0.023	s^{-1}
$k^-_{z^m_8:e^m_{10}}$	1	s^{-1}
$k^+_{z^m_8:e^m_{10}}$	$(k^-_{z^m_8:e^m_{10}} + k^{\text{cat}}_{z^m_8:e^m_{10}})/k^m_{z^m_8:e^m_{10}}$	$M^{-1}s^{-1}$
$k^{\text{cat}}_{e^m_5:APC}$	0.5	s^{-1}
$k^-_{e^m_5:APC}$	1	s^{-1}
$k^m_{e^m_5:APC}$	1.25×10^{-8}	M
$k^+_{e^m_5:APC}$	$(k^-_{e^m_5:APC} + k^{\text{cat}}_{e^m_5:APC})/k^m_{e^m_5:APC}$	$M^{-1}s^{-1}$
$k^{\text{cat}}_{e5:APC}$	0.5	s^{-1}
$k^-_{e5:APC}$	1	s^{-1}
$k^m_{e5:APC}$	1.25×10^{-8}	M
$k^+_{e5:APC}$	$(k^-_{e5:APC} + k^{\text{cat}}_{e5:APC})/k^m_{e5:APC}$	$M^{-1}s^{-1}$
$k^{\text{cat}}_{e^m_8:APC}$	$k^{\text{cat}}_{e^m_5:APC}$	s^{-1}

(continued)

Table 2 (continued)

Parameter	Value	Units
$k^m_{e_8^m:APC}$	$k^m_{e_5^m:APC}$	M
$k^-_{e_8^m:APC}$	$k^-_{e_5^m:APC}$	s^{-1}
$k^+_{e_8^m:APC}$	$k^+_{e_5^m:APC}$	$M^{-1}s^{-1}$
$k^{cat}_{e_8:APC}$	$k^{cat}_{e_5:APC}$	s^{-1}
$k^m_{e_8:APC}$	$k^m_{e_5:APC}$	M
$k^-_{e_8:APC}$	$k^-_{e_5:APC}$	s^{-1}
$k^+_{e_8:APC}$	$k^+_{e_5:APC}$	$M^{-1}s^{-1}$
$k^{in}_{AT:e_{10}}$	0.1	s^{-1}
$k^{in}_{AT:e_9}$	0.1	s^{-1}
$k^{in}_{AT:e_2}$	0.2	s^{-1}
$k^m_{z_7:e_9}$	1.7×10^{-6}	M
$k^{cat}_{z_7:e_9}$	0.32	s^{-1}
$k^-_{z_7:e_9}$	1	s^{-1}
$k^+_{z_7:e_9}$	$(k^-_{z_7:e_9} + k^{cat}_{z_7:e_9})/k^m_{z_7:e_9}$	$M^{-1}s^{-1}$
$k^+_{z_9:e_{11}}$	0.6×10^7	$M^{-1}s^{-1}$
$k^-_{z_9:e_{11}}$	1	s^{-1}
$k^{cat}_{z_9:e_{11}}$	0.21	s^{-1}
$k^+_{z_{11}:e_2}$	2×10^7	$M^{-1}s^{-1}$
$k^-_{z_{11}:e_2}$	1	s^{-1}
$k^{cat}_{z_{11}:e_2}$	1.3×10^{-4}	s^{-1}
k^+_{TM}	10^8	$M^{-1}s^{-1}$
k^-_{TM}	5×10^{-2}	s^{-1}
$k^{cat}_{PC:TM:e_2}$	1/6	s^{-1}
$k^m_{PC:TM:e_2}$	7×10^{-7}	M
$k^-_{PC:TM:e_2}$	1	s^{-1}
$k^+_{PC:TM:e_2}$	$(k^-_{PC:TM:e_2} + k^{cat}_{PC:TM:e_2})/k^m_{PC:TM:e_2}$	$M^{-1}s^{-1}$
k^{deg}_8	0.005	s^{-1}
$\overline{e_2}$	0.001	M

All values taken from Fogelson et al. [10]

Table 3 Modified list of surface binding (on/off) parameters

Parameter	Value	Units	Parameter	Value	Units
k_{adh}^{+}	2×10^{10}	$M^{-1} s^{-1}$	k_{e2}^{act}	0.1	s^{-1}
k_{adh}^{-}	0	s^{-1}	k_{plt}^{act}	3×10^{8}	$M^{-1} s^{-1}$
k_{2}^{on}	10^{7}	$M^{-1} s^{-1}$	k_{2s}^{on}	10^{7}	$M^{-1} s^{-1}$
k_{2}^{off}	5.9	s^{-1}	k_{2s}^{off}	0.2	s^{-1}
k_{5}^{on}	5.7×10^{7}	$M^{-1} s^{-1}$	k_{8}^{on}	5×10^{7}	$M^{-1} s^{-1}$
k_{5}^{off}	0.17	s^{-1}	k_{8}^{off}	0.17	s^{-1}
k_{9}^{on}	10^{7}	$M^{-1} s^{-1}$	k_{10}^{on}	10^{7}	$M^{-1} s^{-1}$
k_{9}^{off}	0.025	s^{-1}	k_{10}^{off}	0.025	s^{-1}
k_{z11}^{on}	10^{7}	$M^{-1} s^{-1}$	k_{e11}^{on}	10^{7}	$M^{-1} s^{-1}$
k_{z11}^{off}	0.1	s^{-1}	k_{e11}^{off}	0.017	s^{-1}
k_{7}^{on}	5×10^{7}	$M^{-1} s^{-1}$			
k_{7}^{off}	0.005	s^{-1}			

All values taken from Fogelson et al. [10]

Table 4 Modified list of platelet, surface and volume scalings, and flow-related parameters

Parameter	Value	Description
cpl	1	1 corresponds to 250,000 platelets per L
rad	400	Radius of blood vessel (μm)
$width$	40	Width of injury (μm)
nph_{max}	80	Maximum height of clot (μm)
dc	5.0	Diffusion rate ($\mu m^2/s$)
vel	1.0	Default midstream velocity (μm/s)
$tfscale$	1.0	Tissue factor concentration (fmol/cm^2)
nh	$0.01 \cdot 3\sqrt[3]{(rad \cdot width \cdot dc/vel)}/4$	Default boundary layer thickness
k_{flow}	$vel \cdot nh \cdot 10^4/(width \cdot rad)$	Flow term (s^{-1})
k_{flow}^{p}	$5 \cdot vel \cdot 10^4/(width \cdot rad)$	Platelet flow term (s^{-1})
p	$4 \cdot 10^{-13}/cpl$	Platelet concentration
ps	$20/(6 \cdot 10^{23} \cdot nh \cdot 10^{-5}(10^2) \cdot 10^{-10})$	
$leuk$	8.5×10^{-15}	Leukocyte concentration (8.5 fM)
ls	$20/(6 \cdot 10^{23} \cdot nh \cdot 10^{-5}(10^2) \cdot 10^{-10})$	
n_{nhc}	1/8	
v_2	1.3×10^{-6}	1300 nM
v_5	10^{-8}	10 nM
v_7	10^{-8}	10 nM
v_8	10^{-9}	1 nM
v_9	10^{-7}	0.1 μM
v_{10}	1.7×10^{-7}	170 nM
v_{11}	10^{-8}	0.01 μM

(continued)

Table 4 (continued)

Parameter	Value	Description
v_f	2.4×10^{-9}	2.4 nM ([TFPI])
n_5	3×10^3	Number of factor V molecules released per platelet
np_{10}	2700	Number of factor X binding sites per platelet
np_5	5000	Number of factor V binding sites per platelet
np_8	450	Number of factor VIII binding sites per platelet
np_9	250	Number of factor IX binding sites per platelet
np_2	1000	Number of thrombin binding sites per platelet
np_{11}	1500	Number of inactive factor XI binding sites per platelet
np_{111}	250	Number of active factor XI binding sites per platelet
s_2	$p \cdot np_2$	Total number of binding sites for thrombin
s_5	$np_5 \cdot p$	Total number of binding sites for factor V
s_7	$tfscale \cdot 10^{-8}/nh$	Volume conversion actor for $[TF]$
s_8	$p \cdot np_8$	Total number of binding sites for factor VIII
s_9	$p \cdot np_9$	Total number of binding sites for factor IX
s_{10}	$p \cdot np_{10}$	Total number of binding sites for factor X
s_{11}	$p \cdot np_{11}$	Total number of binding sites for inactive factor XI
s_{111}	$p \cdot np_{111}$	Total number of binding sites for active factor XI
s_{TM}	5×10^{-7}	Thrombomodulin concentration
vpc	6.5×10^{-8}	Concentration of protein C
nd_{Fg}	8.95×10^{-6}	8945.5 nM
nd_{Pg}	2.15×10^{-6}	2154.3 nM

Values taken from DePillis et al. [4], Elizondo et al. [6], Fogelson et al. [10], Wajima et al. [26]

Initial Conditions

See Table 5.

Table 5 Modified list of concentration scalings and nondimensional initial values used for all variables

Variable(s)	Scaling	Units	Nondimensional Initial Value(s)
$z_7, e_7, [Z_7 : E_2], [Z_7 : E_{10}], [Z_7 : E_9]$	v_7	M	1, 0.01, 0, 0, 0
$z_7^m, e_7^m, [TFPI : E_{10} : E_7^m], [Z_9 : E_7^m]$	s_7	mol/μm^2	0, 0, 0, 0
$[Z_{10}^m : E_{10}^m], [Z_7^m : E_2^m], [Z_{10} : E_7^m]$	s_7	mol/μm^2	0, 0, 0
$z_{10}, e_{10}, e_{10}^{ec}$	v_{10}	M	1, 0, 0
$z_{10}^m, e_{10}^m, [Z_5^m : E_{10}^m]$	s_{10}	mol/μm^2	0 ,0, 0
$z_5, e_5, [Z_5 : E_2], [APC : e_5]$	v_5	M	1, 0, 0, 0
$z_5^m, e_5^m, [PRO]$	s_5	mol/μm^2	0, 0, 0
$z_8, e_8, [Z_8 : E_2], [APC : e_8]$	v_8	M	1, 0, 0, 0
$z_8^m, e_8^m, [TEN], [Z_{10}^m : TEN]$	s_8	mol/μm^2	0, 0, 0, 0
$[Z_8^m : E_{10}^m], [Z_8^m : E_2^m], [APC : E_8^m]$	s_8	mol/μm^2	0, 0, 0
$[TEN^*], [Z_{10}^m : TEN^*]$	s_8	mol/μm^2	0, 0
$z_9, e_9, e_9^{ec}, [Z_9 : E_{11}], [Z_9 : E_{11}^h]$	v_9	M	1, 0, 0, 0, 0
$z_9^m, e_9^m, e_9^{m*}, [Z_9^m : E_{11}^{m*}], [Z_9^m : E_{11}^{h,m}]$	s_9	mol/μm^2	0, 0, 0, 0, 0
$z_2, e_2, [APC]$	v_2	M	1, 0, 0
$z_2^m, e_2^m, [Z_5^m : E_5^m]$	s_2	mol/μm^2	0, 0, 0
$[P_a^e]$	ps	count/μm^2	0
$[P_u^m], [P_a^m], [P_a^l], [P_a^b]$	p	count/L	1, 0, 0
$[L_a^e]$	ls	count/μm^2	0
$[L_u^m], [L_a^p], [NET]$	$leuk$	count/L	0, 0
$[TFPI], [TFPI : E_{10}]$	v_f	M	1, 0
$[Z_2^m : PRO], [APC : E_5^m]$	s_5	mol/μm^2	0, 0
$[TM : e_2], [TM : e_2 : APC]$	s_{TM}	mol/μm^2	0, 0
$z_{11}, e_{11}, e_{11}^h, e_{11}^{h,m}, [Z_{11} : E_2], [E_{11}^h : E_2]$	v_{11}	M	1, 0, 0, 0, 0, 0
$z_{11}^m, [Z_{11} : E_2^m]$	s_{11}	mol/μm^2	0, 0
$[E_{11}^{h,m*} : E_2^m]$	s_{111}	mol/μm^2	0
z_i^{up}	v_i	M	1
$[Fg], [F]$	nd_{Fg}	M	1, 0
$[Pg], [P]$	nd_{Pg}	M	1, 0
W	1	#	Varied

Unless highlighted, these values are taken from DePillis et al. [4], Fogelson et al. [10], and Wajima et al. [26]

References

1. Johanna Altmann, Smriti Sharma, and Irene M Lang. Advances in our understanding of mechanisms of venous thrombus resolution. *Expert Review of Hematology*, 9(1):69–78, 2016.
2. Volker Brinkmann, Ulrike Reichard, Christian Goosmann, Beatrix Fauler, Yvonne Uhlemann, David S Weiss, Yvette Weinrauch, and Arturo Zychlinsky. Neutrophil extracellular traps kill bacteria. *Science*, 303(5663):1532–1535, 2004.
3. Ivan Budnik and Alexander Brill. Immune factors in deep vein thrombosis. *Trends in Immunology*, 39(8):610–623, August 2018.
4. Lisette de Pillis, Erica J. Graham, Kaitlyn Hood, Yanping Ma, Ami Radunskaya, and Julie Simons. Injury-initiated clot formation under flow: a mathematical model with warfarin treatment. In T. Jackson and A. Radunskaya, editors, *Applications of Dynamical Systems in Biology and Medicine*, volume 158 of *IMA Volumes in Mathematics and Its Applications*, pages 75–98. Springer, 2015.
5. Jose Antonio Diaz and Daniel Durant Myers. Inflammation, thrombogenesis, fibrinolysis, and vein wall remodeling after deep venous thrombosis. In *Inflammatory Response in Cardiovascular Surgery*, pages 175–183. Springer, 2013.
6. Priscilla Elizondo and Aaron L Fogelson. A mathematical model of venous thrombosis initiation. *Biophysical Journal*, 111:2722–2734, December 2016.
7. Bernd Engelmann and Steffen Massberg. Thrombosis as an intravascular effector of innate immunity. *Nature Reviews Immunology*, 13(1):34, 2013.
8. C. Farrera and B. Fadeel. Macrophage clearance of neutrophil extracellular traps is a silent process. *Journal of Immunology*, 191(5):2647–2656, 2013.
9. Antonio Fasano and Adélia Sequeira. *Hemomath: The Mathematics of Blood*, volume 18. Springer, 2017.
10. Aaron L Fogelson, Yasmeen H Hussain, and Karin Leiderman. Blood clot formation under flow: the importance of factor XI depends strongly on platelet count. *Biophysical Journal*, 102(1):10–18, 2012.
11. Centers for Disease Control and Prevention. Data and statistics on venous thromboembolism. https://www.cdc.gov/ncbddd/dvt/data.html. Accessed: 2018-08-28.
12. Tobias A Fuchs, Alexander Brill, Daniel Duerschmied, Daphne Schatzberg, Marc Monestier, Daniel D Myers, Shirley K Wroblewski, Thomas W Wakefield, John H Hartwig, and Denisa D Wagner. Extracellular DNA traps promote thrombosis. *Proceedings of the National Academy of Sciences*, 107(36):15880–15885, 2010.
13. Mehran Ghasemzadeh and Ehteramolsadat Hosseini. Platelet-leukocyte crosstalk: linking proinflammatory responses to procoagulant state. *Thrombosis research*, 131(3):191–197, 2013.
14. Peter Hoang, Alex Wallace, Mark Sugi, Andrew Fleck, Yash Pershad, Nirvikar Dahiya, Hassan Albadawi, Grace Knuttinen, Sailendra Naidu, and Rahmi Oklu. Elastography techniques in the evaluation of deep vein thrombosis. *Cardiovascular diagnosis and therapy*, 7(Suppl 3):S238, 2017.
15. Andrew S Kimball, Andrea T Obi, Jose A Diaz, and Peter K Henke. The emerging role of nets in venous thrombosis and immunothrombosis. *Frontiers in immunology*, 7:236, 2016.
16. Andrew L Kuharsky and Aaron L Fogelson. Surface-mediated control of blood coagulation: The role of binding site densities and platelet deposition. *Biophysical Journal*, 80:1050–1074, 2001.
17. Karin Leiderman and Aaron L Fogelson. Grow with the flow: a spatial–temporal model of platelet deposition and blood coagulation under flow. *Mathematical Medicine and Biology: a journal of the IMA*, 28:47–84, 2011.
18. GD Lowe. Virchow's triad revisited: abnormal flow. *Pathophysiology of Haemostasis and Thrombosis*, 33(5–6):455–457, 2002.
19. Nigel Mackman. New insights into the mechanisms of venous thrombosis. *The Journal of clinical investigation*, 122(7):2331–2336, 2012.

20. Amit A Patel, Yan Zhang, James N Fullerton, Lies Boelen, Anthony Rongvaux, Alexander A Maini, Venetia Bigley, Richard A Flavell, Derek W Gilroy, Becca Asquith, et al. The fate and lifespan of human monocyte subsets in steady state and systemic inflammation. *Journal of Experimental Medicine*, 214(7):1913–1923, 2017.

21. Farbod Nicholas Rahaghi, Jasleen Kaur Minhas, and Gustavo A Heresi. Diagnosis of deep venous thrombosis and pulmonary embolism: New imaging tools and modalities. *Clinics in chest medicine*, 39(3):493–504, 2018.

22. Prakash Saha, Julia Humphries, Bijan Modarai, Katherine Mattock, Matthew Waltham, Colin E Evans, Anwar Ahmad, Ashish S Patel, Sobath Premaratne, Oliver TA Lyons, et al. Leukocytes and the natural history of deep vein thrombosis: current concepts and future directions. *Arteriosclerosis, thrombosis, and vascular biology*, 31(3):506–512, 2011.

23. C Schulz, B Engelmann, and S Massberg. Crossroads of coagulation and innate immunity: the case of deep vein thrombosis. *Journal of Thrombosis and Haemostasis*, 11(s1):233–241, 2013.

24. Charlotte Summers, Sara M Rankin, Alison M Condliffe, Nanak Singh, A Michael Peters, and Edwin R Chilvers. Neutrophil kinetics in health and disease. *Trends in immunology*, 31(8):318–324, 2010.

25. Marie-Luise von Brühl, Konstantin Stark, Alexander Steinhart, Sue Chandraratne, Ildiko Konrad, Michael Lorenz, Alexander Khandoga, Anca Tirniceriu, Raffaele Coletti, Maria Köllnberger, et al. Monocytes, neutrophils, and platelets cooperate to initiate and propagate venous thrombosis in mice in vivo. *Journal of Experimental Medicine*, pages jem–20112322, 2012.

26. T Wajima, GK Isbister, and SB Duffull. A comprehensive model for the humoral coagulation network in humans. *Clinical Pharmacology & Therapeutics*, 86(3):290–298, 2009.

Trigonometric-Type Functions Derived from Polygons Inscribed in the Unit Circle

Torina Lewis

Abstract Given a polygon inscribed inside of a circle with vertices satisfying the equation $z^N = 1$, we introduce a new class of periodic functions called the "geometric polygon functions." Methodology used to construct and analyze the classical circular and elliptic functions is essential for defining the cosine polygon, sine polygon, and dine polygon functions, called the "geometric polygon functions." Dividing N into two sub-cases, odd values and even values, allows mutually exclusive sets. Focusing on even values of N, the square functions are introduced as the smallest most significant case within this sub-case. The square functions provide the building blocks for larger even values of N for which additional analysis is presented. The goals of this research are to introduce the "geometric polygon functions" and compute the Fourier series expansion of the square functions. These findings can be manipulated to prove results in matroid theory. In particular, the construct of the "geometric polygon functions" are representations of graphs whose bicircular matroids have well controlled circuit spectra.

1 Introduction

A function f defined on a set $D \subset \mathbb{R}$ is called periodic if there exists a constant $T > 0$ for which $f(x + T) = f(x)$ for all x in D and for some real number $T \neq 0$. Such a constant T is called the period of the function f. We assert that the period will always be positive. The fundamental period is the smallest positive period of f if such a period exists [1, 2]. Continuous effort has been spent on the development and application of periodic functions, for their critical role in modeling physical phenomena through applications and connections to an array of mathematics and science. Many applications of periodic functions and solutions, such as bloodstain pattern analysis in forensic science [3], predator–prey models

T. Lewis (✉)
Clark Atlanta University, Atlanta, GA, USA
e-mail: tlewis@cau.edu

© The Author(s) and the Association for Women in Mathematics 2019
S. D'Agostino et al. (eds.), *A Celebration of the EDGE Program's Impact on the Mathematics Community and Beyond*, Association for Women in Mathematics Series 18, https://doi.org/10.1007/978-3-030-19486-4_19

in mathematical biology [4, 5], and blood pressure analysis and glucose–insulin interaction in medicine [6, 7], have been useful.

Trigonometric functions, also called circular functions are the first examples of periodic functions in formal educational studies. Let us suppose that θ is an angle in standard position, (x, y) is a point on the terminal side of θ which lies on a circle with Cartesian equation $x^2 + y^2 = r^2$, and r is the radius of the circle. Then the traditional circular functions as defined in trigonometry are $cos(\theta) = \frac{x}{r}$ and $sin(\theta) = \frac{y}{r}$. However, a third function, the dine function, is also derived from the circle. This function is defined as $din(\theta) = r$ [8]. The dine function is typically omitted from trigonometry courses because the radius of a circle is constant. Hence, the dine function on a circle is a constant function and plots a horizontal line. In the next section, the elliptic functions are discussed and the dine function reappears. Unlike the circle, the ellipse has two radius measures, one horizontally along the x-axis, and the other vertically along the y-axis. The elliptic dine function is not constant. Thus the dine function is realized and more significant in the discussion concerning elliptic functions.

2   Periodic Functions Associated with an Ellipse

The elliptical trigonometric unit is an ellipse centered at the origin with radii a and b, and has the equation of the form,

$$\left(\frac{x}{a}\right)^2 + \left(\frac{y}{b}\right)^2 = 1. \tag{1}$$

Here, the y-intercept is normalized to $b = 1$ and the x-intercept $a > 1$, so that the eccentricity or the modulus of the ellipse is defined as, $k^2 = 1 - \frac{1}{a^2}$ (Fig. 1).

Fig. 1 The elliptic cosine, sine, and dine functions, see Eqs. (2)–(4)

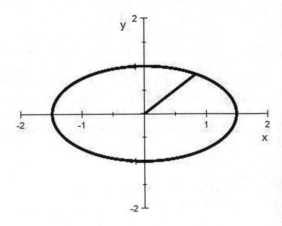

Suppose θ is an angle in standard position, (x, y) is a point on the terminal side of θ which lies on the normalized ellipse, and r is the distance from the origin to a point (x, y) on the ellipse. Hence the elliptic cosine, "$cn(\theta)$," elliptic sine, "$sn(\theta)$," and elliptic dine, "$dn(\theta)$," functions are given by the following ratios [11–13]:

$$cn(\theta) = \frac{x(\theta)}{a}, \tag{2}$$

$$sn(\theta) = y(\theta), \tag{3}$$

$$dn(\theta) = \frac{r(\theta)}{a}. \tag{4}$$

If radii a and b are equal, then Eq. (1) reduces to $x^2 + y^2 = r^2$. Thus the properties for the circular functions hold, and the elliptic functions as written in Eqs. (2)–(4) reduce to the equations,

$$\cos(\theta) = \frac{x}{r}, \ \sin(\theta) = \frac{y}{r}, \ din(\theta) = r. \tag{5}$$

On the normalized ellipse, the radii are not equal. Hence equation $x^2 + y^2 = r^2$ is replaced by the relation,

$$\left(\frac{x}{a}\right)^2 + y^2 = 1. \tag{6}$$

3 The "Geometric Polygon Functions"

For each positive integer $N \geq 3$, the Nth roots of unity satisfying the equation $z^N = 1$ have solutions $z_k = \cos\left(\frac{2\pi k}{N}\right) + i\sin\left(\frac{2\pi k}{N}\right)$, where k is an integer. The N solutions plotted in a complex plane correspond to the number of vertices of a regular polygon inscribed in a circle centered at the origin with the Cartesian equation $x^2 + y^2 = 1$. Each pair of adjacent vertices is equidistant apart with a central angle that measures $\frac{2\pi}{N}$ because the polygon is regular.

A regular polygon inscribed inside of a unit circle is a closed, simple, and convex curve. Thus it gives rise to periodic curves. To give an explicit representation of these curves in the plane, concepts in planar geometry, and motivation from the circular and elliptic functions provide a foundation. An integration of theories facilitates the derivation of the radius of a polygon inscribed inside of a unit circle. Let $v_i = (a, b)$ be a vertex, and m be the slope between adjacent vertices v_i and v_{i+1} on the outer edge of a polygon inscribed inside of the unit circle, then

$$r \equiv \frac{b - ma}{\sin(\theta) - m(\cos(\theta))}. \tag{7}$$

Given a construction of a regular polygon inscribed inside of a unit circle with vertices satisfying the equation $z^N = 1$, and the radius r, we define three associated periodic functions, the cosine polygon, "$C_N(\theta)$," sine polygon, "$S_N(\theta)$," and dine polygon, "$D_N(\theta)$," functions. The three functions are called the "geometric polygon functions." The definitions of these functions are written as,

$$C_N(\theta) \equiv r(\theta)cos(\theta), \tag{8}$$

$$S_N(\theta) \equiv r(\theta)sin(\theta), \tag{9}$$

$$D_N(\theta) \equiv r(\theta), \tag{10}$$

where N is the number of vertices on a regular polygon.

There are infinitely many regular polygons; therefore, the naming convention used for the cosine polygon, sine polygon, and dine polygon functions is the trigonometric function followed by the name of the polygon. For example, consider the regular polygon with four sides and vertices satisfying the equation $z^4 = 1$. We call the explicit functions obtained, the cosine square, sine square, and dine square functions, denoted by $C_4(\theta)$, $S_4(\theta)$, and $D_4(\theta)$, respectively. These functions are the square functions.

The "geometric polygon functions" are a new class of periodic functions. While we have generalized the Nth order, regular polygons inscribed inside of the unit circle, details are presented regarding the case $N = 4$. The square functions are periodic functions derived by Lewis [9] and Mickens [10] in separate works that utilize contrasting methods. The graphical representation of the square functions yields plots that are periodic. Radio and sound waves are modeled by periodic curves. Decomposing waves using Fourier series have been extensively studied and shown to affect signal transmission. The main theorem shows the computation while decomposing the square functions into a sum of sine waves.

4   Geometric Features for $N =$ Odd and $N =$ Even

The construction of the "geometric polygon functions" follows techniques used to establish the circular and elliptic functions and the methods are transferable for each value of N. Before performing a deeper analysis of the "geometric polygon functions," N is subdivided into two fundamental cases having some similar properties, but a unique difference that guarantees distinction between cases. The two cases explored consider whether N is odd or even.

4.1   The Triangle Functions and Odd N

The triangle functions derived by inscribing a triangle into a unit circle possess a challenge that is absent from square functions or more generally, functions with even N values. When the triangle functions are constructed, the triangle is invariant only about the x-axis. Thus the triangle functions and furthermore all "geometric polygon functions" with odd values of N require more attentive methods because each symmetry transformation is not present. We expect that the methods used for analyzing the triangle functions will be similar for any odd value N because of the existing consistent symmetry property among the functions.

4.2   The Square Functions and Even N

The square functions are explicitly defined by inscribing a square inside of a unit circle (see Fig. 2) with vertices satisfying $z^4 = 1$, and manipulation of the equation,

$$| x | + | y | = 1 \tag{11}$$

using trigonometric properties. This construction gives the square functions; the core case when N is even. Note that the independent variable is the angle θ. Further, the square functions are periodic functions of θ, with the following indicated periods:

Fig. 2 The construction of D_4

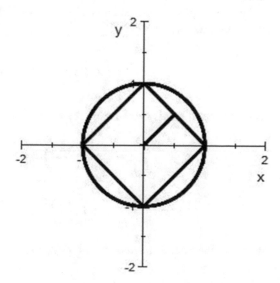

$$C_4(\theta + 2\pi) = C_4(\theta), \ S_4(\theta + 2\pi) = S_4(\theta), \ and \ D_4\left(\theta + \frac{\pi}{2}\right) = D_4(\theta). \quad (12)$$

One point of significance is that the three systems, circular, elliptic, and square, satisfy the following symmetry transformations:

$$T_1 : x \rightarrow -x, \quad y \rightarrow y, \quad (13)$$

$$T_2 : x \rightarrow x, \quad y \rightarrow -y, \quad (14)$$

$$T_3 = T_1 T_2 = T_2 T_1 : x \rightarrow -x, \quad y \rightarrow -y. \quad (15)$$

All "geometric polygon functions" with even N values are symmetric as described in Eqs. (13), (14), and (15). Thus solutions on the interval $0 < \theta \leq 2\pi$ are determined from solutions found in the first quadrant by the application of shift and reflection properties in trigonometry [13]. The "geometric polygon functions" with an even number of vertices have similar properties, and an analogous approach is used to investigate the functions. As a result of our approach, knowledge is discovered regarding the "geometric polygon functions" with larger even N values.

5 The Fourier Series Expansion of the Square Functions

Since $N = 4$ is the smallest most interesting even case, it is examined in more detail. In particular, we show by explicit calculation the Fourier coefficients for the square functions.

Let f be a piecewise continuous function on $[-l, l]$. Then the Fourier expansion of f is the expression,

$$\frac{a_0}{2} + \sum_{k=1}^{\infty} \left(a_k cos\left(\frac{\pi k x}{l}\right) + b_k sin\left(\frac{\pi k x}{l}\right) \right), \quad (16)$$

where the coefficients a_k and b_k in the series defined by

$$a_k = \frac{1}{l} \int_{-l}^{l} f(x) cos\left(\frac{\pi k x}{l}\right) dx \qquad (k = 0, 1, 2, \ldots), \quad (17)$$

$$b_k = \frac{1}{l} \int_{-l}^{l} f(x) sin\left(\frac{\pi k x}{l}\right) dx \qquad (k = 1, 2, \ldots) \quad (18)$$

are called the Fourier coefficients of f [2, 14].

5.1 Preliminary Information

To prove the main theorem of this work, we introduce the following result. This result is derived from unpublished work completed by Ronald E. Mickens and is essential for computing the Fourier coefficients for the square functions.

Lemma 1 *Let $m \in \mathbb{N}$, $\theta \in \mathbb{R}$, and $\theta \neq \frac{\pi}{2}, \frac{3\pi}{2}$, then*

$$\frac{1}{cos(\theta)} = 2 \sum_{m=0}^{\infty} (-)^m cos[(2m + 1)\theta]. \tag{19}$$

Solution This result is obtained by straightforward calculation. Let us begin with the equation,

$$1 = 1. \tag{20}$$

By adding and subtracting cosine functions, we write the following equation:

$$1 = 1 + cos\,2\theta - cos\,2\theta + cos\,4\theta - cos\,4\theta + cos\,6\theta - cos\,6\theta + \cdots \quad . \tag{21}$$

Rewriting the equation using a particular grouping of the functions, the equation is acquired,

$$1 = (1 + cos\,2\theta) - (cos\,2\theta + cos\,4\theta) + (cos\,4\theta + cos\,6\theta) + \cdots$$
$$+ (-)^m [cos\,(2m\theta) + cos[2(m + 1)\theta]. \tag{22}$$

Using trigonometric properties, the previous equation simplifies to

$$1 = 2cos\theta\,cos\theta - 2cos\theta\,cos\,3\theta + 2cos\theta\,cos\,5\theta + \cdots$$
$$+ 2(-)^m cos\theta\,cos[(2m + 1)\theta]. \tag{23}$$

Factoring $2 \cos \theta$ from each term and dividing by $cos\theta$ gives the equation,

$$\frac{1}{cos\theta} = 2[cos\theta - cos\,3\theta + cos\,5\theta + \ldots + (-)^m cos[(2m + 1)\theta]]. \tag{24}$$

Therefore the desired result is obtained,

$$\frac{1}{cos(\theta)} = 2 \sum_{m=0}^{\infty} (-)^m cos[(2m + 1)\theta]. \tag{25}$$

$\square$

5.2 The Fourier Coefficients for D_4

The dine square function is represented in the definitions of the cosine and sine square functions. Therefore, we determine the Fourier coefficients for the dine square function which enables us to write a Fourier series expansion of the square functions. It is indeed worthwhile to note that the coefficients b_k are zero due to the orthogonality condition.

Theorem 1 *In the first quadrant, $0 < \theta < \frac{\pi}{2}$, suppose that $D_4 = r(\theta) = \frac{1}{cos(\theta)+sin(\theta)}$ is a piecewise periodic function such that $D_4(\theta) = D_4\left(\theta + \frac{\pi}{2}\right)$. Then D_4 has a Fourier series expansion of the form,*

$$D_4(\theta) = \frac{a_0}{2} + \sum_{k=1}^{\infty}\left(\sum_{m=0}^{\infty}(-)^m(-)^k\left(\frac{4\sqrt{2}}{\pi}\right)\left[\frac{sin[(4k+2m+1)\frac{\pi}{4}]}{4k+2m+1}\right.\right.$$
$$\left.\left.+\frac{sin[(4k-2m-1)\frac{\pi}{4}]}{4k-2m-1}\right]\right)cos(4k\theta). \tag{26}$$

Proof Since $D_4(\theta)$ is periodic with period $T = \frac{\pi}{2}$, then $l = \frac{\pi}{4}$. Hence $cos\left(\frac{k\pi\theta}{\frac{\pi}{4}}\right) = cos(4k\theta)$ and its Fourier coefficients are given by the equation,

$$a_k = \left(\frac{4}{\pi}\right)\int_{\frac{-\pi}{4}}^{\frac{\pi}{4}} D_4(\theta)\,cos(4k\theta)\,d\theta. \tag{27}$$

By symmetry, the equation reduces to

$$a_k = \left(\frac{4}{\pi}\right)(2)\int_0^{\frac{\pi}{4}} D_4(\theta)\,cos(4k\theta)\,d\theta. \tag{28}$$

Substituting $D_4 = \frac{1}{sin(\theta)+cos(\theta)}$ leads to

$$a_k = \left(\frac{8}{\pi}\right)\int_0^{\frac{\pi}{4}} \frac{cos(4k\theta)}{sin(\theta)+cos(\theta)}\,d\theta. \tag{29}$$

Let $y = \frac{\pi}{4}$, then $cos(y) = sin(y) = \frac{1}{\sqrt{2}}$. The use of a trigonometric relation determines $sin(\theta) + cos(\theta) = \sqrt{2}cos(\theta - \frac{\pi}{4})$ and we get

$$a_k = \left(\frac{4\sqrt{2}}{\pi}\right)\int_0^{\frac{\pi}{4}} \frac{cos(4k\theta)}{cos(\theta - \frac{\pi}{4})}\,d\theta. \tag{30}$$

A Ψ transformation, $\Psi = \theta - \frac{\pi}{4}$, gives $d\Psi = d\theta$. This transformation and a change of integration limits are reflected in the following equation:

$$a_k = \left(\frac{4\sqrt{2}}{\pi}\right) \int_{-\frac{\pi}{4}}^{0} \frac{cos(4k\Psi + k\pi)}{cos(\Psi)} d\Psi. \qquad (31)$$

Using a Ptolemy's identity, we find $cos(4k\Psi + k\pi) = (-)^k cos(4k\Psi)$. Thus the series becomes

$$a_k = (-)^k \left(\frac{4\sqrt{2}}{\pi}\right) \int_{-\frac{\pi}{4}}^{0} \frac{cos(4k\Psi)}{cos(\Psi)} d\Psi. \qquad (32)$$

Let $\Psi = -\phi$. Hence $d\Psi = -d\phi$. Making the substitutions and simplifying the equation gives

$$a_k = (-)^k \left(\frac{4\sqrt{2}}{\pi}\right) \int_{0}^{\frac{\pi}{4}} \frac{cos(4k\phi)}{cos(\phi)} d\phi. \qquad (33)$$

By applying the result from Lemma 1,

$$\frac{1}{cos(\phi)} = 2 \sum_{m=0}^{\infty} (-)^m cos[(2m+1)\phi], \qquad (34)$$

to (33) we obtain,

$$a_k = (-)^k \left(\frac{8\sqrt{2}}{\pi}\right) \sum_{m=0}^{\infty} (-)^m \int_{0}^{\frac{\pi}{4}} cos(4k\phi) cos[(2m+1)\phi] d\phi. \qquad (35)$$

A product-to-sum trigonometric identity and simplification transform the integrand $cos(4k\phi)cos[(2m+1)\phi]$ into

$$a_k = (-)^k \left(\frac{4\sqrt{2}}{\pi}\right) \sum_{m=0}^{\infty} (-)^m \int_{0}^{\frac{\pi}{4}} (cos[(4k+2m+1)\phi] + cos[(4k-2m-1)\phi]) d\phi. $$
$$(36)$$

Using the substitution method for integration, we evaluate the integral and hence receive the related coefficients,

$$a_k = (-)^k \left(\frac{4\sqrt{2}}{\pi}\right) \sum_{m=0}^{\infty} (-)^m \left[\frac{sin[(4k+2m+1)\frac{\pi}{4}]}{4k+2m+1} + \frac{sin[(4k-2m-1)\frac{\pi}{4}]}{4k-2m-1}\right]. $$
$$(37)$$

Therefore, the Fourier coefficients of the dine square function on the interval $0 < \theta < \frac{\pi}{2}$ are

$$D_4 = \frac{a_0}{2} + \sum_{k=1}^{\infty} a_k \cos(4k\theta). \tag{38}$$

□

5.3 Fourier Coefficients for $C_4(\theta)$ and $S_4(\theta)$

In Sect. 5.2, the Fourier coefficients for the dine square function were determined. Now the Fourier series expansion is introduced for the square functions. First, let us recall the definitions of the square functions,

$$C_4(\theta) \equiv \cos(\theta) \, D_4(\theta),$$

$$S_4(\theta) \equiv \sin(\theta) \, D_4(\theta),$$

$$D_4(\theta) \equiv r(\theta).$$

These definitions and results from Sect. 3 are used to write the Fourier series expansion for C_4 and S_4. The expansions are as follows:

$$C_4 = \left(\frac{a_0}{2}\right)\cos(\theta) + \left(\frac{a_1}{2}\right)(\cos 3\theta + \cos 5\theta) + \left(\frac{a_2}{2}\right)(\cos 7\theta + \cos 9\theta) + \cdots$$
$$+ \left(\frac{a_k}{2}\right)\Big[\cos[(4k-1)\theta] + \cos[(4k+1)\theta]\Big], \tag{39}$$

$$S_4 = \left(\frac{a_0}{2}\right)\sin(\theta) + \left(\frac{a_1}{2}\right)(-\sin 3\theta + \sin 5\theta) + \left(\frac{a_2}{2}\right)(-\sin 7\theta + \cos 9\theta) + \cdots$$
$$+ \left(\frac{a_k}{2}\right)\Big[-\sin[(4k-1)\theta] + \sin[(4k+1)\theta]\Big]. \tag{40}$$

6 Discussion

In this work, a new class of periodic functions is introduced, the "geometric polygon functions." Methods involving the circular and elliptic functions are used to give explicit definitions for the "geometric polygon functions." The "geometric polygon functions" are constructed by inscribing a regular polygon with N vertices satisfying the equation $z^N = 1$ inside of a unit circle. Subdividing the vertex set N into two fundamental cases, (1) N is odd and (2) N is even, allows information extraction from an analysis of the triangle and square functions. However, additional results center on the case where N is even. Consistent symmetry transformations among the "geometric polygon functions" with an even number of vertices allow investigation of the smallest most significant case $N = 4$ (square functions). Hence we also derive explicit analytic formulas for the Fourier coefficients of the square functions and show their related Fourier series expansion.

The construction of the "geometric polygon functions" provokes future questions in the area of matroid theory as the graphs of the functions represent bicircular matroids. An article published by Lewis et al. [15] investigates when bicircular matroids have circuit spectra of one or two. While the "geometric polygon functions" have larger circuit spectra, the bicycles have very controlled sizes. Relating the graphs generated by the "geometric polygon functions" to current theoretical results involving bicircular matroids can lead to theoretical expansion and development of conjectures in matroid theory.

Acknowledgements The author's research is supported by the National Science Foundation, Human Resources Division, grant number 1700408; Dr. Ronald Mickens serves as senior collaborator on the project. The author thanks Springer for providing a catalyst that displays the talent of the EDGE women and their colleagues. Support from the EDGE organizers, editors, and reviewers is greatly appreciated as their efforts help to showcase the scholarly work of women studying Mathematics. Special thanks to family and friends that sacrificed time to ensure the completion of the project. Lastly, the author sincerely thanks Michelle Craddock for introducing her to the EDGE program in 2008.

References

1. Mirotin, A. R., Mirotin E. A. On sums and products of periodic functions. Real Analysis Exchange. **34**, 347–358 (2009)
2. Silverman, R., Tolstov, G.P. Fourier Series. Dover Publications. (1976)
3. Huang, N., Huang, W., Fung, Y., Shen, Z. Engineering analysis of biological variables: An example of blood pressure over 1 day. Real Analysis Exchange. **95**, 4816–4821 (1998)
4. Fan, M., Wang, K. Periodic solutions of a discrete time nonautonomous ratio-dependent predator-prey system. Mathematical and Computational Modeling. **34**, 951–961 (2002)
5. Wang, J., Zhang, J. Periodic solutions for discrete predator-prey systems with the Beddington-DeAngelis functional response. Applied Mathematics Letters. **19**, 1361–1366 (2006)
6. Germano, G and et al. Fourier and related analyses: An approach to study blood pressure nyctohemeral cycle. Blood Pressure and Heart Rate Variability: Computer Analysis, Methodology and Clinical Applications. 85–93 (1992)
7. Jiaxu, L., Yang, K. Analysis of a model of the glucose-insulin regulatory system with two delays. Society for Industrial and Applied Mathematics. **67** 757–776 (2007)
8. Schwalm, W. Elliptic functions as trigonometry. pp. 1–1 to 1–10. Morgan and Claypool Publishers (2002)
9. Lewis, T., Mickens, R.E. Square functions as a dynamic system. Proceedings of Dynamic Systems and Applications. **7** 268–274 (2016)
10. Mickens, R.E. Some properties of square (periodic) functions. Proceedings of Dynamic Systems and Applications. **7** 282–286 (2016)
11. Bayeh, C., Bernard, M., Moubayed, N. Introduction to the elliptical trigonometry. WSEAS Transactions on Mathematics. **8** 282–286 (2009)

12. Johannessen, K. A nonlinear differential equation related to the Jacobi elliptic functions. International Journal of Differential Equations. **2012** 282–286 (2012)
13. Milne-Thomson, L.M. Handbook of Mathematical Functions, Dover Publications, New York, NY (1972), Edited by: M. Abramowitz, M., Stegun, I.A.
14. Stewart, J. Essential Calculus: Early Transcendentals 6th. Thomson Higher Education, Belmont CA (2007)
15. Lewis, T., McNulty, J., Neudauer, N.A., Reid, T.J., Sheppardson, L. Bicircular matroid designs. Ars Combinatoria. **110** 513–523 (2013)

An Extension of Wolfram's Rule 90 for One-Dimensional Cellular Automata over Non-Abelian Group Alphabets

Erin Craig and Eirini Poimenidou

Abstract We study one-dimensional cellular automata with local update rule defined by an extension of Wolfram's Rule 90 over non-abelian group alphabets. In particular we develop necessary and sufficient conditions for a state in such an automaton to have a predecessor. We apply our results to compute the fraction of states that are reachable through evolution of an automaton over a finite dihedral group.

1 Introduction

In the study of cellular automata, one is interested in predicting the long term behavior of an automaton based on its local update rules. In the classical case of additive cellular automata, one relies heavily on the fact that an update rule can be thought of as a group homomorphism in the following sense: Given a finite cellular automaton of length W with periodic boundary conditions over an abelian group G, we may think of a particular state of the automaton $S = (d_0, \ldots, d_{W-1})$ with $d_i \in G$ as an element of $G \times G \times \cdots \times G = G^W$ and the update rule as a homomorphism $\phi : G^W \to G^W$. Rule 90 is then given as follows: let $S = (d_0, \ldots, d_{W-1})$ be a state over $(\mathbb{Z}/2\mathbb{Z})^W$, and define

$$\phi : (\mathbb{Z}/2\mathbb{Z})^W \to (\mathbb{Z}/2\mathbb{Z})^W$$

$$\phi(S) = ((d_{W-1} + d_1) \bmod 2, (d_0 + d_2) \bmod 2, \ldots, (d_{W-2} + d_0) \bmod 2).$$

We wish to consider an extension of Rule 90 over a non-abelian group, G. Letting $S = (d_0, \ldots, d_{W-1})$ be a state in G^W, we study the following update rule:

E. Craig · E. Poimenidou (✉)
New College of Florida, Sarasota, FL, USA
e-mail: erin.craig@ncf.edu; poimenidou@ncf.edu

© The Author(s) and the Association for Women in Mathematics 2019
S. D'Agostino et al. (eds.), *A Celebration of the EDGE Program's Impact on the Mathematics Community and Beyond*, Association for Women in Mathematics Series 18, https://doi.org/10.1007/978-3-030-19486-4_20

$$\phi : G^W \to G^W$$

$$\phi(S) = (d_{W-1}d_1, d_0d_2, d_1d_3, d_2d_4, \dots, d_{W-2}d_0).$$

It is clear that, because the d_is do not necessarily commute, ϕ is not in general a homomorphism. As a result, we cannot apply the techniques that have been used in the past, see [1] or [2]. In our study of cellular automata, we develop new techniques to study the behavior of cellular automata over non-abelian groups. In particular, we study the fraction of states in a cellular automaton that have predecessors, that is, we study the fraction of states that are reachable through evolution of the automaton. Our strategy is to develop the necessary and sufficient conditions for a state to have a predecessor. We show that, given a state $S = (d_0, \dots, d_{W-1})$ and at most *two* entries from its predecessor state S', namely c_0 and c_1, for c_i and d_i in a multiplicative group G, we can obtain all entries in S'. We will then show that there are requirements on c_0 and c_1 depending on the entries of S. Hence, given S, we can determine properties that c_0 and c_1 must have, and given c_0 and c_1, we can determine S' in its entirety. If no such c_0 or c_1 exists, then S cannot have a predecessor state and so is unreachable through evolution.

Our main object of study is a one-dimensional finite cellular automaton with periodic boundary conditions over a multiplicative non-abelian group, G. As necessary throughout the paper, we will understand the automaton as being on a length W (discrete) circle or as being periodic on the entire discrete line with period W. As such, given a state $S' = (c_0, \dots, c_{W-1})$ and its successor, $S = (d_0, \dots, d_{W-1})$, we may write our local update rule as $d_{i+1 \bmod W} = c_{i \bmod W} c_{i+2 \bmod W}$ for all $i \in \mathbb{Z}$. This is the rule which will be studied throughout the paper unless otherwise specified.

2   Cells Determined by Their Successors

This section is motivated by the desire to write the entries of a state in terms of the entries in its successor state. This can be thought of as an attempt to move backwards one time step in the evolution of the automaton defined.

Let $S = (d_0, \dots, d_{W-1})$ be a state in the automaton and suppose that its predecessor, $S' = (c_0, \dots, c_{W-1})$, exists. Then, given our choice of update rule, we can write

$$c_a c_{a+2} = d_{a+1} \Rightarrow c_{a+2} = c_a^{-1} d_{a+1}$$

and

$$c_{a+2} c_{a+4} = d_{a+3} \Rightarrow c_{a+4} = c_{a+2}^{-1} d_{a+3}.$$

c_0		c_2			
d_0	d_1	d_2	d_3	d_4	d_5

c_0		$c_2 = c_0^{-1}d_1$		c_4	
d_0	d_1	d_2	d_3	d_4	d_5

c_0		$c_2 = c_0^{-1}d_1$		$c_4 = d_1^{-1}c_0d_3$	
d_0	d_1	d_2	d_3	d_4	d_5

$c_0 = d_0^{-1}d_3^{-1}c_0^{-1}d_1d_5$		$c_2 = c_0^{-1}d_1$		$c_4 = d_1^{-1}c_0d_3$	
d_0	d_1	d_2	d_3	d_4	d_5

Fig. 1 Given c_i from a predecessor state, and its successor state $(d_0, d_1, \ldots, d_5)$, we use our update rule to find an equation for c_0, which, once solved, can help us find c_2 and c_4. In this case, we note that we have no information about c_1, c_3, and c_5; however, we could write an equation for c_1 as we have done for c_0

c_0		c_2		
d_0	d_1	d_2	d_3	d_4

c_0		$c_2 = c_0^{-1}d_1$		c_4
d_0	d_1	d_2	d_3	d_4

c_0	c_1	$c_2 = c_0^{-1}d_1$		$c_4 = d_1^{-1}c_0d_3$
d_0	d_1	d_2	d_3	d_4

c_0	$c_1 = d_3^{-1}c_0^{-1}d_1d_0$	$c_2 = c_0^{-1}d_1$	c_3	$c_4 = d_1^{-1}c_0d_3$
d_0	d_1	d_2	d_3	d_4

c_0	$c_1 = d_3^{-1}c_0^{-1}d_1d_0$	$c_2 = c_0^{-1}d_1$	$c_3 = d_0^{-1}d_1^{-1}c_0d_3d_2$	$c_4 = d_1^{-1}c_0d_3$
d_0	d_1	d_2	d_3	d_4

$c_0 = d_2^{-1}d_3^{-1}c_0^{-1}d_1d_0d_4$	$c_1 = d_3^{-1}c_0^{-1}d_1d_0$	$c_2 = c_0^{-1}d_1$	$c_3 = d_0^{-1}d_1^{-1}c_0d_3d_2$	$c_4 = d_1^{-1}c_0d_3$
d_0	d_1	d_2	d_3	d_4

Fig. 2 Given c_i from a predecessor state, and its successor state $(d_0, d_1, \ldots, d_4)$, we use our update rule to write an equation for c_0. In this case, we have information about all of the elements of the predecessor state

Combining the two, we have $c_{a+4} = d_{a+1}^{-1}c_ad_{a+3}$. In this way, we are able to write $c_{(a+2j) \bmod w}$ in terms of $c_a^{\pm 1}$ and the entries in S for any $j \in \mathbb{N}$. This is illustrated in Figs. 1 and 2.

In Lemmas 1 and 2, we will explicitly describe this property. To prove these lemmas, we will consider the automaton on a length W (discrete) circle as an automaton on the entire discrete line which is periodic with period W.

Lemma 1 *Let c_i, $d_i \in G$, a multiplicative group, and let $S = (\ldots, d_j, d_{j+1}, \ldots)$ and $S' = (\ldots, c_j, c_{j+1}, \ldots)$ be states defined on the entire discrete line in the automaton given, such that S' evolves to S in one time step under the update rule defined. Then, for $a \in \mathbb{Z}$ and $j \in \mathbb{N}$ such that j **odd** and $j \geq 3$,*

$$P(j) : c_{a+2j} = \left(\prod_{l=0}^{\frac{j-1}{2}-1} d_{a+2j-3-4l}^{-1} \right) c_a^{-1} \left(\prod_{l=0}^{\frac{j+1}{2}-1} d_{a+1+4l} \right). \qquad (1.1)$$

Proof We will use induction on j to prove the result. Without loss of generality, we prove only the case when $a = 0$ and we note that adding $\pm a$ to every index shifts each entry a cells to the right or left and does not change the proof. Let S' and S be as given in the statement of the lemma. To see that $P(3)$ is true, we apply the update rule to S':

$$c_0 c_2 = d_1 \Rightarrow c_2 = c_0^{-1} d_1$$

$$c_2 c_4 = d_3 \Rightarrow c_4 = c_2^{-1} d_3 = d_1^{-1} c_0 d_3$$

$$c_4 c_6 = d_5 \Rightarrow c_6 = c_4^{-1} d_5 = d_3^{-1} c_0^{-1} d_1 d_5.$$

And so $P(3)$ is true. Suppose now that $P(k)$ is true for k odd. Then,

$$c_{2k} c_{2(k+1)} = d_{2k+1} \Rightarrow c_{2(k+1)} = c_{2k}^{-1} d_{2k+1}$$

$$c_{2(k+1)} c_{2(k+2)} = d_{2k+3} \Rightarrow c_{2(k+2)} = c_{2(k+1)}^{-1} d_{2k+3} = d_{2k+1}^{-1} c_{2k} d_{2k+3}.$$

By the inductive hypothesis,

$$c_{2(k+2)} = d_{2k+1}^{-1} \left(\prod_{l=0}^{\frac{k-1}{2}-1} d_{2k-3-4l}^{-1} \right) c_0^{-1} \left(\prod_{l=0}^{\frac{k+1}{2}-1} d_{1+4l} \right) d_{2k+3}$$

$$= d_{2(k+2)-3}^{-1} \left(\prod_{l=0}^{\frac{k+1}{2}-2} d_{2(k+2)-3-4l}^{-1} \right) c_0^{-1} \left(\prod_{l=0}^{\frac{k+1}{2}-1} d_{1+4l} \right) d_{1+4\left(\frac{k+3}{2}-1\right)}$$

$$= \left(\prod_{l=0}^{\frac{(k+2)-1}{2}-1} d_{2k-3-4l}^{-1} \right) c_0^{-1} \left(\prod_{l=0}^{\frac{(k+2)+1}{2}-1} d_{1+4l} \right).$$

Hence, $P(3)$ is true and $P(k) \Rightarrow P(k+2)$ so $P(j)$ is true for all $j \in \mathbb{N}$, j odd, and $j \geq 3$.

A nearly identical proof gives the following lemma.

Lemma 2 *Let c_i, $d_i \in G$, a multiplicative group, and let $S = (\ldots, d_j, d_{j+1}, \ldots)$ and $S' = (\ldots, c_j, c_{j+1}, \ldots)$ be states defined on the entire discrete line in the automaton given, such that S' evolves to S in one time step under the update rule defined. Then, for $a \in \mathbb{Z}$ and $j \in \mathbb{N}$ such that j **even** and $j \geq 2$,*

$$P(j): c_{a+2j} = \left(\prod_{l=0}^{\frac{j}{2}-1} d_{a+2j-3-4l}^{-1} \right) c_a \left(\prod_{l=0}^{\frac{j}{2}-1} d_{a+3+4l} \right). \tag{1.2}$$

Recalling that for $j \in \mathbb{N}$,

$$\left\lfloor \frac{j}{2} \right\rfloor = \begin{cases} \frac{j-1}{2} & j \text{ odd} \\ \frac{j}{2} & j \text{ even} \end{cases}$$

$$\left\lceil \frac{j}{2} \right\rceil = \begin{cases} \frac{j+1}{2} & j \text{ odd} \\ \frac{j}{2} & j \text{ even} \end{cases}$$

we combine Lemmas 1 and 2 to prove the following theorem.

Theorem 1 *Let c_i, $d_i \in G$, a multiplicative group, and let $S = (\ldots, d_j, d_{j+1}, \ldots)$ and $S' = (\ldots, c_j, c_{j+1}, \ldots)$ be states defined on the entire discrete line. Then S' evolves to S in one time step iff for all $a \in \mathbb{Z}$ and $j \in \mathbb{N}$ $(j \geq 2)$,*

$$P(j): c_{a+2j} = \left(\prod_{l=0}^{\lfloor \frac{j}{2} \rfloor - 1} d_{a-3+2j-4l}^{-1} \right) c_a^{(-1)^j} \left(\prod_{l=0}^{\lceil \frac{j}{2} \rceil - 1} d_{a+1+2(j+1 \bmod 2)+4l} \right).$$

Proof We can see that S' evolves to S in one time step using Lemmas 1 and 2. Conversely, suppose that $P(j)$ is always satisfied. We want to show that $c_{a+2k} c_{a+2(k+1)} = d_{a+2k+1}$. Without loss of generality, we will choose $a = 0$ and suppose that k is even. Then

$$c_{2k} = \left(\prod_{l=0}^{\frac{k}{2}-1} d_{2k-3-4l}^{-1} \right) c_0 \left(\prod_{l=0}^{\frac{k}{2}-1} d_{3+4l} \right),$$

and

$$c_{2(k+1)} = \left(\prod_{l=0}^{\frac{k}{2}-1} d^{-1}_{2(k+1)-3-4l} \right) c_0^{-1} \left(\prod_{l=0}^{\frac{k+2}{2}-1} d_{1+4l} \right).$$

So,

$$c_{2k} c_{2(k+1)} = \left(\prod_{l=0}^{\frac{k}{2}-1} d^{-1}_{2k-3-4l} \right) c_0 \left(\prod_{l=0}^{\frac{k}{2}-1} d_{3+4l} \right) \left(\prod_{l=0}^{\frac{k}{2}-1} d^{-1}_{2(k+1)-3-4l} \right) c_0^{-1} \left(\prod_{l=0}^{\frac{k+2}{2}-1} d_{1+4l} \right).$$

We will show first that

$$\left(\prod_{l=0}^{\frac{k}{2}-1} d_{3+4l} \right) = \left(\prod_{l=0}^{\frac{k}{2}-1} d^{-1}_{2(k+1)-3-4l} \right)^{-1},$$

by showing that their product is equal to 1

$$\left(\prod_{l=0}^{\frac{k}{2}-1} d_{3+4l} \right) \left(\prod_{l=0}^{\frac{k}{2}-1} d^{-1}_{2(k+1)-3-4l} \right) = \left(d_3 d_7 \dots d_{3+4\left(\frac{k}{2}-1\right)} \right)$$

$$\times \left(d^{-1}_{2(k+1)-3} d^{-1}_{2(k+1)-7} \dots d^{-1}_{2(k+1)-3-4\left(\frac{k}{2}-1\right)} \right)$$

$$= (d_3 d_7 \dots d_{2k-1}) \left(d^{-1}_{2k-1} d^{-1}_{2k-5} \dots d^{-1}_3 \right)$$

$$= 1$$

as desired. It remains to show that

$$\left(\prod_{l=0}^{\frac{k}{2}-1} d^{-1}_{2k-3-4l} \right) \left(\prod_{l=0}^{\frac{k+2}{2}-1} d_{1+4l} \right) = d_{2k+1}.$$

We expand the products and compute to find

$$\left(\prod_{l=0}^{\frac{k}{2}-1} d^{-1}_{2k-3-4l} \right) \left(\prod_{l=0}^{\frac{k+2}{2}-1} d_{1+4l} \right) = \left(d^{-1}_{2k-3} \dots d^{-1}_{2k-3-4\left(\frac{k}{2}-2\right)} d^{-1}_{2k-3-4\left(\frac{k}{2}-1\right)} \right)$$

$$\times \left(d_1 d_5 \dots d_{1+4\left(\frac{k+2}{2}-2\right)} d_{1+4\left(\frac{k+2}{2}-1\right)} \right)$$

$$= \left(d_{2k-3}^{-1} \ldots d_5^{-1} d_1^{-1}\right) (d_1 d_5 \ldots d_{2k-3} d_{2k+1})$$

$$= d_{2k+1}$$

as desired. This completes the proof.

Knowing now how to describe specific entries in a state using entries from its successor state, we move forward to describe an entire state in terms of its successor.

3 States Determined by Their Successors

As before, let $S' = (c_0, \ldots, c_{W-1})$ evolve to $S = (d_0, \ldots, d_{W-1})$ in one time step. The value of Theorem 1 lies in the fact that it allows us to write c_{a+2j} in terms only of c_a and the entries in the successor state, S. Taking the indices of the elements of S and S' modulo W, we note that, when $j = W$ or $2j = W$, we can write c_a nontrivially in terms of itself and the entries of S. Assuming the entries of S are known and c_a is unknown, then Theorem 1 gives a formula for c_a. If it can be solved, then c_a and every entry of S' indexed by $a + 2k$ for $k \in \mathbb{N}$ exist. If not, then there is no solution for c_a or any entry of S' indexed by $a + 2k$ for $k \in \mathbb{N}$. Hence, for W odd, the existence of c_a guarantees the existence of every element of S'. For W even, the existence of c_a guarantees only the existence of the elements of S' that have the same parity as a. This property is illustrated in Figs. 1 and 2. Below, we take a closer look at the case when $W = 6$.

Example 1 $W = 6, a = 0$

$$c_0 c_2 = d_1 \Rightarrow c_2 = c_0^{-1} d_1$$

$$c_2 c_4 = d_3 \Rightarrow c_4 = c_2^{-1} d_3 = d_1^{-1} c_0 d_3$$

$$c_4 c_0 = d_5 \Rightarrow c_0 = c_4^{-1} d_5 = d_3^{-1} c_0^{-1} d_1 d_5.$$

As before, we take the indices in Theorem 1 modulo $W = 6$. In this example, we see that c_0 can only describe c_0, c_2, and c_4, and the only elements from S necessary to do so are d_1, d_3, and d_5. To describe c_1, c_3, and c_5, we proceed as before

$$c_1 c_3 = d_2 \Rightarrow c_3 = c_1^{-1} d_2$$

$$c_3 c_5 = d_4 \Rightarrow c_5 = c_3^{-1} d_4 = d_2^{-1} c_1 d_4$$

$$c_5 c_1 = d_0 \Rightarrow c_1 = c_5^{-1} d_0 = d_4^{-1} c_1^{-1} d_2 d_0.$$

Similarly, c_1 can only describe c_1, c_3, and c_5, and the only elements from S necessary to do so are d_0, d_2, and d_4.

The example above motivates the development of notation to discuss which elements from S are needed to describe the elements of S'. We note that $S' = (c_0, \ldots, c_5)$ can be understood as two sub-states, $S'_0 = (c_0, c_2, c_4)$ and $S'_1 = (c_1, c_3, c_5)$, that evolve independent of each other and we can define $S_0 = (d_1, d_3, d_5)$ and $S_1 = (d_0, d_2, d_4)$ to correspond accordingly.[1] With a slight abuse of terminology, we will refer to the two independently evolving sub-states that arise when the automaton has even length, as "states ". We define a reduced state as follows.

Definition 1 A reduced state, T_a, is a state such that its predecessor state, T'_a, if it exists, can be completely generated by the elements of T_a and the ath entry of T'_a.

Definition 2 A reduced predecessor state, T'_a, is a state whose successor state, T_a, is a reduced state.

For simplicity, we will always take $a = 0$. For automata of odd length, all states are reduced: as noted, their predecessor states can be generated completely by the entries of the state itself and c_0. For automata of even length, a state $S = (d_0, \ldots, d_{W-1})$ is a combination of two reduced states, $S_0 = (d_1, d_3, \ldots, d_{W-1})$ with reduced predecessor $S'_0 = (c_0, c_2, \ldots, c_{W-2})$ and $S_1 = (d_0, d_2, \ldots, d_{W-2})$ with reduced predecessor $S'_1 = (c_1, c_3, \ldots, c_{W-1})$.

Example 2 Let $G = D_6 = \langle r, s \mid r^3 = s^2 = 1, srs = r^2 \rangle$, and let $S' = (s, rs, r, r)$ be a state in the automaton defined such that

$$S' = (s, rs, r, r) \to (r^2s, r^2s, s, rs) = S.$$

The corresponding reduced states are as follows: $S'_0 = (s, r)$, with $S_0 = (r^2s, rs)$, and $S'_1 = (rs, r)$, with $S_1 = (r^2s, s)$.

4 Summary

The notation developed in Sects. 2 and 3 gives us a way to understand a state of the automaton defined in terms only of one of its entries and the entries of its successor. Corollaries 1, 2, and 3 describe when a state has a predecessor using the work and notation developed in Sects. 2 and 3.

---

[1]It is important to mention that S'_0 evolves to S_0 under the update rule which maps

$$(c_0, c_1, c_2) \to (c_0c_1, c_1c_2, c_2c_0).$$

Hence, the study of automata with even length, W, under the original update rule corresponds to study of automata of length $\frac{W}{2}$ under the update rule written above. Corresponding results are inherent within the paper. While we note this here, we do not discuss it again explicitly.

Corollary 1 *Let $W \in \mathbb{N}$ such that $2 \nmid W$. Let $S = (d_0, \ldots, d_{W-1})$ be a state in the automaton defined over a multiplicative group, G. Then S has a predecessor $S' = (c_0, \ldots, c_{W-1})$ iff there exists $c_0 \in G$ satisfying*

$$c_0 \left(\prod_{k=0}^{\frac{W-1}{2}-1} d_{(3+4k) \bmod W} \right) c_0 = \prod_{k=0}^{\frac{W+1}{2}-1} d_{(1 \mid 4k) \bmod W}. \tag{1.3}$$

Proof For this proof, we will understand S and S' as states in the automaton defined on the entire discrete line that are periodic with period W. For simplicity, we will write the states as having length W.

Suppose first that S has a predecessor $S' = (c_0, \ldots, c_{W-1})$. Then, by application of Lemma 1 for $j = W$, we see that c_0 satisfies Eq. (1.3).

Conversely, suppose that there exists $c_0 \in G$ satisfying Eq. (1.3) and let $S = (d_0, \ldots, d_{W-1})$. Construct $S' = (c_0, \ldots, c_{W-1})$ such that $c_{i \bmod W} c_{i+2 \bmod W} = d_{i+1 \bmod W}$ using the following method:

$$c_0 c_2 = d_1 \Rightarrow c_2 = c_0^{-1} d_1$$

$$c_2 c_4 = d_3 \Rightarrow c_4 = d_1^{-1} c_0 d_3$$

$$\vdots$$

$$c_{W-1} c_1 = d_0 \Rightarrow c_1 = c_{W-1}^{-1} d_0$$

$$\vdots$$

and so on. Because we have already chosen c_0, we can work this way only until $c_{W-2} c_0 = d_{W-1} \Rightarrow c_0 = c_{W-2} d_{W-1}$. All that remains is to show that the c_0 chosen satisfies $c_0 = c_{W-2} d_{W-1}$. To do so, we will show that $c_{W-2} c_0 = d_{W-1}$. We rewrite Eq. (1.3) to find

$$c_0 = \left(\prod_{k=0}^{\frac{W-1}{2}-1} d_{(2W-3-4k) \bmod W}^{-1} \right) c_0^{-1} \left(\prod_{k=0}^{\frac{W+1}{2}-1} d_{(1+4k) \bmod W} \right).$$

By construction of $c_{(W-2)}$, we can use Lemma 2 to see that

$$c_{(W-2)} = c_{2(W-1) \bmod W} = \left(\prod_{k=0}^{\frac{W-1}{2}-1} d_{(2(W-1)-3-4k) \bmod W}^{-1} \right) c_0 \left(\prod_{k=0}^{\frac{W-1}{2}-1} d_{(3+4k) \bmod W} \right).$$

Hence,

$$
\begin{aligned}
c_{W-2}c_0 &= \left(d_{2W-5}^{-1}\dots d_5^{-1}d_1^{-1}c_0 d_3 d_7 \dots d_{2W-3}\right) \\
&\quad \times \left(d_{2W-3}^{-1}\dots d_7^{-1}d_3^{-1}c_0^{-1}d_1 d_5 \dots d_{2W-5}d_{2W-1}\right) \\
&= \left(d_{W-5}^{-1}\dots d_5^{-1}d_1^{-1}c_0 d_3 d_7 \dots d_{W-3}\right) \\
&\quad \times \left(d_{W-3}^{-1}\dots d_7^{-1}d_3^{-1}c_0^{-1}d_1 d_5 \dots d_{W-5}d_{W-1}\right) \\
&= d_{W-1}
\end{aligned}
$$

as desired.

A similar proof reveals the following two corollaries.

Corollary 2 *Let $W \in \mathbb{N}$ such that $W = 2m$ and $2 \nmid m$. Let $S = (d_1, d_3, \dots, d_{W-1})$ be a reduced state in the automaton defined over a multiplicative group, G. Then S has a predecessor $S' = (c_0, c_2, \dots, c_{W-2})$ iff there exists $c_0 \in G$ satisfying*

$$
c_0 \left(\prod_{k=0}^{\frac{m-1}{2}-1} d_{(3+4k)\bmod W} \right) c_0 = \prod_{k=0}^{\frac{m+1}{2}-1} d_{(1+4k)\bmod W}. \tag{1.4}
$$

We note here that, for a state of the form given in Corollary 2 to have a predecessor, there must also exist a $c_1 \in G$ satisfying

$$
c_1 \left(\prod_{k=0}^{\frac{m-1}{2}-1} d_{(4+4k)\bmod W} \right) c_1 = \prod_{k=0}^{\frac{m+1}{2}-1} d_{(2+4k)\bmod W}. \tag{1.5}
$$

This equation is in accordance with Theorem 1 for $a = 1$.

Corollary 3 *Let $W \in \mathbb{N}$ such that $W = 2m$ and $2 \mid m$. Let $S = (d_1, d_3, \dots, d_{W-1})$ be a reduced state in the automaton defined over a multiplicative group, G. Then S has a predecessor $S' = (c_0, c_2, \dots, c_{W-2})$ iff there exists $c_0 \in G$ satisfying*

$$
\left(\prod_{k=0}^{\frac{m}{2}-1} d_{(1+4k)\bmod W} \right) c_0 = c_0 \left(\prod_{k=0}^{\frac{m}{2}-1} d_{(3+4k)\bmod W} \right). \tag{1.6}
$$

As in Corollary 2, a state of the form given in Corollary 3 has a predecessor if there also exists a $c_1 \in G$ satisfying

$$\left(\prod_{k=0}^{\frac{m}{2}-1} d_{(2+4k) \bmod W} \right) c_1 = c_1 \left(\prod_{k=0}^{\frac{m}{2}-1} d_{(4+4k) \bmod W} \right). \tag{1.7}$$

Again, this equation is in accordance with Theorem 1 for $a = 1$.

5   An Automaton Over the Dihedral Group

In this section, we study the automaton defined in Sects. 3 and 4 over the dihedral group of order $2n$ generated by a rotation r and a flip s, namely $D_{2n} = \langle r, s \mid r^n = s^2 = 1, srs = r^{-1} \rangle$. By writing solutions for the equations appearing in Corollaries 1, 2, and 3 with entries from D_{2n}, we describe the fraction of states that have at least one predecessor.

We note here that the equations from Corollaries 1 and 2 have the same structure and so we will study all reduced states of odd length concurrently. The equation for a reduced state of even length seen in Corollary 3 will be treated separately. However, to simplify all the aforementioned equations, we will use the following property about the elements of D_{2n}.

Lemma 3 *Let D_{2n} be generated by a rotation, r, and flip, s, such that $r^n = 1 = s^2$ and $srs = r^{-1}$. Then for $a \in \mathbb{Z}/n\mathbb{Z}$ and $b \in \mathbb{Z}/2\mathbb{Z}$,*

$$s^b r^a = r^{(a-2ab) \bmod n} s^b.$$

Proof Either $b = 0$ or $b = 1$. Suppose first that $b = 0$. Then $r^a = s^0 r^a = r^{a-2a(0)} s^0 = r^a$. If $b = 1$, then $s^1 r^a = r^{-a \bmod n} s^1 = r^{a-2a(1) \bmod n} s^1$ as expected.

We proceed now to simplify Eqs. (1.3) and (1.4). It is important first to develop notation that will be used throughout the paper.

Note 1 Throughout the paper, we use the following to define i_1, i_2, j_1, and j_2. For a reduced state of odd width, W,

$$r^{i_1} s^{j_1} = \prod_{k=0}^{\frac{W-1}{2}-1} d_{(3+4k) \bmod W}$$

$$r^{i_2} s^{j_2} = \prod_{k=0}^{\frac{W+1}{2}-1} d_{(1+4k) \bmod W}.$$

When the state has even width, $W = 2m$ such that $2 \nmid m$, we use

$$r^{i_1} s^{j_1} = \prod_{k=0}^{\frac{m-1}{2}-1} d_{(3+4k) \bmod W}$$

$$r^{i_2} s^{j_2} = \prod_{k=0}^{\frac{m+1}{2}-1} d_{(1+4k) \bmod W}.$$

Finally, when a state has width $W = 2m$ such that $2 \mid m$, we use

$$r^{i_1} s^{j_1} = \prod_{k=0}^{\frac{m}{2}-1} d_{(1+4k) \bmod W}$$

$$r^{i_2} s^{j_2} = \prod_{k=0}^{\frac{m}{2}-1} d_{(3+4k) \bmod W}.$$

Using notation from Note 1, we simplify Eqs. (1.3) and (1.4). We will write them as

$$r^x s^y r^{i_1} s^{j_1} r^x s^y = r^{i_2} s^{j_2},$$

where $c_0 = r^x s^y$. Using Lemma 3, we rewrite the above as

$$r^{(x+i_1-2i_1y) \bmod n} s^{(j_1+y) \bmod 2} r^x s^y = r^{i_2} s^{j_2}$$

$$r^{(x+i_1-2i_1y+x-2x(j_1+y \bmod 2)) \bmod n} s^{(j_1+2y) \bmod 2} = r^{i_2} s^{j_2}$$

$$r^{(x+i_1-2i_1y+x-2x(j_1+y \bmod 2)) \bmod n} s^{j_1 \bmod 2} = r^{i_2} s^{j_2}$$

$$r^{(i_1+2x-2x(j_1+y \bmod 2)-2i_1y) \bmod n} s^{j_1 \bmod 2} = r^{i_2} s^{j_2}.$$

So, for a reduced state of odd length to have a corresponding reduced predecessor, its elements must satisfy the requirement

$$j_1 \equiv j_2 \bmod 2. \tag{1.8}$$

Additionally, there must exist a pair $x \in \mathbb{Z}/n\mathbb{Z}$ and $y \in \mathbb{Z}/2\mathbb{Z}$ satisfying

$$2x(1 - (j_1 + y \bmod 2)) - 2i_1 y \equiv i_2 - i_1 \bmod n. \tag{1.9}$$

We now repeat the above for a reduced state of even length. We write Eq. (1.6) as

$$r^{i_1}s^{j_1}r^x s^y = r^x s^y r^{i_2} s^{j_2}.$$

Again, $c_0 = r^x s^y$. We use Lemma 3 to see that

$$r^{i_1+x-2xj_1 \bmod n} s^{j_1+y \bmod 2} = r^{x+i_2-2i_2y \bmod n} s^{j_2+y \bmod 2}$$

$$r^{2i_2y-2xj_1 \bmod n} s^{j_1} = r^{i_2 \ i_1 \bmod n} s^{j_2}.$$

As before, for a reduced state S to have a corresponding reduced predecessor, its elements must satisfy $j_1 \equiv j_2 \bmod 2$ (Eq. (1.8)). There must also exist a pair $x \in \mathbb{Z}/n\mathbb{Z}$ and $y \in \mathbb{Z}/2\mathbb{Z}$ satisfying

$$2i_2y - 2xj_1 \equiv i_2 - i_1 \bmod n. \qquad (1.10)$$

Examples illustrating the above calculations can be found in Sects. 6 of the paper.

Having addressed all cases, we begin to interpret these requirements with the goal of stating the fraction of states that are reachable through evolution of the automaton over D_{2n}. For the reader's convenience, we restate and rephrase Wolfram, Martin, and Odlyzko's theorem for the automaton over $\mathbb{Z}/2\mathbb{Z}$ before stating our analogous theorem for the automaton over D_{2n}.

Theorem 2 (Theorem 3.1 in [2]) *The fraction of the 2^N possible configurations of a size N cellular automaton defined [by the update rule given] which can be reached by evolution is $1/2$ for N odd and $1/4$ for N even.*

Theorem 3 *The fraction of the $(2n)^W$ possible states of a size W cellular automaton defined in Sect. 5 over the dihedral group of $2n$ elements which can be reached through evolution of the automaton is given in the table below (Table 1).*

Knowing now what requirements a reduced state must satisfy to have a predecessor, we count the fraction of states that satisfy these requirements. Because both the W odd and W even case require a state S to satisfy Eq. (1.8), we interpret this requirement first. Some preliminary lemmas are needed before we can prove Theorem 3.

Lemma 4 *The product of $2k + 1$ elements in $D_{2n} - \langle r \rangle$ does not lie in $\langle r \rangle$.*

Proof Let

$$\phi : D_{2n} \to D_{2n}/\langle r \rangle$$

Table 1 Fraction of states with predecessors in automata of length W over finite dihedral groups

	n odd	n even
W odd	$\frac{1}{2}$	$\frac{1}{4}$
$W = 2m$, m odd	$\frac{1}{4}$	$\frac{1}{16}$
$W = 2m$, m even	$\frac{(n(n+2)-1)^2}{16n^4}$	$\frac{(n(n+4)-4)^2}{64n^4}$

be the natural projection homomorphism so that $\phi(r) = 0$ and $\phi(s) = 1$. Define $\{a_i\}_{i=1}^{2k+1}$ such that $a_i \in D_{2n} - \langle r \rangle$ for all i. Then

$$\phi\left(\prod_{i=1}^{2k+1} a_i\right) = \phi(a_1) + \phi(a_2) + \cdots + \phi(a_{2k+1}) = 1 + 1 + \cdots + 1 = 1.$$

It follows that $\prod_{i=1}^{2k+1} a_i$ lies in $D_{2n} - \langle r \rangle$.

Lemma 5 *Let* $S = (d_0, \ldots, d_{W-1})$ *be a reduced state. Then, using the notation from Note 1,* $j_1 = j_2$ *iff* S *has an even number of terms that lie in* $D_{2n} - \langle r \rangle$.

Proof Suppose first that S is a reduced state and $j_1 = j_2$. Then using the notation from Note 1, we consider the product of all elements of S as $r^{i_1} s^{j_1} r^{i_2} s^{j_2}$. Using Lemma 3, we write this product as $r^{i_1 + i_2 - 2i_2 j_1 \bmod n} s^{j_1 + j_2 \bmod 2}$. Since $j_1 = j_2$, this product lies in $\langle r \rangle$, and by Lemma 4, the product can be written only as products of an even number of terms in $D_{2n} - \langle r \rangle$. Hence, $j_1 = j_2 \Rightarrow S$ has an even number of terms in $D_{2n} - \langle r \rangle$.

Conversely, suppose that S has an even number of terms in $D_{2n} - \langle r \rangle$. Then any product of all of its elements lies in $\langle r \rangle$; specifically, the product $r^{i_1} s^{j_1} r^{i_2} s^{j_2} = r^{i_1 + i_2 - 2i_2 j_1 \bmod n} s^{j_1 + j_2 \bmod 2}$ lies in $\langle r \rangle$. From this, we see clearly that $j_1 + j_2 \equiv 0 \bmod 2$ and so $j_1 \equiv j_2 \bmod 2$. Since j_1 and j_2 are in $\mathbb{Z}/2\mathbb{Z}$, $j_1 \equiv j_2 \bmod 2 \Rightarrow j_1 = j_2$. Hence, if S has an even number of terms in $D_{2n} - \langle r \rangle$, then $j_1 = j_2$. This completes the proof.

So, if a reduced state does not have an even number of terms from $D_{2n} - \langle r \rangle$, then it cannot have a corresponding predecessor state. The lemma below counts the number of states which have an even number of terms from $D_{2n} - \langle r \rangle$.

Lemma 6 *For* $W \in \mathbb{N}$, *the fraction of the* $(2n)^W$ *states in the cellular automaton defined in Sect. 5 over* D_{2n} *which have an even number of terms in* $D_{2n} - \langle r \rangle$ *is* $1/2$.

Proof Let $W \in \mathbb{N}$, and let $S = (d_0, \ldots, d_{W-1})$ denote a reduced state in the automaton with an even number of terms in $D_{2n} - \langle r \rangle$. Define E_W as the set of all states of length W with an even number of terms in $D_{2n} - \langle r \rangle$ and O_W as the set of all states of length W with an odd number of terms in $D_{2n} - \langle r \rangle$ and let

$$\phi : E_W \to O_W$$

$$\phi(S) = (sd_0, d_1, \ldots, d_{W-1}).$$

It is easy to see that ϕ is well defined. We show that ϕ is a bijection. Let

$$\psi : O_W \to E_W$$

$$\psi(S) = (sd_0, d_1, \ldots, d_{W-1}).$$

Then $\psi(\phi(S)) = \psi(s * d_0, d_1, \ldots, d_{W-1}) = (ssd_0, d_1, \ldots, d_{W-1}) = S$, so ϕ has an inverse and hence is a bijection. It follows that the fraction of the $(2n)^W$ states in the automaton defined in Sect. 5 over D_{2n} which have an even number of terms in $D_{2n} - \langle r \rangle$ is $1/2$.

Hence, exactly $1/2$ of states do not satisfy $j_1 = j_2$, and so at least $1/2$ of states cannot have predecessors. It is important now to examine the previous statement in a different light. We can understand this statement by recalling that D_{2n} is the semidirect product of $\mathbb{Z}/n\mathbb{Z}$ and $\mathbb{Z}/2\mathbb{Z}$, and so we can gain insight from the study of the automata over $\mathbb{Z}/2\mathbb{Z}$. In [2], Wolfram, Martin, and Odlyzko prove the following lemma:

Lemma 7 (Lemma 3.1 in [2]) *Configurations containing an odd number of sites with value 1 can never be generated in the evolution of the cellular automaton defined [by the update rule given], and can occur only as initial states.*

To make the connection between the automaton over $\mathbb{Z}/2\mathbb{Z}$ and the automaton over D_{2n}, we take a reduced state S over D_{2n} and write each entry modulo $\langle r \rangle$ to obtain a new state, $S_{\mathbb{Z}/2\mathbb{Z}}$, which lies in the automata over $\mathbb{Z}/2\mathbb{Z}$. Hence, if $S_{\mathbb{Z}/2\mathbb{Z}}$ does not have a predecessor, neither does S. Given Lemma 7, it follows that S can only have a predecessor if it has an even number of sites with value 1 from the $\mathbb{Z}/2\mathbb{Z}$ component of D_{2n}. This correlates to a state having an even number of sites with value from $D_{2n} - \langle r \rangle$. Hence, at most, $1/2$ of states in the automaton defined have predecessors.

We now address the other requirement for a reduced state S to have a predecessor: the necessity for the existence of a pair $x \in \mathbb{Z}/n\mathbb{Z}$ and $y \in \mathbb{Z}/2\mathbb{Z}$ to satisfy Eqs. (1.9) or (1.10). To solve these equations, we will call upon a well-known result from number theory.

Theorem 4 (Linear Congruence) *Let $a, b \in \mathbb{Z}$, let $n \in \mathbb{N}$, and let $\gcd(a, n) = d$. Then*

$$ax \equiv b \bmod n$$

has a solution for x iff $d \mid b$. If $d \mid b$ and x_0 is a solution of $ax \equiv b \bmod n$, then the set

$$\left\{ x_0 + k\frac{n}{d} \mid k \in \mathbb{Z} \right\}$$

is the set of all solutions for x. This set will reduce to d solutions modulo n.

Before proceeding with the proof of Theorem 3 we need one more lemma.

Lemma 8 *For $n \in \mathbb{N}$ such that $2 \mid n$, the fraction of pairs (i_1, i_2) such that $i_1, i_2 \in \mathbb{Z}/n\mathbb{Z}$ and $i_1 \equiv i_2 \bmod 2$ is $\frac{1}{2}$.*

Proof Let $n \in \mathbb{N}$ such that $2 \mid n$. Let A be the set of all pairs (a, b) such that $a, b \in \mathbb{Z}/n\mathbb{Z}$ and $a \equiv b \bmod 2$, and let B be the set of pairs (c, d) such that $c, d \in \mathbb{Z}/n\mathbb{Z}$ and $c \not\equiv d \bmod 2$. Define

$$\phi : A \to B$$

$$\phi(a, b) = (a + 1 \bmod n, b).$$

We show that ϕ is a bijection. Define

$$\psi : B \to A$$

$$\psi(c, d) = (c - 1 \bmod n, d).$$

Then $\psi(\phi(a, b)) = \psi(a + 1 \bmod n, b) = (a, b)$ so ϕ has an inverse and hence is a bijection. It follows that the fraction of pairs (i_1, i_2) such that $i_1, i_2 \in \mathbb{Z}/n\mathbb{Z}$ and $i_1 \equiv i_2 \bmod 2$ is $\frac{1}{2}$.

We begin now the proof of Theorem 3.

Proof (of Theorem 3) Using notation from Note 1, we know that a reduced state must satisfy $j_1 = j_2$ in order to have a predecessor. Additionally, we must be able to solve Eqs. (1.9) and (1.10) for x and y.

Supposing that $j_1 = j_2$ is satisfied, we proceed by solving Eq. (1.9)

$$2x(1 - (y + j_1) \bmod 2) - 2yi_1 \equiv i_2 - i_1 \bmod n$$

for $x \in \mathbb{Z}/n\mathbb{Z}$ and $y \in \mathbb{Z}/2\mathbb{Z}$.

To find x and y that satisfy the above, we note that there are two cases: $j_1 = 0$ and $j_1 = 1$. First, let $j_1 = 0$. Then Eq. (1.9) becomes

$$2x(1 - y) - 2yi_1 \equiv i_2 - i_1 \bmod n.$$

Because $y \in \mathbb{Z}/2\mathbb{Z}$, all solutions are of the form $(x, 0)$ or $(x, 1)$. This allows us to solve for x by letting $y = 0$ and then letting $y = 1$. Suppose $y = 0$. Then Eq. (1.9) becomes

$$2x \equiv i_2 - i_1 \bmod n. \tag{1.11}$$

For n such that $2 \nmid n$, 2 has a multiplicative inverse modulo n, and so Eq. (1.11) reduces to and has the solution $x \equiv 2^{-1}(i_2 - i_1) \bmod n$. By the Linear Congruence Theorem, this is the unique solution. When $2 \mid n$, then 2 has no multiplicative inverse modulo n and Eq. (1.11) cannot be reduced further. Applying the Linear Congruence Theorem, Eq. (1.11) has a solution iff $2 \mid i_2 - i_1$. If $2 \mid i_2 - i_1$, then Eq. (1.11) has two solutions, namely $x \equiv \frac{i_2 - i_1}{2} \bmod n$ and $x \equiv \frac{i_2 - i_1}{2} + \frac{n}{2} \bmod n$.

When $y = 1$, then Eq. (1.9) becomes

$$-2i_1 \equiv i_2 - i_1 \bmod n$$

$$0 \equiv i_2 + i_1 \bmod n. \tag{1.12}$$

In this case, there is no dependence on x. Hence, if $0 \equiv i_2 + i_1 \bmod n$, then the set

$$\{x = k \mid k \in \mathbb{Z}/n\mathbb{Z}\}$$

is the set of solutions for x. This completes the case for $j_1 = 0$.

Letting $j_1 = 1$, Eq. (1.9) reduces to

$$2x(1 - (y + 1) \bmod 2) - 2yi_1 \equiv i_2 - i_1 \bmod n.$$

Proceeding as before, we suppose that $y = 0$. Then we can reduce the above to

$$0 \equiv i_2 - i_1 \bmod n. \tag{1.13}$$

As in Eq. (1.12), there is no dependence on x. Hence, if $0 \equiv i_2 - i_1 \bmod n$, then the set

$$\{x = k \mid k \in \mathbb{Z}/n\mathbb{Z}\}$$

is the set of solutions for x.

When $y = 1$, Eq. (1.9) becomes

$$2x - 2i_1 \equiv i_2 - i_1 \bmod n$$

$$2x \equiv i_2 + i_1 \bmod n. \tag{1.14}$$

For n such that $2 \nmid n$, 2 has a multiplicative inverse modulo n, and so Eq. (1.14) reduces to and has the solution $x \equiv 2^{-1}(i_2 + i_1) \bmod n$. By the Linear Congruence Theorem, this is the unique solution. When $2 \mid n$, then 2 has no multiplicative inverse modulo n and Eq. (1.14) cannot be reduced further. Applying the Linear Congruence Theorem, Eq. (1.14) has a solution iff $2 \mid i_2 + i_1$. If $2 \mid i_2 + i_1$, then Eq. (1.14) has two solutions, namely $x \equiv \frac{i_2+i_1}{2} \bmod n$ and $x \equiv \frac{i_2+i_1}{2} + \frac{n}{2} \bmod n$.

For the reader's convenience, these solutions are summarized in Table 2.

With the goal of describing the fraction of states in this automaton that have predecessors, we use Table 2 to count how often Eq. (1.9) has a solution. Having already counted the fraction of states that satisfy $j_1 = j_2$, we can now complete the study of states of length W such that $4 \nmid W$.

Suppose first that W is odd. Then for n odd, all states satisfying $j_1 = j_2$ have at least one predecessor and so $\frac{1}{2}$ of states have at least one predecessor. When n is even, only the states which satisfy $j_1 = j_2$ and $i_1 \equiv i_2 \bmod 2$ have at least one predecessor. By Lemma 8, $\frac{1}{2}$ of states satisfy $i_1 \equiv i_2 \bmod 2$. So, when W is odd and n is even, the fraction of all states that have at least one predecessor is $1/4$.

When W is even such that $4 \nmid W$, we recall that a state S is composed of two reduced states that need to concurrently satisfy the same requirements as those in the case for W odd. Hence, we square the probabilities above to see that when W is

Table 2 $2x(1 - (y + j_1) \bmod 2) - 2yi_1 \equiv i_2 - i_1 \bmod n$

$j_1 = 1$	$2 \nmid n$	$i_2 \equiv i_1 \bmod n$	$i_2 \equiv i_1 \bmod 2$	$x \equiv 2^{-1}(i_2 + i_1) \bmod n, y = 1$ $\{x = k \mid k \in \mathbb{Z}/n\mathbb{Z}\}, y = 0$
			$i_2 \not\equiv i_1 \bmod 2$	$x \equiv 2^{-1}(i_2 + i_1) \bmod n, y = 1$ $\{x = k \mid k \in \mathbb{Z}/n\mathbb{Z}\}, y = 0$
		$i_2 \not\equiv i_1 \bmod n$	$i_2 \equiv i_1 \bmod 2$	$x \equiv 2^{-1}(i_2 + i_1) \bmod n, y = 1$
			$i_2 \not\equiv i_1 \bmod 2$	$x \equiv 2^{-1}(i_2 + i_1) \bmod n, y = 1$
	$2 \mid n$	$i_2 \equiv i_1 \bmod n$	$i_2 \equiv i_1 \bmod 2$	$x \equiv \frac{i_2+i_1}{2} \bmod n, \quad y = 1;$ $x \equiv (\frac{i_2+i_1}{2} + \frac{n}{2}) \bmod n, y = 1$ $\{x = k \mid k \in \mathbb{Z}/n\mathbb{Z}\}, y = 0$
			$i_2 \not\equiv i_1 \bmod 2$	No solutions
		$i_2 \not\equiv i_1 \bmod n$	$i_2 \equiv i_1 \bmod 2$	$x \equiv \frac{i_2+i_1}{2} \bmod n, \quad y = 1;$ $x \equiv (\frac{i_2+i_1}{2} + \frac{n}{2}) \bmod n, y = 1$ $\{x = k \mid k \in \mathbb{Z}/n\mathbb{Z}\}, y = 0$
			$i_2 \not\equiv i_1 \bmod 2$	No solutions
$j_1 = 0$	$2 \nmid n$	$i_2 \equiv -i_1 \bmod n$	$i_2 \equiv i_1 \bmod 2$	$x \equiv 2^{-1}(i_2 - i_1) \bmod n, y = 0$ $\{x = k \mid k \in \mathbb{Z}/n\mathbb{Z}\}, y = 1$
			$i_2 \not\equiv i_1 \bmod 2$	$x \equiv 2^{-1}(i_2 - i_1) \bmod n, y = 0$ $\{x = k \mid k \in \mathbb{Z}/n\mathbb{Z}\}, y = 1$
		$i_2 \not\equiv -i_1 \bmod n$	$i_2 \equiv i_1 \bmod 2$	$x \equiv 2^{-1}(i_2 - i_1) \bmod n, y = 0$
			$i_2 \not\equiv i_1 \bmod 2$	$x \equiv 2^{-1}(i_2 - i_1) \bmod n, y = 0$
	$2 \mid n$	$i_2 \equiv -i_1 \bmod n$	$i_2 \equiv i_1 \bmod 2$	$x \equiv \frac{i_2-i_1}{2} \bmod n, \quad y = 0;$ $x \equiv (\frac{i_2-i_1}{2} + \frac{n}{2}) \bmod n, y = 0$ $\{x = k \mid k \in \mathbb{Z}/n\mathbb{Z}\}, y = 1$
			$i_2 \not\equiv i_1 \bmod 2$	No solutions
		$i_2 \not\equiv -i_1 \bmod n$	$i_2 \equiv i_1 \bmod 2$	$x \equiv \frac{i_2-i_1}{2} \bmod n, y = 0;$ $x \equiv (\frac{i_2-i_1}{2} + \frac{n}{2}) \bmod n, y = 0$
			$i_2 \not\equiv i_1 \bmod 2$	No solutions

even $(4 \nmid W)$ and n is odd, the fraction of states which have at least one predecessor is $1/4$. When W is even $(4 \nmid W)$ and n is even, this fraction is $1/16$.

Having completed the case for automata of length W such that $4 \nmid W$, we proceed to study the case when $4 \mid W$. To do so, we suppose that $j_1 = j_2$ is satisfied and we look for solutions of Eq. (1.10)

$$2i_2 y - 2x j_1 \equiv i_2 - i_1 \bmod n.$$

Suppose first that $j_1 = 0$. Then, Eq. (1.10) becomes

$$2i_2 y \equiv i_2 - i_1 \bmod n.$$

No matter the choice of y, there is no dependence on x in this case. When $y = 0$, if $0 \equiv i_2 - i_1 \bmod n$, the set of solutions for x is

$$\{x = k \mid k \in \mathbb{Z}/n\mathbb{Z}\}.$$

When $y = 1$, if $0 \equiv -i_2 - i_1$, the set of solutions for x is as before,

$$\{x = k \mid k \in \mathbb{Z}/n\mathbb{Z}\}.$$

Now, let $j_1 = 1$. Then Eq. (1.10) becomes

$$2i_2 y - 2x \equiv i_2 - i_1 \bmod n.$$

When $y = 0$, this simplifies to

$$-2x \equiv i_2 - i_1 \bmod n. \tag{1.15}$$

For n such that $2 \nmid n$, 2 has a multiplicative inverse modulo n, and so Eq. (1.15) reduces to and has the solution $x \equiv 2^{-1}(i_2 + i_1) \bmod n$. By the Linear Congruence Theorem, this is the unique solution. When $2 \mid n$, then 2 has no multiplicative inverse modulo n and Eq. (1.15) cannot be reduced further. Applying the Linear Congruence Theorem, Eq. (1.15) has a solution iff $2 \mid i_2 - i_1$. If $2 \mid i_2 - i_1$, then Eq. (1.15) has two solutions, namely $x \equiv -\frac{i_2-i_1}{2} \bmod n$ and $x \equiv -\frac{i_2-i_1}{2} + \frac{n}{2} \bmod n$.

We now let $y = 1$ and simplify Eq. (1.10) to

$$2i_2 - 2x \equiv i_2 - i_1 \bmod n$$

$$-2x \equiv -i_2 - i_1 \bmod n$$

$$2x \equiv i_2 + i_1 \bmod n. \tag{1.16}$$

For n such that $2 \nmid n$, 2 has a multiplicative inverse modulo n, and so Eq. (1.16) reduces to and has the solution $x \equiv 2^{-1}(i_2 + i_1) \bmod n$. By the Linear Congruence Theorem, this is the unique solution. When $2 \mid n$, then 2 has no multiplicative inverse modulo n and Eq. (1.16) cannot be reduced further. Applying the Linear Congruence Theorem, Eq. (1.16) has a solution iff $2 \mid i_2 + i_1$. If $2 \mid i_2 + i_1$, then Eq. (1.16) has two solutions, namely $x \equiv \frac{i_2+i_1}{2} \bmod n$ and $x \equiv \frac{i_2+i_1}{2} + \frac{n}{2} \bmod n$.

These solutions are summarized in Table 3.

Using Table 3, we count how often Eq. (1.10) has at least one solution. Having already counted the fraction of states that satisfy $j_1 = j_2$, we can now complete the study of states of length W such that $4 \mid W$. We write W as $W = 2m$ and recall that we are studying reduced states of length m.

The fraction of the states of length m that satisfy $j_1 = j_2$ is $1/2$, and $1/2$ of these satisfy $j_1 = j_2 = 1$. For n odd, all of these states have predecessors; this accounts for $1/4$ of the $(2n)^m$ total states of length m. For n even and $j_1 = j_2 = 1$, a state also has to satisfy $i_1 \equiv i_2 \bmod 2$ in order to have a predecessor; hence, of the states

Table 3 $2yi_2 - 2xj_1 \equiv i_2 - i_1 \bmod n$

$j_1 = 1$	$2 \nmid n$	$i_2 \equiv i_1 \bmod 2$	$x \equiv 2^{-1}(i_2 + i_1) \bmod n,\ y = 1$
			$x \equiv 2^{-1}(i_1 - i_2) \bmod n,\ y = 0$
		$i_2 \not\equiv i_1 \bmod 2$	$x \equiv 2^{-1}(i_2 + i_1) \bmod n,\ y = 1$
			$x \equiv 2^{-1}(i_1 - i_2) \bmod n,\ y = 0$
	$2 \mid n$	$i_2 \equiv i_1 \bmod 2$	$x \equiv \frac{i_2 + i_1}{2} \bmod n,\ y = 1;\ x \equiv (\frac{i_2 + i_1}{2} + \frac{n}{2}) \bmod n,\ y = 1$
			$x \equiv \frac{i_1 - i_2}{2} \bmod n,\ y = 0$
		$i_2 \not\equiv i_1 \bmod 2$	No solutions
$j_1 = 0$	$2 \nmid n$	$i_2 \equiv i_1 \bmod n$	$\{x \equiv k \mid k \in \mathbb{Z}/n\mathbb{Z}\},\ y = 0$
		$i_2 \equiv -i_1 \bmod n$	$\{x \equiv k \mid k \in \mathbb{Z}/n\mathbb{Z}\},\ y = 1$
		$i_2 \not\equiv \pm i_1 \bmod n$	No solutions
	$2 \mid n$	$i_2 \equiv i_1 \bmod n$	$\{x \equiv k \mid k \in \mathbb{Z}/n\mathbb{Z}\},\ y = 0$
		$i_2 \equiv -i_1 \bmod n$	$\{x \equiv k \mid k \in \mathbb{Z}/n\mathbb{Z}\},\ y = 1$
		$i_2 \not\equiv \pm i_1 \bmod n$	No solutions

satisfying $j_1 = j_2 = 1, 1/2$ have predecessors. This accounts for $1/8$ of the $(2n)^m$ total states of length m.

Now we address the case when $j_1 = j_2 = 0$. To have a predecessor, a state must satisfy $i_1 \equiv \pm i_2 \bmod n$; we now count the number of states which satisfy this requirement. We begin by noting that there are n choices for i_1 and once chosen, i_2 is fixed as $\pm i_1$. Recall that i_1 comes from an ordered product of $m/2$ elements. The first element of this product can be one of $2n$ choices and we have the same freedom in choice for all the other elements until the final one. The final element in this product is uniquely determined by the choices made before it and the choice for i_1. Hence, there are $(2n)^{m/2-1}$ products which give r^{i_1} and so there are $(2n)^{m/2-1}$ products which give $r^{i_2} = r^{i_1}$. Hence, including the choice of n, there are $n(2n)^{m-2}$ ways to satisfy $i_2 \equiv i_1 \bmod n$. By a similar argument, there are $n(2n)^{m-2}$ ways to satisfy $i_2 \equiv -i_1 \bmod n$. We add the two to see that there are $2n(2n)^{m-2}$ ways to satisfy $i_2 \equiv \pm i_1 \bmod n$. However, for n odd, we have counted the case $i_1 = i_2 = 0$ twice, so we subtract $(2n)^{m-2}$ from our total count. For n even, we have counted both $i_1 = i_2 = 0$ and $i_1 = i_2 = n/2$ twice and so we subtract $2(2n)^{m-2}$ from our count.

This gives, for n odd, the number of the $(2n)^m$ total reduced states of length m ($2 \mid m$) which have predecessors is $\frac{(2n)^m}{4} + (2n)^{m-1} - (2n)^{m-2}$. For n even, this number is $\frac{(2n)^m}{8} + (2n)^{m-1} - 2(2n)^{m-2}$. We recall that a state S of length W ($4 \mid W$) is composed of two reduced states that need to concurrently satisfy the same requirements, and so we square these fractions to arrive at the following: for W even ($4 \mid W$) and n odd, the number of states which is reachable in the evolution of the automata is $(\frac{(2n)^m}{4} + (2n)^{m-1} - (2n)^{m-2})^2$, when n is even, this number is $(\frac{(2n)^m}{8} + (2n)^{m-1} - 2(2n)^{m-2})^2$. This completes the proof of Theorem 3.

The proof of Theorem 3 yields the following corollary regarding the in-degree of a given state.

Corollary 4 *The following table describes the number of predecessors possible for a state of length W in the automaton defined over D_{2n}.*

	$2 \nmid n$	$2 \mid n$
$2 \nmid W$	$0, 1, n+1$	$0, 2, n+2$
$2 \mid W, 4 \nmid W$	$0, 1, n+1, (n+1)^2$	$0, 4, 2(n+2), (n+2)^2$
$4 \mid W$	$0, 4, 2n, 4n, n^2, 2n^2$	$0, 9, 3n, 6n, n^2, 2n^2$

Proof The proof of Corollary 4 is seen by application of Tables 2 and 3. The application of Tables 2 and 3 will be illustrated here, though we will not do all cases explicitly.

Suppose first that $2 \nmid W$, n is odd. Then, suppose a state satisfies $j_1 = j_2 = 1$. If the state also satisfies $i_2 \equiv i_1 \bmod n$, then it has $n + 1$ predecessors. The predecessors are determined by $x \equiv 2^{-1}(i_2 + i_1) \bmod n$, $y = 1$ and $\{x = k \mid k \in \mathbb{Z}/n\mathbb{Z}\}$ and $y = 0$. If a state does not satisfy $i_2 \equiv i_1 \bmod n$, then it has only one predecessor, determined by $x \equiv 2^{-1}(i_2 + i_1) \bmod n$ and $y = 1$.

If a state satisfies $j_1 = j_2 = 0$ and additionally, the state satisfies $i_2 \equiv -i_1 \bmod n$, then it has $n + 1$ predecessors. They are determined by $x \equiv 2^{-1}(i_2 - i_1) \bmod n$ and $y = 0$, and $\{x = k \mid k \in \mathbb{Z}/n\mathbb{Z}\}$ and $y = 1$. If the state does not satisfy $i_2 \equiv -i_1 \bmod n$, then it has only the predecessor determined by $x \equiv 2^{-1}(i_2 - i_1) \bmod n$ and $y = 0$.

Now, suppose that $2 \nmid W$ and n is even. Then, if a state satisfies $j_1 = j_2 = 1$, $i_2 \equiv i_1 \bmod n$, and $i_2 \equiv i_1 \bmod 2$, it has $n + 2$ predecessors, determined by $x \equiv \frac{i_2+i_1}{2} \bmod n$ and $y = 1$, $x \equiv \left(\frac{i_2+i_1}{2} + \frac{n}{2}\right) \bmod n$ and $y = 1$, and the set $\{x = k \mid k \in \mathbb{Z}/n\mathbb{Z}\}$ and $y = 0$. If a state of the form $j_1 = j_2 = 1$ does not satisfy $i_2 \equiv i_1 \bmod 2$, then it has no predecessors.

If a state satisfies $j_1 = j_2 = 0$, $i_2 \equiv -i_1 \bmod n$, and $i_2 \equiv i_1 \bmod 2$, then it has $n + 2$ predecessors, determined by $x \equiv \frac{i_2-i_1}{2} \bmod n$ and $y = 0$, $x \equiv \left(\frac{i_2-i_1}{2} + \frac{n}{2}\right) \bmod n$ and $y = 0$, and the set $\{x = k \mid k \in \mathbb{Z}/n\mathbb{Z}\}$ and $y = 1$. If the state satisfies $j_1 = j_2 = 0$ and $i_2 \equiv i_1 \bmod 2$, it has only two predecessors, given by $x \equiv \frac{i_2-i_1}{2} \bmod n$ and $y = 0$, $x \equiv \left(\frac{i_2-i_1}{2} + \frac{n}{2}\right) \bmod n$ and $y = 0$. If a state does not satisfy $i_2 \equiv i_1 \bmod n$, then it has no predecessors.

The rest of the proof for this corollary lies in Tables 2 and 3.

6 Examples

Using the methods described in Sect. 5, we will give examples of states in the automaton defined over D_{2n} to determine if they have predecessors and if so, what those predecessors are.

Example 1 Let $S = (s, r, rs, r^2s, s, s)$ be a state in the automaton defined over D_8. Then S has no predecessor because it has an odd number of terms in $D_8 - \langle r \rangle$.

Example 2 Let $S = (s, r, rs, r^2s, s)$ be a state in the automaton defined in Sect. 5 over D_{10}. Because S has an even number of terms in $D_{10} - \langle r \rangle$, we compute i_1, i_2, j_1, and j_2 by application of Corollary 1 for automata of odd length over a multiplicative group G. This corollary states that $S = (d_0, \ldots, d_{W-1})$ has a predecessor $S' = (c_0, \ldots, c_{W-1})$ iff there exists $c_0 \in G$ satisfying

$$c_0 \left(\prod_{k=0}^{\frac{W-1}{2}-1} d_{(3+4k) \bmod W} \right) c_0 = \prod_{k=0}^{\frac{W+1}{2}-1} d_{(1+4k) \bmod W}.$$

In this case, S has a predecessor iff there exists $c_0 \in D_{10}$ satisfying

$$c_0 d_3 d_2 c_0 = d_1 d_0 d_4,$$

that is,

$$c_0 (r^2 s)(rs) c_0 = (r)(s)(s).$$

This simplifies to

$$c_0 r c_0 = r.$$

Hence, $i_1 = i_2 = 1$ and $j_1 = j_2 = 0$. Using Table 2, we see that S has exactly one predecessor. This predecessor is determined by $x \equiv 2^{-1}(i_2 - i_1) \bmod n$ and $y = 0$. Explicitly, we see that the solution for c_0 is

$$c_0 = r^0 s^0 = 1.$$

Now that c_0 is fixed, we compute the other terms using Theorem 1. Because Theorem 1 does not offer a solution for c_2 in terms of c_0, we compute this term explicitly

$$c_1 = \prod_{l=0}^{\frac{3-1}{2}-1} d_{6-3-4l \bmod 5}^{-1} c_0^{-1} \prod_{l=0}^{\frac{j+1}{2}-1} d_{1+4l \bmod 5}$$

$$= d_3^{-1} c_0^{-1} d_1 d_0$$

$$= (r^2 s)^{-1}(1)(r)(s)$$

$$= r$$

$$c_0 c_2 = d_1$$

$$c_2 = c_0^{-1}d_1$$
$$= (1)(r)$$
$$= r$$
$$c_1c_3 = d_2$$
$$c_3 = c_1^{-1}d_2$$
$$c_3 = r^{-1}(rs)$$
$$= s$$
$$c_2c_4 = d_3$$
$$c_4 = c_2^{-1}d_3$$
$$= (r^{-1})(r^2s)$$
$$= rs.$$

Finally, we check to see that the state obtained, $S' = (1, r, r, s, rs)$, updates to S in one time step:

$$(1, r, r, s, rs) \rightarrow (rsr, r, rs, rrs, s) = (s, r, rs, r^2s, s),$$

as desired.

Example 3 Let $S = (s, r, r, rs, rs, s)$ be a state in the automaton defined in Sect. 5 over D_8. Because S has an even number of terms in $D_8 - \langle r \rangle$, we compute i_1, i_2, j_1, and j_2 by application of Corollary 2 for automata of even length over a multiplicative group G. This corollary states that $S = (d_0, \ldots, d_{W-1})$ has a predecessor $S' = (c_0, \ldots, c_{W-1})$ iff there exist $c_0, c_1 \in G$ satisfying

$$c_0 \left(\prod_{k=0}^{\frac{m-1}{2}-1} d_{(3+4k) \bmod W} \right) c_0 = \prod_{k=0}^{\frac{m+1}{2}-1} d_{(1+4k) \bmod W}$$

$$c_1 \left(\prod_{k=0}^{\frac{m-1}{2}-1} d_{(4+4k) \bmod W} \right) c_1 = \prod_{k=0}^{\frac{m+1}{2}-1} d_{(2+4k) \bmod W}.$$

In this case, S has a predecessor iff there exists $c_0 \in D_8$ satisfying

$$c_0 d_3 c_0 = d_1 d_5,$$

that is,

$$c_0(rs)c_0 = (r)(s)$$

and $c_1 \in D_8$ satisfying

$$c_1 d_4 c_1 = d_2 d_0,$$

that is,

$$c_1(rs)c_1 = (r)(s).$$

These simplify to

$$c_0 rs c_0 = rs$$

and

$$c_1 rs c_1 = rs.$$

Because these equations are identical, we solve only for c_0. We see that $i_1 = i_2 = 1$ and $j_1 = j_2 = 1$. Using Table 2, we know that there are 6 predecessors for this reduced state. Using the formulas given, the predecessors for the reduced state are determined by $1, r^3, s, rs, r^2 s$, and $r^3 s$.

Suppose that $c_0 = 1$ and $c_1 = rs$. We now compute the other terms in the predecessor state using Theorem 1 as before

$$c_0 c_2 = d_1$$
$$c_2 = c_0^{-1} d_1$$
$$= r$$
$$c_1 c_3 = d_2$$
$$c_3 = c_1^{-1} d_2$$
$$= (rs)(r) = s$$
$$c_2 c_4 = d_3$$
$$c_4 = c_2^{-1} d_3$$
$$= r^{-1}(rs)$$
$$= s$$
$$c_3 c_5 = d_4$$

$$c_5 = c_3^{-1} d_4$$
$$= s^{-1}(rs) = r^3.$$

Finally, we check to see that the state obtained, $S' = (1, rs, r, s, s, r^3)$, updates to S in one time step:

$$(1, rs, r, s, s, r^3) \rightarrow (r^3rs, r, rss, rs, sr^3, s) = (s, r, r, rs, rs, s),$$

as desired.

Example 4 Let $S = (1, r, s, rs, s, s, r, r)$ be a state in the automaton defined in Sect. 5 over D_6. Because S has an even number of terms in $D_6 - \langle r \rangle$, we compute i_1, i_2, j_1, and j_2 by application of Corollary 3 for automata of even length over a multiplicative group G. This corollary states that for $W = 2m$, $S = (d_0, \ldots, d_{W-1})$ has a predecessor $S' = (c_0, \ldots, c_{W-1})$ iff there exist $c_0, c_1 \in G$ satisfying

$$\left(\prod_{k=0}^{\frac{m}{2}-1} d_{(1+4k) \bmod W} \right) c_0 = c_0 \left(\prod_{k=0}^{\frac{m}{2}-1} d_{(3+4k) \bmod W} \right)$$

and

$$\left(\prod_{k=0}^{\frac{m}{2}-1} d_{(2+4k) \bmod W} \right) c_0 = c_0 \left(\prod_{k=0}^{\frac{m}{2}-1} d_{(4+4k) \bmod W} \right).$$

In this case, S has a predecessor iff there exists $c_0 \in D_6$ satisfying

$$d_1 d_5 c_0 = c_0 d_3 d_7,$$

that is,

$$(r)(s)c_0 = c_0(rs)(r)$$

and $c_1 \in D_8$ satisfying

$$d_2 d_6 c_1 = c_1 d_4 d_0,$$

that is,

$$(s)(r)c_1 = c_1(s)(1).$$

These simplify to

$$rsc_0 = c_0 s$$

and

$$r^2 s c_1 = c_1 s.$$

We solve first for c_0. In this case, $i_1 = 1$, $i_2 = 0$, and $j_1 = j_2 = 1$. Using Table 3, we see that there are two solutions for x and y, determined by $x \equiv 2^{-1}(1) \bmod 3$, $y = 0$ and $x \equiv 2^{-1}(1) \bmod 3$, $y = 1$. We will choose the latter.

To solve for c_1, we see that $i_1 = 2$, $i_2 = 0$, and $j_1 = j_2 = 1$. Using Table 3, we see that there are two solutions for x and y, namely $x \equiv 2^{-1}(2) \bmod 3$, $y = 0$ and $x \equiv 2^{-1}(2) \bmod 3$, $y = 1$. Again, we choose the latter. We solve now for the remaining entries in S'. When $c_0 = r^2 s$ and $c_1 = rs$, we have

$$c_0 c_2 = d_1$$

$$c_2 = c_0^{-1} d_1$$

$$= (r^2 s)^{-1}(r) = rs$$

$$c_1 c_3 = d_2$$

$$c_3 = c_1^{-1} d_2$$

$$= (rs)^{-1}(s) = r$$

$$c_4 = \prod_{l=0}^{\frac{2}{2}-1} d_{4-3-4l \bmod 8}^{-1} c_0 \prod_{l=0}^{\frac{2}{2}-1} d_{3+4l \bmod 8}$$

$$= d_1^{-1} c_0 d_3$$

$$= (r^{-1})(r^2 s)(rs)$$

$$= 1$$

$$c_5 = \prod_{l=0}^{\frac{2}{2}-1} d_{1+4-3-4l \bmod 8}^{-1} c_1 \prod_{l=0}^{\frac{2}{2}-1} d_{1+3+4l \bmod 8}$$

$$= d_2^{-1} c_1 d_4$$

$$= (s)^{-1}(rs)(s)$$

$$= r^2 s$$

$$c_6 = \prod_{l=0}^{\frac{2}{2}-1} d_{6-3-4l \bmod 8}^{-1} c_0 \prod_{l=0}^{\frac{4}{2}-1} d_{1+4l \bmod 8}$$

$$= d_3^{-1} c_0^{-1} d_1 d_5$$

$$= (rs)^{-1}(r^2s)^{-1}(r)(s)$$

$$= s$$

$$c_7 = \prod_{l=0}^{\frac{2}{2}-1} d_{1+6-3-4l \bmod 6}^{-1} c_1^{-1} \prod_{l=0}^{\frac{4}{2}-1} d_{1+1+4l \bmod 6}$$

$$= d_4^{-1} c_1^{-1} d_2 d_6$$

$$= (s)^{-1}(rs)^{-1}(s)(r)$$

$$= rs.$$

Finally, we check to see that the state obtained, $S' = (r^2s, rs, rs, r, 1, r^2s, s, rs)$, updates to S in one time step:

$$(r^2s, rs, rs, r, 1, r^2s, s, rs) \rightarrow (rsrs, r^2srs, rsr, rs, rr^2s, s, r^2srs, sr^2s),$$

$$= (1, r, s, rs, s, s, r, r)$$

as desired.

7 Further Study

In this work we focused on studying cellular automata over non-abelian group alphabets with Wolfram's Rule 90 as the update rule. We applied our finding to automata over dihedral groups. By studying different rules over other non-abelian groups we have a rich source of problems particularly suited for undergraduates. The program GAP [3] was used extensively to conjecture and verify results in this paper. The PascGalois [4] is a great resource for studying and visualizing automata evolution.

In the study of cellular automata, one wishes to draw the state transition diagram (STD), that is, one wishes to draw the map of all states in the automaton as nodes mapping to each other under the update rule for the automaton. On the STD, all nodes with in-degree zero represent states that have no predecessors and these nodes map into the remaining states, transients, and states on cycles. This diagram is useful in interpreting the long term behavior of automata. In our study of the automaton over D_{2n}, we did not focus on drawing the state transition diagram in favor of finding the fraction of states which have predecessors. We note that this is an important piece of data for the STD; this fraction gives us the number of nodes that have in-degree of at least 1. Figuring out the STD for given automata over given groups is another fruitful area of research particularly suited for undergraduates.

In the interest of brevity, we also omitted any historical elements on cellular automata and their applications. Interested readers can find some historical elements in [5–7].

References

1. Bardzell, M. and Miller, N. *The Evolution Homomorphism and Permutation Actions on Group Generated Cellular Automata.* Complex Systems, Volume 15, Issue 2. 2004.
2. Wolfram, S., Martin, O., and Odlyzko, A.M. *Algebraic Properties of Cellular Automata.* Communications in Mathematical Physics, 93. 1984.
3. *GAP - Groups, Algorithms, Programming - a System for Computational Discrete Algebra.* http://www.gap-system.org/. 1995.
4. *The PascGalois Project: Visualizing Abstract Mathematics.* pascgalois.org. 2008.
5. Rennard, J. *Introduction to Cellular Automata.* http://www.rennard.org/alife/english/acintrogb01.html. 2006.
6. Neumann, J. von. *Re-evaluation of the problems of complicated automata: Problems of hierarchy and evolution.* Papers of John von Neumann on Computing and Computer Theory. W. Aspray and A. Burks, eds., MIT Press. 1987.
7. Neumann, J. von. *General and Logical Theory of Automata.* Cerebral Mechanisms of Behavior. Lloyd A. Jeffress, ed. New York: John Wiley and Sons, Inc. 1951.

A Preliminary Exploration of the Professional Support Networks the EDGE Program Creates

Candice R. Price and Nina H. Fefferman

Abstract Programs such as the Enhancing Diversity in Graduate Education (EDGE) Program for women focus on improving outcomes for women and minorities in postgraduate degree programs in mathematics. One of the functions of these programs is increasing the size and reach of professional networks, including both mentors and peers. This is, in part, based on research showing a direct correlation between the strength of professional networks and individual professional success. However, little work has been done to analyze networks created by such programs. We extracted the network connections between EDGE participants from their undergraduate and graduate training in addition to their EDGE co-participants. We then examined these connections according to some frequently explored metrics of network organization to ascertain whether features of participation within the EDGE network were critical to individual success. Unfortunately, for the purpose of this work, EDGE was a victim of its own success: nearly all the past participants of the program were successfully employed in roles that utilized their postgraduate degrees in mathematics! We subsequently chose a more restrictive definition for success and explored factors within the available EDGE network that might predict the completion of a doctoral degree in mathematics. Our preliminary study lays the groundwork for future efforts to understand the true impact of EDGE-like programs and how to design purposefully targeted, efficient, and effective interventions.

C. R. Price (✉)
University of San Diego, San Diego, CA, USA
e-mail: cprice@sandiego.edu

N. H. Fefferman
University of Tennessee, Knoxville, TN, USA

© The Author(s) and the Association for Women in Mathematics 2019
S. D'Agostino et al. (eds.), *A Celebration of the EDGE Program's Impact on the Mathematics Community and Beyond*, Association for Women in Mathematics Series 18, https://doi.org/10.1007/978-3-030-19486-4_21

1 The EDGE Program

Underrepresented students, women of color specifically, are more likely to be retained in STEM programs if they have opportunities to engage in course content with peers, participate in undergraduate research, and join clubs and organizations. The authors in [1] discuss how belongingness is crucial as students learn about the people who pursue a STEM career, the values in STEM workplaces, and whether they can see themselves in those workplaces or graduate programs. Programs successful in recruiting and retaining women in science note the importance of opportunities to develop deeper understandings and connections to science through living and learning programs, broad mentoring, and opportunities to work closely with faculty in hands-on science experiences [1]. Networking opportunities provide more than social connections while students pursue a professional position—they provide images of who the students can be in the future [7]. As a result, some programs specifically have among their goals the generation of ongoing social and mentoring networks. One such program is the Enhancing Diversity in Graduate Education (EDGE) Program.

The EDGE Program was launched in 1998 by Dr. Sylvia Bozeman and Dr. Rhonda Hughes. As faculty at women's colleges, Spelman College and Bryn Mawr College, respectively, Bozeman and Hughes found that attrition takes its toll on women mathematician at the graduate level. While there are programs supporting the pipeline for women at the K-16 level, women and minority students were particularly adversely affected by the change in culture in the transition from undergraduate to graduate study [13]. The EDGE program is designed to strengthen the ability of women and minority students to successfully complete graduate programs in the mathematical sciences. In its 20 year tenure, over 275 women have participated in EDGE either as summer participants, mentors, instructors, or local organizers. Because mentoring is a critical component of the EDGE experience, EDGE participants find mentors among the EDGE summer faculty, the advanced graduate student assistants, and the directors. The structure of the program involves two basic components: an intensive summer program, and a follow-up mentoring program [13]. The program creates a strong network between its participants and thus we set out to explore the types of connections made by this highly successful program and which of these connects lead to success.

Since social psychology and educational theory suggest that strong networks can be important factors in succeeding in graduate programs in STEM fields [6, 8], we consider that the networks EDGE creates may be largely responsible for the acknowledged success of the program overall [9]. We therefore analyze the professional social network that EDGE creates among its participants to explore which features of this network might be most effective in support of the success of the participants of EDGE. Once this is understood, our hope is that it may be possible to purposefully design the types of networks programs like EDGE generates to improve outcomes of those in STEM fields.

One challenge we face in accomplishing this goal is that the idea of success can mean different things. We may conclude that success of the EDGE program includes securing the pipeline for women and minorities in STEM. Yet, complications arise when the metric for this success is discussed. Does it include finding employment in STEM? Achieving a terminal master's degree? Graduating with a Ph.D. in a mathematical science? If outcomes and goals can vary, this becomes a complicated problem. Since the ultimate goal of the EDGE program is to enhance the professional success of women with advanced degrees in mathematical sciences, our first inclination was to consider "success" as simply being employed in a role that would utilize such a degree. To our joy and consternation, we found this to provide no means for analysis of predictors within the EDGE participant network, since 97% of participants for whom there was information on current employment (86%) had achieved this outcome. There was therefore an insufficient set of examples of negative outcomes against which to correlate potential indicators. We therefore made the choice to restrict our definition of success solely to completing a Ph.D. in mathematical sciences.

2   Social Network Analysis

A social network approach to discussing the impact of EDGE on the success of it participants can assist in understanding what aspect of the network is most influential. Although there are multiple network centrality measures [4], we focus on and measure the following four: degree, betweenness, closeness, and eigenvector.

Degree centrality is historically the first, and conceptually simplest centrality measure. The degree of a node can be interpreted as the number of direct connections node n_i has, see Eq. (1)

$$D_i = \sum_{j=1}^{n} a_{ij} \tag{1}$$

where $a_{ij} = \begin{cases} 1 & \text{if node } n_i \text{ is connected to node } n_j \\ 0 & \text{if node } n_i \text{ is not connected to node } n_j. \end{cases}$

Individuals or organizations with high degree are those who are "in the know," i.e., who are connected to many others in the network.

Closeness centrality describes a measure of the average path length between node n_i and each of the other nodes in the graph. This measure expresses the average social distance from each individual to every other individual in the network. The concept of social distance is easily understood by considering the concept of an "Erdös number," calculated by finding the shortest set of connections from any oneself to Paul Erdös based on "collaborative distance" (authorship of mathematical papers) [12]. To calculate closeness centrality, we use the following algorithm:

$$C_i = \left[\sum_{j=1}^{n} b_{ij} \right]^{-1}$$

where b_{ij} represents the number of links in the shortest path connecting nodes n_i and n_j. In this way an individual with a direct tie to everyone else ends up with the largest closeness value.

Betweenness provides a mathematical compromise between degree and closeness. Betweenness of a node n_i measures the percentage of shortest paths from one node in the network to another node in which node n_i is included. To calculate betweenness centrality, start by finding all the shortest paths between any two nodes in the network. Then, count the number of these shortest paths that go through each node (Eq. (2)). This number is betweenness centrality

$$B_i = \frac{\sum_{i<j} g_{jk}(n_i)}{g_{jk}} \tag{2}$$

where g_{jk} represents the number of links in the shortest path connecting agents n_j and n_k and $g_{jk}(n_i)$ represents the number of these paths that contain agent n_i.

Eigenvector centrality, in contrast, measures how well connected a node is to other well connected nodes. To calculate eigenvector centrality, first construct an adjacency matrix, M, that describes who is connected to whom in the social network. We use equation a_{ij} for our entries in the matrix. Then, we calculate the eigenvalues λ of this matrix and chose the largest eigenvalue. After making this choice, then find the associated eigenvector. This eigenvector, v, provides the eigenvector centrality measure for each agent in the network. The largest component in the eigenvector corresponds to the agent with the highest eigenvector centrality.

2.1   Extraction of EDGE Data

Although the data is publicly available via the EDGE website, for ease of analysis, we obtained information from the EDGE-maintained database by direct request to the program administrators. This database extraction provided us with the following information for each participant: name, year of participation, the undergraduate institution they attended, the graduate program they attended immediately following participation in EDGE, whether or not a doctoral degree had been awarded at time of extraction (to the best of the program's knowledge), and current place of employment (if known; 86% self-reported). The database extraction also provided the names of the mentors and instructors for each year of the program. To allow for students who may still be successful in completing a doctoral program, but have not yet completed their graduate degrees within a standard 6 years [10], we analyzed the provided data for individuals who participated in EDGE only between 1998 (the first year of the program) and 2011. We then built the contact network by including an edge between any two individuals who (a) participated in the same EDGE cohort year, (b) attended the same undergraduate institution, (c) matriculated the same graduate institution directly after participation in EDGE, or (d) were a mentor/instructor to participant pair. This extraction was done by hand, with spot checking between the two co-authors to attempt to minimize errors. It should be noted that this preliminary analysis omits some very likely huge aspects of the EDGE participant network: the reunion meetings at which individuals from multiple cohorts interact with each other, and participation in EDGE mentoring clusters. These activities most certainly act to both broaden and strengthen the professional network fostered by the EDGE program. Information on attendance of or participation in these activities was not publicly available at the time of analysis. This missing data is clearly crucially important to any effort to fully characterize the professional social network fostered by the EDGE program, and therefore also critical to any analysis that would understand which factors in such networks are the most important drivers of individual success. We therefore present our findings as a proposal for how future efforts might best evaluate a more complete description of the interactions and activities EDGE creates and supports, rather than as any definitive finding of current impact.

2.2 Analysis of Extracted Network

We used the included network analysis tools in Matlab 2017a to calculate the individual centrality metrics (degree, closeness, betweenness, and eigenvector) for our extracted network (Fig. 1).

We then calculated the pairwise correlations between these metrics and the Boolean "success" outcome from the initial database of completion of a doctoral degree. To determine whether or not there were meaningful associations between the network metrics and the success outcomes, we considered coefficient of correlation, R^2 fit of the linear approximation, and the F−statistic associated with the analysis of variance (ANOVA) performed [11]. The statistical methods themselves conformed with standard practice in the social network analysis literature, details of which are beautifully explained in the original introductory paper on centrality analysis [5]. (Note that while individuals who already had their degrees before participating in the program, e.g., instructors, were included in the extracted network in order to capture the potential for social connectivity mediated by knowing them, they were excluded from the correlation analysis since their "success," based on our definition, cannot be attributed to participation in EDGE.)

To consider whether the impacts of network position may be less nuanced than continuous centrality measures, we also transformed each centrality metric into a Boolean version of either greater than or equal to the median centrality outcome, or less than the median centrality outcome for each of the four measures and considered the correlation between these Boolean versions of the centralities with the success outcome. These, however, were less successful in prediction than their more exact, continuous counterparts.

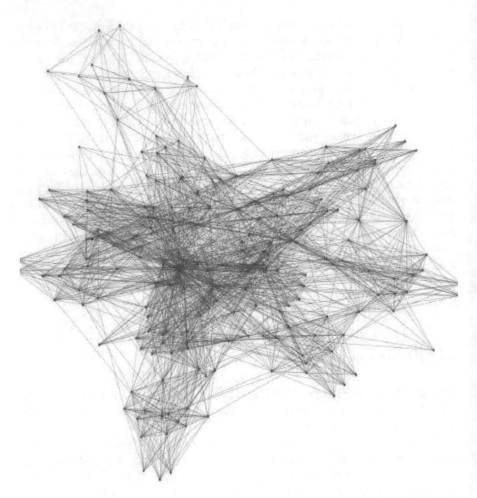

Fig. 1 This is a visualization of the extracted network of EDGE participants from 1998 through 2011, including mentors and instructors. Connections were included if two individuals shared the same EDGE cohort, if two individuals attended the same undergraduate or graduate institution, and between instructors/mentors and participants within the same year of the EDGE program

3   Observed Correlations of Network Metrics with EDGE Participant Success

Based on these fairly straightforward analyses of the most commonly analyzed metrics of social network organization, we see that the feature of individual centrality that correlated at all with the successful completion of a Ph.D. in STEM is degree centrality. However, the explanatory power even of this most correlated metric is extremely low (coefficient of correlation of 0.008 with an $R^2 = 0.02$, $p \leq 0.05$). Though a weak correlation, this is consistent with other studies [3] that have shown that increases in numbers of connections (in our case, even just within the EDGE participant cohort) might support long-term career success. Of course, to fully understand the impact of the EDGE network, data from the additional professional network development activities fostered by the EDGE program (e.g., reunion gatherings, etc.) would need to be included in this type of analysis.

It is important to note that the observable impact from the EDGE dataset is limited to exploring the internal determinants of which factors indicated likelihood of success among EDGE participants. We are not making conclusions about the impact of the EDGE program itself. While data exist as published statistics about the increase in successful completion of graduate programs for EDGE participants relative to comparable students who did not participate in EDGE [2, 13] that data does not include sufficient detail about professional networks for us to have analyzed the same network metrics to determine the specifics of *how* EDGE effects change in supportive professional networks. We strongly support any future endeavors to capture these types of comparative data so we may begin to understand which features of social and professional support may be most important in improving career outcomes for women and minorities in mathematics.

It is also important to revisit the discussion on the definition of "success" itself. The stated goal of the EDGE program for women is "to strengthen the ability of women and minority students to successfully complete graduate programs in the mathematical sciences" [13]. Due to the focus of the program's activities and website highlighting the completion of a Ph.D., we did not include the completion of a terminal master's degree as a "successful" outcome in our analyses even though this clearly constitutes a successful completion of a graduate program in cases in which it was the student's intended goal. Conversely, as mentioned above, it is also important to reiterate that the broader outcome of supporting individuals pursuing careers in STEM had to be rejected as an analyzed program outcome because too few of the EDGE participants were not successfully working in STEM fields as of the time of this analysis (of the 86% of past participants for whom employment status was known, only 3% were employed outside of a STEM-focused position). That by itself is a nearly unheard of rate of professional success in STEM for any group of women and/or minority students.

4 Motivation for Further Analysis

As already mentioned above, the strong evidence that improved professional networks in STEM increase rates of success for women and minorities is just a first step at understanding how these goals are being achieved. Advances in the field of social network analysis over the past few decades have highlighted the diversity of mechanisms and impacts of different structures and actions of network types. This understanding has led the field to the insight that simply increasing the number of connections or numbers of participants in a supportive professional network may be correlates, rather than causal drivers, of successful outcomes. Until we begin to analyze the comparative networks of individuals, we will not truly be able to understand *how* interventions such as the successful EDGE program are achieving their goals. However, once analyzed, our hope is that the field will be able to design efficient, targeted strategies to improve exactly those aspects which most directly impact career outcomes for students. To that end, our efforts here presented provide the extracted network for the first 13 years of EDGE and the analyses that can act as a comparative dataset for a "known successful" intervention. (Note: The authors will happily share the de-identified EDGE participant network.) It is our hope that future work will shortly enable these comparisons and insights.

5 Conclusions

The EDGE program has been shown to be remarkably effective in helping students become STEM professionals, but thus far, it remains unknown which aspects of the support EDGE provides are the most important drivers of those benefits. EDGE may therefore not only act to support their participants directly, but may also benefit the broader scientific community by providing a test bed for understanding how supportive interventions work in STEM fields. This understanding could then not only help refine and improve EDGE activities, but could aid in the targeted design of future programs, to more efficiently help students through their graduate programs and launch their professional STEM careers. Discovering these features may also generate testable hypotheses for the types of social support that can best address even broader pipeline issues, beyond preparation for graduate work in mathematics among women and minority students. While our current analyses were limited by the data available, we believe the methods here presented may prove valuable tools for future use in understanding how the support of programs like EDGE can shape effective professional networks that support and enhance our community. We eagerly anticipate the expansion of these efforts to include further input from network and education scientists and sociologists and are grateful for the opportunity to analyze even these preliminary social networks of EDGE participants in aid of advancing this agenda.

Acknowledgements The authors both wish to thank the National Institute for Mathematical and Biological Synthesis (NIMBioS) for visitor support to CRP.

References

1. Addressing stem culture and climate to increase diversity in stem disciplines. *Higher Education Today*, May 2018.
2. American Mathematical Society. The American Mathematical Society, Annual Survey. http://www.ams.org/profession/data/annual-survey/annual-survey. [Online; accessed 27-July-2018].
3. T. T. Baldwin, M. D. Bedell, and J. L. Johnson. The social fabric of a team-based M.B.A. program: Network effects on student satisfaction and performance. *Academy of management journal*, 40(6):1369–1397, 1997.

4. L. C. Freeman.   Centrality in social networks conceptual clarification.   *Social networks*, 1(3):215–239, 1978.
5. L. C. Freeman, D. Roeder, and R. R. Mulholland. Centrality in social networks: ii. experimental results. *Social networks*, 2(2):119–141, 1979.
6. D. M. Merolla and R. T. Serpe. Stem enrichment programs and graduate school matriculation: the role of science identity salience. *Social Psychology of Education*, 16(4):575–597, 2013.
7. B. W.-L. Packard and N. L. Fortenberry. *Successful STEM mentoring initiatives for underrepresented students: a research-based guide for faculty and administrators*. Stylus, 2016.
8. R. T. Palmer, D. C. Maramba, and T. E. Dancy. A qualitative investigation of factors promoting the retention and persistence of students of color in stem. *The Journal of Negro Education*, pages 491–504, 2011.
9. The EDGE Program for women. EDGE Awards. http://www.ams.org/profession/data/annual-survey/annual-survey, 1998. [Online; accessed 27-July-2018].
10. The Mathematical Association of America. A graduate school primer. https://www.maa.org/a-graduate-school-primer. [Online; accessed 27-July-2018].
11. S. Weerahandi. ANOVA under unequal error variances. *Biometrics*, pages 589–599, 1995.
12. Wikimedia Foundation.   Erds number.   https://en.wikipedia.org/wiki/Erds_number, 1998. [Online; accessed 2-July-2018].
13. www.edgeforwomen.org. The EDGE program for women. https://www.edgeforwomen.org/about-edge/, 1998. [Online; accessed 27-July-2018].

A Model for Three-Phase Flow in Porous Media with Rate-Dependent Capillary Pressure

Kimberly Spayd and Ellen R. Swanson

Abstract As a contaminant, such as oil, travels through a porous media, such as soil, there is contact between the contaminant, groundwater, and the intermediate gas, air. At this interface there is a pressure difference, capillary pressure, which impacts the flow of the contaminant through the porous media. We derive a model for three-phase flow in porous media with the inclusion of capillary pressure, as given by thermodynamically constrained averaging theory (TCAT). Starting with conservation of mass, an incompressibility condition, and Darcy's law, we include constitutive equations that extend a strictly hyperbolic system analyzed by Juanes and Patzek. In the absence of gravity and capillarity, they show that solutions include rarefaction waves and shocks which satisfy the Liu entropy criterion. By incorporating capillary pressure, we show that the model gains dissipation and dispersion terms, the latter of which is rate-dependent. This extends the framework developed by Hayes and LeFloch in which there are solutions involving shocks which do not satisfy the Liu entropy criterion.

1 Introduction

The installation and use of underground pipelines to transport oil and other fluid mixtures has been a controversial topic of late, particularly in regard to accidental leaks contaminating the surrounding soil and water supply [16, 26, 31]. Such events are of concern to petroleum engineers and environmentalists alike. Much work has been done to optimize remediation after a chemical spill or pipeline malfunction (see [8] and the references therein), but there remains much to understand about

K. Spayd
Gettysburg College, Gettysburg, PA, USA
e-mail: kspayd@gettysburg.edu

E. R. Swanson (✉)
Centre College, Danville, KY, USA
e-mail: ellen.swanson@centre.edu

© The Author(s) and the Association for Women in Mathematics 2019
S. D'Agostino et al. (eds.), *A Celebration of the EDGE Program's Impact on the Mathematics Community and Beyond*, Association for Women in Mathematics Series 18, https://doi.org/10.1007/978-3-030-19486-4_22

how the contaminant, water, and air underground behave in time and space. In this paper, we assume there is a point-source leakage of an insoluble contaminant into a porous medium; the fluid is a nonaqueous phase liquid (NAPL) and its source is above the water table, in a region of the porous medium where air and water are both present. For simplicity, we assume that the horizontal flow of the NAPL is limited in scope so that a model with only one spatial dimension is sufficient to capture the evolution of the contaminant's position. Our interest is in modeling the changing saturations, i.e., volume fractions, of all three phases through time.

To do this, we derive a system of two one-dimensional partial differential equations which arise from physical principles including conservation of mass, an incompressibility condition, Darcy's law, and many constitutive equations that provide specific functional forms for quantities of interest. Section 2 presents these equations in detail. Previous work in this area has neglected the important contributions of dissipation and dispersion terms that appear in the system through the capillary pressure equation; see [3, 20, 22] as examples. Capillary pressure, P_c, is defined to be the difference of two-phase pressures at their interface and is given at the microscale by the Young–Laplace equation

$$P_c = P_n - P_w = \frac{2\gamma}{R}, \tag{1}$$

where P_n is the pressure of the nonwetting phase (for example, oil), P_w is the pressure of the wetting phase (groundwater), γ is the surface tension between the two phases, and R is the radius of curvature for a spherically shaped interface. Historically, Eq. (1) has been thought to accurately model the equilibration of the two-phase pressures as an instantaneous event dependent only on the saturation of the wetting phase, u. Its contribution to the larger model was insignificant as the resulting terms only smoothed solutions whose structure was captured by the more substantial governing equations.

In recent years, Gray and Miller introduced *thermodynamically constrained averaging theory* (TCAT), including a representation of capillary pressure at the macroscale [10–12]:

$$P_c(u, u_t) = P_n - P_w = P_c^e(u) - \tau\phi u_t - \tau\gamma\hat{k}\frac{\epsilon - \epsilon^{eq}}{P_c^e(u)} \tag{2}$$

in which P_c^e is the equilibrium capillary pressure, τ is a relaxation time, ϕ is the porosity of the medium, ϵ and ϵ^{eq} are interfacial areas, the latter at equilibrium, and $\hat{k}$ is a generation rate coefficient for the interfacial area. The presence of the second and third terms on the right-hand side of Eq. (2) signifies that capillary pressure between two phases does not equilibrate instantaneously; rather there is a time dependence in the physical event, which is not represented in Eq. (1). Prior work by Hassanizadeh and Gray introduced the importance of this rate dependence through a simpler first-order correction to Eq. (1) known as *dynamic capillary pressure* [13, 14], in which Eq. (2) is reduced to the first two terms on the right-hand side. A rigorous model based on TCAT principles for two-phase flow at the macroscale is discussed in [9]; an intense effort can be made to extend this work to three phases. With our simpler model developed below, our ultimate interest is in potential solution structures rather than a high level of physical applicability.

The mathematical implications of dynamic and TCAT capillary pressure on multiphase flow models have been considered extensively, for example, in [5, 19, 29, 32, 33, 35]. In particular, traveling wave analysis uncovers the presence of *undercompressive shocks* in these models when rate-dependent capillary pressure equations are incorporated. An undercompressive shock is a jump discontinuity which has characteristics converging only on one side; they arise as limits of traveling wave solutions for regularized equations with nonconvex fluxes. Their consideration has expanded the catalog of possible solution profiles for multiphase flow models through combinations with rarefaction and other shock waves.

Experimental results support that capillary pressure depends on more than just the saturation of the wetting phase. Nonmonotonic saturation profiles observed in [4, 5, 34] for single-phase flow cannot be adequately modeled with only the expression of capillary pressure given in Eq. (1); when water infiltrates columns of dry sand, the fluid can exhibit a phenomenon known as saturation overshoot in which water saturation is higher at the leading edge. Eckberg and Sunada describe their experimental results for three-phase flow, specifically with oil spills in mind, in [6] and note that static equations for capillary pressure are inadequate to capture the observed saturation behavior. Two-dimensional contaminant transport in different porous media, saturated and unsaturated, is described in [18] and nonmonotonic saturation profiles appear as well. More recently, two-phase flow experiments in which NAPL displaces water in a saturated medium have been analyzed in the context of dynamic effects [25]. O'Carroll et al. find an improved fit to the data when their model incorporates dynamic capillary pressure but not relative permeability constitutive relationships.

In this paper, we develop a rudimentary model for three-phase flow in a porous medium using TCAT capillary pressure; see Sect. 2. In Sect. 3, we describe how the model fits a framework in which undercompressive shocks develop as solutions to the associated Riemann problem [15]. In Sect. 4, we summarize our work and provide further lines of inquiry into this model.

2 Model

We consider the saturations of three fluid phases: water, air, and a nonaqueous phase liquid (NAPL), represented by $u(x, t)$, $v(x, t)$, $s(x, t)$, respectively. The NAPL contaminant is treated as a singular phase although in practice it may have multiple constituents. Our model represents the transport of the contaminant in a porous medium above the water table, where air is present. Thus, $u + v + s = 1$. The NAPL phase is assumed to be nonreactive, so microbial remediation is not represented in the model. Each phase has density ρ_i and moves with velocity V_i, $i = w$ (water), a (air), and n (NAPL).

Conservation of mass for each phase gives

$$\frac{\partial}{\partial t}(\phi \rho_w u) + \frac{\partial}{\partial x}(V_w) = 0, \tag{3a}$$

$$\frac{\partial}{\partial t}(\phi \rho_a v) + \frac{\partial}{\partial x}(V_a) = 0, \tag{3b}$$

$$\frac{\partial}{\partial t}(\phi \rho_n s) + \frac{\partial}{\partial x}(V_n) = 0. \tag{3c}$$

Following the development of similar multiphase flow models, for example [20, 32], we assume incompressibility of the three phases, unrealistic as it is with air present in the system. This implies that the total fluid velocity $V_t = V_a + V_w + V_n$ is constant. Darcy's law for each phase, neglecting gravitational terms, relates the phase velocity to the pressure gradient:

$$V_i = -\rho_i \frac{K k_i}{\mu_i} \frac{\partial P_i}{\partial x}, \tag{4}$$

where K represents the absolute permeability of the porous medium, k_i the relative permeability, μ_i the viscosity, and P_i the pressure of phase i. For ease of notation, let

$$\lambda_i = \rho_i \frac{K k_i}{\mu_i} \tag{5}$$

and $\lambda_t = \lambda_a + \lambda_w + \lambda_n$, so that $\sum_i \frac{\lambda_i}{\lambda_t} = 1$.

The NAPL phase is typically treated as the intermediary wetting phase between air and water [8] as the contact angle it makes with the medium is smaller than air and larger than water. Then the TCAT capillary pressures between NAPL and water, P_{cnw}, as well as air and NAPL, P_{can}, are functions of the respective wetting phase saturations. Using the rate-dependent capillary pressure forces of Eq. (2), we obtain:

$$P_{cnw} = P_n - P_w = P_{cnw}^{eq}(u) - \tau_{nw}\phi\frac{\partial u}{\partial t} - \tau_{nw}\gamma_{nw}\hat{k}_{nw}\frac{\epsilon_{nw} - \epsilon_{nw}^{eq}}{P_{cnw}^{eq}(u)}, \tag{6a}$$

$$P_{can} = P_a - P_n = P_{can}^{eq}(s) - \tau_{an}\phi\frac{\partial s}{\partial t} - \tau_{an}\gamma_{an}\hat{k}_{an}\frac{\epsilon_{an} - \epsilon_{an}^{eq}}{P_{can}^{eq}(s)}. \tag{6b}$$

Following [2], we express $V_w = \sum_i \frac{\lambda_i}{\lambda_t}V_w \pm \frac{\lambda_w}{\lambda_t}V_n \pm \frac{\lambda_w}{\lambda_t}V_a$. Then reorganizing terms and incorporating Darcy's law for each phase gives

$$V_w = \frac{\lambda_w}{\lambda_t}V_t + \frac{\lambda_n}{\lambda_t}V_w + \frac{\lambda_a}{\lambda_t}V_w - \frac{\lambda_w}{\lambda_t}V_n - \frac{\lambda_w}{\lambda_t}V_a \tag{7}$$

$$= \frac{\lambda_w}{\lambda_t} V_t - \frac{\lambda_w \lambda_n}{\lambda_t} \frac{\partial P_w}{\partial x} - \frac{\lambda_w \lambda_a}{\lambda_t} \frac{\partial P_w}{\partial x} + \frac{\lambda_w \lambda_n}{\lambda_t} \frac{\partial P_n}{\partial x} + \frac{\lambda_w \lambda_a}{\lambda_t} \frac{\partial P_a}{\partial x}. \quad (8)$$

Capillary pressure is introduced into Eq. (8) by grouping terms and distributing negative signs:

$$V_w = \frac{\lambda_w}{\lambda_t} V_t - \frac{\lambda_w \lambda_n}{\lambda_t} \frac{\partial}{\partial x} (P_w - P_n) - \frac{\lambda_w \lambda_a}{\lambda_t} \frac{\partial}{\partial x} (P_w - P_a) \quad (9)$$

$$= \frac{\lambda_w}{\lambda_t} V_t + \frac{\lambda_w \lambda_n}{\lambda_t} \frac{\partial}{\partial x} (P_n - P_w) + \frac{\lambda_w \lambda_a}{\lambda_t} \frac{\partial}{\partial x} (P_a - P_n + P_n - P_w) \quad (10)$$

$$= \frac{\lambda_w}{\lambda_t} V_t + \frac{\lambda_w \lambda_n}{\lambda_t} \frac{\partial P_{cnw}}{\partial x} + \frac{\lambda_w \lambda_a}{\lambda_t} \frac{\partial P_{can}}{\partial x} + \frac{\lambda_w \lambda_a}{\lambda_t} \frac{\partial P_{cnw}}{\partial x}. \quad (11)$$

We introduce the following functions to streamline the expanding notation:

$$f_w(x, t) = \frac{\lambda_w}{\lambda_t}, \quad (12a)$$

$$H(x, t) = \frac{\lambda_w(\lambda_n + \lambda_a)}{\lambda_t}, \quad (12b)$$

$$G(x, t) = \frac{\lambda_w \lambda_a}{\lambda_t}. \quad (12c)$$

Then Eq. (11) becomes $V_w = f_w(x, t)V_t + H(x, t)\dfrac{\partial P_{cnw}}{\partial x} + G(x, t)\dfrac{\partial P_{can}}{\partial x}$.

Similarly, we let $V_a = \sum_i \dfrac{\lambda_i}{\lambda_t} V_a \pm \dfrac{\lambda_a}{\lambda_t} V_w \pm \dfrac{\lambda_a}{\lambda_t} V_n$ so that

$$V_a = \frac{\lambda_a}{\lambda_t} V_t + \frac{\lambda_w}{\lambda_t} V_a + \frac{\lambda_n}{\lambda_t} V_a - \frac{\lambda_a}{\lambda_t} V_w - \frac{\lambda_a}{\lambda_t} V_n \quad (13)$$

$$= \frac{\lambda_a}{\lambda_t} V_t - \frac{\lambda_w \lambda_a}{\lambda_t} \frac{\partial P_a}{\partial x} - \frac{\lambda_n \lambda_a}{\lambda_t} \frac{\partial P_a}{\partial x} + \frac{\lambda_w \lambda_a}{\lambda_t} \frac{\partial P_w}{\partial x} + \frac{\lambda_n \lambda_a}{\lambda_t} \frac{\partial P_n}{\partial x} \quad (14)$$

$$= \frac{\lambda_a}{\lambda_t} V_t + \frac{\lambda_a \lambda_n}{\lambda_t} \frac{\partial}{\partial x} (P_n - P_a) + \frac{\lambda_w \lambda_a}{\lambda_t} \frac{\partial}{\partial x} (P_w - P_n + P_n - P_a) \quad (15)$$

$$= \frac{\lambda_a}{\lambda_t} V_t - \frac{\lambda_a \lambda_n}{\lambda_t} \frac{\partial P_{can}}{\partial x} - \frac{\lambda_w \lambda_a}{\lambda_t} \frac{\partial P_{cnw}}{\partial x} - \frac{\lambda_w \lambda_a}{\lambda_t} \frac{\partial P_{can}}{\partial x}, \quad (16)$$

$$= f_a(x, t)V_t - G(x, t)\frac{\partial P_{cnw}}{\partial x} - F(x, t)\frac{\partial P_{can}}{\partial x} \quad (17)$$

where

$$f_a(x, t) = \frac{\lambda_a}{\lambda_t}, \quad (18a)$$

$$F(x, t) = \frac{\lambda_a (\lambda_w + \lambda_n)}{\lambda_t}. \tag{18b}$$

The equilibrium capillary pressure between water and NAPL is taken to be a decreasing linear function of water saturation, while the equilibrium capillary pressure between air and NAPL is taken to be an increasing linear function of air saturation [2]. Substituting Eq. (11) into Eq. (3a), we have that the conservation of mass for the water phase is

$$
\phi \rho_w \frac{\partial u}{\partial t} + V_t \frac{\partial f_w(x, t)}{\partial x}
$$
$$
= \frac{\partial}{\partial x} \left[-H(x, t) \frac{\partial}{\partial x} \left(-u - \tau_{nw} \phi \frac{\partial u}{\partial t} - \tau_{nw} \gamma_{nw} \hat{k}_{nw} \frac{\epsilon_{nw} - \epsilon_{nw}^{eq}}{-u} \right) \right.
$$
$$
\left. + G(x, t) \frac{\partial}{\partial x} \left(v - \tau_{an} \phi \frac{\partial s}{\partial t} - \tau_{an} \gamma_{an} \hat{k}_{an} \frac{\epsilon_{an} - \epsilon_{an}^{eq}}{v} \right) \right]. \tag{19}
$$

Then substituting Eq. (17) into Eq. (3b), conservation of mass for the air phase becomes

$$
\phi \rho_a \frac{\partial v}{\partial t} + V_t \frac{\partial f_a(x, t)}{\partial x}
$$
$$
= \frac{\partial}{\partial x} \left[G(x, t) \frac{\partial}{\partial x} \left(-u - \tau_{nw} \phi \frac{\partial u}{\partial t} - \tau_{nw} \gamma_{nw} \hat{k}_{nw} \frac{\epsilon_{nw} - \epsilon_{nw}^{eq}}{-u} \right) \right.
$$
$$
\left. + F(x, t) \frac{\partial}{\partial x} \left(v - \tau_{an} \phi \frac{\partial s}{\partial t} - \tau_{an} \gamma_{an} \hat{k}_{an} \frac{\epsilon_{an} - \epsilon_{an}^{eq}}{v} \right) \right]. \tag{20}
$$

We make the substitution $s = 1 - u - v$ in Eqs. (19, 20) so that the two equations only depend on water and air saturations. Reorganizing the right-hand sides of Eqs. (19, 20) into dissipation and dispersion terms gives

$$
\phi \rho_w \frac{\partial u}{\partial t} + V_t \frac{\partial f_w(x, t)}{\partial x} = \frac{\partial}{\partial x} \left[H(x, t) \left(1 + \tau_{nw} \gamma_{nw} \hat{k}_{nw} \frac{\epsilon_{nw} - \epsilon_{nw}^{eq}}{u^2} \right) \frac{\partial u}{\partial x} \right.
$$
$$
\left. + G(x, t) \left(1 + \tau_{an} \gamma_{an} \hat{k}_{an} \frac{\epsilon_{an} - \epsilon_{an}^{eq}}{v^2} \right) \frac{\partial v}{\partial x} \right]
$$
$$
+ \frac{\partial}{\partial x} \left[\phi \left(H(x, t) \tau_{nw} + G(x, t) \tau_{an} \right) \frac{\partial^2 u}{\partial x t} \right.
$$
$$
\left. + \phi \tau_{an} G(x, t) \frac{\partial^2 v}{\partial x t} \right] \tag{21}
$$

and

$$\phi\rho_a\frac{\partial v}{\partial t} + V_t\frac{\partial f_a(x,t)}{\partial x} = \frac{\partial}{\partial x}\left[-G(x,t)\left(1 + \tau_{nw}\gamma_{nw}\hat{k}_{nw}\frac{\epsilon_{nw}-\epsilon_{nw}^{eq}}{u^2}\right)\frac{\partial u}{\partial x}\right.$$

$$+ F(x,t)\left(1 + \tau_{an}\gamma_{an}\hat{k}_{an}\frac{\epsilon_{an}-\epsilon_{an}^{eq}}{v^2}\right)\frac{\partial v}{\partial x}\right]$$

$$+ \frac{\partial}{\partial x}\left[\phi\left(F(x,t)\tau_{an} - G(x,t)\tau_{nw}\right)\frac{\partial^2 u}{\partial xt}\right.$$

$$+ \phi\tau_{an}F(x,t)\frac{\partial^2 v}{\partial xt}\right]. \tag{22}$$

To nondimensionalize Eqs. (21, 22), we let $t = T\bar{t}$ and $x = L\bar{x}$, where T and L are constant characteristic time and length scales, respectively, and the bar notation represents the nondimensional variable. Substituting these into Eqs. (21, 22) and balancing coefficients, we obtain the relation $\frac{\phi}{T} = \frac{V_t}{L} = \frac{1}{L^2}$. Then Eqs. (21, 22) become

$$u_t + (f_w(u,v))_x = \left[(1 + \frac{\tau_1}{u^2})H(u,v)u_x + (1 + \frac{\tau_2}{v^2})G(u,v)v_x\right]_x$$

$$+ [(\tau_3 H(u,v) + \tau_4 G(u,v))u_{tx} + \tau_4 G(u,v)v_{tx}]_x \tag{23a}$$

$$v_t + (f_a(u,v))_x = \left[-(1 + \frac{\tau_1}{u^2})G(u,v)u_x + (1 + \frac{\tau_2}{v^2})F(u,v)v_x\right]_x$$

$$+ [(-\tau_3 G(u,v) + \tau_4 F(u,v))u_{tx} + \tau_4 F(u,v)v_{tx}]_x \tag{23b}$$

in which we drop the bar notation and introduce the following notation for simplicity:

$$\tau_1 = \tau_{nw}\gamma_{nw}\hat{k}_{nw}(\epsilon_{nw} - \epsilon_{nw}^{eq}), \tag{24a}$$

$$\tau_2 = \tau_{an}\gamma_{an}\hat{k}_{an}(\epsilon_{an} - \epsilon_{an}^{eq}), \tag{24b}$$

$$\tau_3 = \frac{\tau_{nw}\phi}{T} = \frac{\tau_{nw}}{L^2}, \tag{24c}$$

$$\tau_4 = \frac{\tau_{an}\phi}{T} = \frac{\tau_{an}}{L^2}. \tag{24d}$$

Finally, we scale x and t by a small parameter $\epsilon > 0$ in order to control the dissipative and dispersive effects. Then, system (23) becomes

$$\frac{\partial}{\partial t}\begin{pmatrix} u_\epsilon \\ v_\epsilon \end{pmatrix} + \frac{\partial}{\partial x}\begin{pmatrix} f_w \\ f_a \end{pmatrix}$$

$$= \epsilon \frac{\partial}{\partial x}\left[\begin{pmatrix} (1+\frac{\tau_1}{u^2})H(u,v) & (1+\frac{\tau_2}{v^2})G(u,v) \\ -(1+\frac{\tau_1}{u^2})G(u,v) & (1+\frac{\tau_2}{v^2})F(u,v) \end{pmatrix} \frac{\partial}{\partial x}\begin{pmatrix} u_\epsilon \\ v_\epsilon \end{pmatrix}\right]$$

$$+ \epsilon^2 \frac{\partial}{\partial x}\left[\begin{pmatrix} \tau_3 H(u,v)+\tau_4 G(u,v) & \tau_4 G(u,v) \\ -\tau_3 G(u,v)+\tau_4 F(u,v) & \tau_4 F(u,v) \end{pmatrix} \frac{\partial^2}{\partial x \partial t}\begin{pmatrix} u_\epsilon \\ v_\epsilon \end{pmatrix}\right]. \qquad (25)$$

Note that when TCAT capillary pressure is reduced to dynamic capillary pressure, so that $\tau_1 = \tau_2 = 0$, the dissipation coefficient matrix in system (25) simplifies but the dispersion coefficient matrix is unaffected. Moreover, when only Eq. (1) is used in the derivation, the regularization terms in system (25) are limited to the same simplified dissipation terms but now the dispersion terms vanish. It is precisely the inclusion of rate-dependent capillary pressure that allows for undercompressive shocks in the solution structure.

3   Nonclassical Solutions: Undercompressive Shocks

System (25) approaches a strictly hyperbolic system of conservation laws as $\epsilon \to 0$ when

$$\lambda_w = \frac{1}{\mu_w} u^2, \qquad (26a)$$

$$\lambda_a = \frac{1}{\mu_a}\left(\beta_a v + (1-\beta_a)v^2\right), \qquad (26b)$$

$$\lambda_n = \frac{1}{\mu_n}(1-u-v)(1-u)(1-v), \qquad (26c)$$

with $0 < \dfrac{\mu_a}{\sqrt{\mu_n \mu_w}} < \beta_a$ and $\mu_w < 2\mu_n$ [21]. Juanes and Patzek show that Eq. (26), based on typical assumptions in hydrology and petroleum engineering, allows for the unregularized system to be classified as strictly hyperbolic for all u, $v \in (0, 1)$ instead of elliptic for some saturations u, v [2, 17, 27, 28]. Figure 1 illustrates the flux functions (12a, 18a) when $\mu_w = 0.875$, $\mu_a = 0.03$, $\mu_n = 2$, $\beta_a = 0.1$ as in [21]. The surface plot of f_w has a shape analogous to the typical S-shaped flux curve in the Buckley–Leverett equation for two-phase flow; see [32, 33].

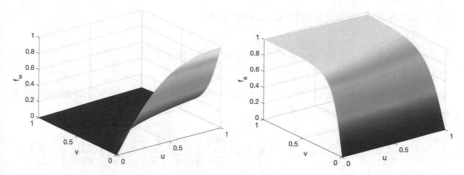

Fig. 1 Flux functions f_w (left) and f_a (right), given by Eqs. (12a, 18a) with mobility functions (26) and constants $\mu_w = 0.875$, $\mu_a = 0.03$, $\mu_n = 2$, $\beta_a = 0.1$ from [21]

Under these conditions, Juanes and Patzek determine that solutions of the associated Riemann problem in $\mathbf{U} = (u, v)$

$$\mathbf{U}_t + \mathbf{f}(\mathbf{U})_x = \mathbf{0}, \quad -\infty < x < \infty, \ t > 0 \tag{27a}$$

$$\mathbf{U}(x, 0) = \begin{cases} \mathbf{U}_\ell & \text{if } x < 0 \\ \mathbf{U}_r & \text{if } x > 0 \end{cases} \tag{27b}$$

are combinations of rarefaction and shock waves [20]. Further, the characteristic fields are not genuinely nonlinear, so that the shocks found in [20], traveling with speed σ, all satisfy the *Liu entropy criterion* [23, 24]:

$$\sigma_j(\mathbf{U}, \mathbf{U}_-) > \sigma_j(\mathbf{U}_+, \mathbf{U}_-) \tag{28}$$

for all $\mathbf{U}$ lying on the j-shock curve that passes through the point $\mathbf{U}_-$ between left and right states $\mathbf{U}_\pm$. See [7, 30] for further background reading.

While Juanes considers the impact of non-equilibrium constitutive equations in [19], he uses a model from Barenblatt and Vinnichenko [1] that incorporates time dependency into the relative permeability functions through an evolution equation for water saturation rather than a rate-dependent capillary pressure equation. The dynamic system in [19], then, is different from system (23) and does not reveal nonclassical solutions because capillarity is still neglected. The main conclusion in [19] is that the regularization terms only smooth the solutions; undercompressive shocks are not considered to be attainable by including the regularization terms in the analysis and numerics.

Together, system (25) and the work of Juanes and Patzek in [20, 21] match the framework developed by Hayes and LeFloch in [15] for nonclassical shock solutions of strictly hyperbolic systems of conservation laws. In [15], the authors consider a system of N strictly hyperbolic conservation laws

$$\frac{\partial \mathbf{U}}{\partial t} + \frac{\partial \mathbf{f}(\mathbf{U})}{\partial x} = 0, \ \mathbf{U}(x,t) \in \mathscr{U}, \tag{29}$$

for an open and convex $\mathscr{U} \subset \mathbb{R}^N$ and flux function $\mathbf{f} : \mathscr{U} \to \mathbb{R}^N$. The regularized approximation to Eq. (29) is given as

$$\frac{\partial \mathbf{U}_\epsilon}{\partial t} + \frac{\partial \mathbf{f}(\mathbf{U}_\epsilon)}{\partial x} = \epsilon \frac{\partial}{\partial x} \left[B_1(\mathbf{U}_\epsilon) \frac{\partial \mathbf{U}_\epsilon}{\partial x} \right] + \epsilon^2 \frac{\partial}{\partial x} \left[B_2(\mathbf{U}_\epsilon) \frac{\partial^2 \mathbf{U}_\epsilon}{\partial x^2} \right] \tag{30}$$

in which $\epsilon > 0$ and B_1, B_2 are $N \times N$ matrices. Hayes and LeFloch rely on the use of an entropy pair (Φ, Ψ) such that $\Phi, \Psi : \mathbb{R}^N \to \mathbb{R}$ are smooth functions that satisfy

$$\nabla \Psi^T = \nabla \Phi^T D\mathbf{f} \tag{31}$$

$$\nabla^2 \Phi \geq cI \tag{32}$$

for some positive constant c. Then a *nonclassical shock* is defined to be a shock that satisfies the entropy inequality $\Phi(\mathbf{U})_t + \Psi(\mathbf{U})_x \leq 0$ but not the Liu entropy criterion (28); this is equivalent to an undercompressive shock [15].

While the results in [15] assume that dispersion is represented by spatial derivatives only, the justifications are unaffected for system (25) with its rate-dependent dispersive term. Thus, by Theorem 2.6B from [15], the j-wave fan for Eq. (27) may contain a combined rarefaction-undercompressive shock or a leading classical shock trailed by a nonclassical shock.

4 Discussion

In this paper, we have derived a simple three-phase flow model with TCAT capillary pressure and described how this full system fits within the context of general results from Hayes and LeFloch [15]. The inclusion of a rate-dependent capillary pressure constitutive equation is essential in this process. The resultant dissipation and dispersion terms are the keys to unlocking previously neglected nonclassical solutions for this model.

While the work of Hayes and LeFloch in [15] demonstrates the existence of undercompressive shocks in the solution profile for system (25), applying their methods in this case study remains an open line of inquiry. The complexity of Eqs. (12, 18), with the specific functional forms given by Eq. (26), makes determining the entropy pair (Φ, Ψ) for system (25) exceedingly difficult, even with software. Once found, an appropriate entropy pair dictates an entropy dissipation function [15] which can be used to determine specific initial conditions $(\mathbf{U}_-, \mathbf{U}_+)$

for which undercompressive shocks arise. Explicitly identifying such initial conditions, along with numerical simulations to visualize corresponding solutions, would be an interesting and substantial addition to the work of Juanes and Patzek in [20].

References

1. Barenblatt GI, Vinnichenko AP (1980) Non-equilibrium seepage of immiscible fluids. Adv Mech 3(3):35–50
2. Bell JB, Trangenstein JA, Shubin GR (1986) Conservation laws of mixed type describing three-phase flow in porous media. SIAM J Appl Math 46(6):1000–1017
3. Corapcioglu MY, Hossain MA (1990) Ground-water contamination by high-density immiscible hydrocarbon slugs in gravity-driven gravel aquifers. Ground Water 28(3):403–412
4. DiCarlo DA (2004) Experimental measurements of saturation overshoot on infiltration. Water Resour Res 40: W04215
5. DiCarlo DA, Juanes R, LaForce T, Witelski TP (2008) Nonmonotonic traveling wave solutions of infiltration into porous media. Water Resour Res 44:W02406
6. Eckberg DK, Sunada DK (1984) Nonsteady three-phase immiscible fluid distribution in porous media. Water Resour Res 20(12):1891–1897
7. Evans, LC (1998) Partial Differential Equations. AMS, Providence, Rhode Island
8. Essaid HI, Bekins BA, Cozzarelli IM (2015) Organic contaminant transport and fate in the subsurface: Evolution of knowledge and understanding. Water Resour Res 51:4861–4902
9. Gray WG, Dye AL, McClure JE, Pyrak-Nolte LJ, Miller CT (2015) On the dynamics and kinematics of two-fluid-phase flow in porous media. Water Resour Res 51:5365–5381
10. Gray WG, Miller CT (2011) TCAT analysis of capillary pressure in non-equilibrium, two-fluid-phase, porous medium systems. Adv Water Resour 34:770–778
11. Gray WG, Miller CT (2014) Introduction to the Thermodynamically Constrained Averaging Theory for porous media systems. Springer, New York
12. Gray WG, Miller CT, Schrefler BA (2013) Averaging theory for description of environmental problems: What have we learned?. Adv Water Resour 51:123–138
13. Hassanizadeh SM, Gray WG (1990) Mechanics and thermodynamics of multiphase flow in porous media including interphase boundaries. Adv Water Resour 13:169–186
14. Hassanizadeh SM, Gray WG (1993) Thermodynamic basis of capillary pressure in porous media. Water Resour Res 29:3389–3405
15. Hayes BT, LeFloch PG (2000) Nonclassical shocks and kinetic relations: strictly hyperbolic systems. SIAM J Math Anal 31(5):941–991
16. Hersher R (2017) Key moments in the Dakota Access Pipeline fight. NPR. https://www.npr.org/sections/thetwo-way/2017/02/22/514988040/key-moments-in-the-dakota-access-pipeline-fight. Accessed 12 June 2018
17. Holden L (1990) On the strict hyperbolicity of the Buckley-Leverett equation for three-phase flow in a porous medium. SIAM J Appl Math 50(3):667–682
18. Illangasekare TH, Ramsey Jr. JL, Jensen KH, Butts MB (1995) Experimental study of movement and distribution of dense organic contaminants in heterogeneous aquifers. J Contaminant Hydrol 20:1–25
19. Juanes R (2009) Nonequilibrium effects in models of three-phase flow in porous media. Adv Water Resour 31:661–673
20. Juanes R, Patzek TW (2004) Analytical solution to the Riemann problem of three-phase flow in porous media. Trans Porous Med 55:47–70
21. Juanes R, Patzek TW (2004) Relative permeabilities for strictly hyperbolic models of three-phase flow in porous media. Trans Porous Med 57:125–152

22. LaForce T, Johns RT (2005) Analytical solutions for surfactant-enhanced remediation of nonaqueous phase liquids. Water Resour Res 41:W10420
23. Liu TP (1974) The Riemann problem for general 2×2 conservation laws. Trans Amer Math Soc 199:89–112
24. Liu TP (1975) The Riemann problem for general systems of conservation laws. J Diff Eq 18:218–234
25. O'Carroll DM, Phelan TJ, Abriola LM (2005) Exploring dynamic effects in capillary pressure in multistep outflow experiments. Water Resour Res 41:W11419
26. Peralta, E (2016) Dakota Access Pipeline protests in North Dakota turn violent. NPR. https://www.npr.org/sections/thetwo-way/2016/09/04/492625850/dakota-access-pipeline-protests-in-north-dakota-turn-violent. Accessed 12 June 2018
27. Shearer M (1988) Loss of strict hyperbolicity of the Buckley-Leverett equations for three phase flow in a porous medium. In: Wheeler MF (ed) Numerical Simulation in Oil Recovery, pp. 263–283. Springer-Verlag, New York
28. Shearer M, Trangenstein JA (1989) Loss of real characteristics for models of three-phase flow in a porous medium. Trans Porous Med 4:499–525
29. Shearer M, Spayd K, Swanson E (2015) Traveling waves for conservation laws with cubic nonlinearity and BBM type dispersion. J Diff Eq 259:3216–3232
30. Shearer M, Levy R (2015) Partial Differential Equations: An Introduction to Theory and Applications. Princeton Univ. Press, Princeton, New Jersey
31. Smith M, Bosman J (2017) Keystone Pipeline leaks 210,000 gallons of oil in South Dakota. New York Times. https://www.nytimes.com/2017/11/16/us/keystone-pipeline-leaks-south-dakota.html. Accessed 12 June 2018
32. Spayd K (2018) Generalizing the modified Buckley-Leverett equation with TCAT capillary pressure. Euro J Appl Math 29(2):338–351
33. Spayd K, Shearer M (2011) The Buckley-Leverett equation with dynamic capillary pressure. SIAM J Appl Math 71(4):1088–1108
34. Stonestrom DA, Akstin KC (1994) Nonmonotonic matric pressure histories during constant flux infiltration into homogeneous profiles. Water Resour Res 30(1): 81–91
35. van Duijn CJ, Peletier LA, Pop IS (2007) A new class of entropy solutions of the Buckley-Leverett equation. SIAM J Math Anal 39(2):507–536

An Invitation to Noncommutative Algebra

Chelsea Walton

Abstract This is a brief introduction to the world of Noncommutative Algebra aimed at advanced undergraduate and beginning graduate students.

1 Introduction

The purpose of this note is to invite you, the reader, into the world of Noncommutative Algebra. What is it? In short, it is the study of algebraic structures that have a noncommutative multiplication. One's first encounter with these structures occurs typically with matrices. Indeed, given two n-by-n matrices X and Y with $n > 1$, we get that $XY \neq YX$ in general. But this simple observation motivates a deeper reason why Noncommutative Algebra is ubiquitous...

Let's consider two basic transformations of images in real 2-space: Rotation by $90°$ clockwise and Reflection about the vertical axis. As we see in the figures below, the *order* in which these transformations are performed *matters*.

Since these transformations are *linear* (i.e., in $\mathbb{R}^2$, lines are sent to lines), they can be encoded by 2-by-2 matrices with entries in $\mathbb{R}$ [2, Section 3.C]. Namely

- $90°$ CW Rotation corresponds to $\left(\begin{smallmatrix} 0 & 1 \\ -1 & 0 \end{smallmatrix}\right)$, which sends vector $\left(\begin{smallmatrix} v_1 \\ v_2 \end{smallmatrix}\right)$ to $\left(\begin{smallmatrix} v_2 \\ -v_1 \end{smallmatrix}\right)$;
- Reflection about the y-axis is encoded by $\left(\begin{smallmatrix} -1 & 0 \\ 0 & 1 \end{smallmatrix}\right)$, which sends $\left(\begin{smallmatrix} v_1 \\ v_2 \end{smallmatrix}\right)$ to $\left(\begin{smallmatrix} -v_1 \\ v_2 \end{smallmatrix}\right)$.

The composition of linear transformations is then encoded by matrix multiplication. So, the first row in Fig. 1 corresponds to $\left(\begin{smallmatrix} -1 & 0 \\ 0 & 1 \end{smallmatrix}\right)\left(\begin{smallmatrix} 0 & 1 \\ -1 & 0 \end{smallmatrix}\right) = \left(\begin{smallmatrix} 0 & -1 \\ -1 & 0 \end{smallmatrix}\right)$, which sends $\left(\begin{smallmatrix} v_1 \\ v_2 \end{smallmatrix}\right)$ to $\left(\begin{smallmatrix} -v_2 \\ -v_1 \end{smallmatrix}\right)$. Yet the second row is given by $\left(\begin{smallmatrix} 0 & 1 \\ -1 & 0 \end{smallmatrix}\right)\left(\begin{smallmatrix} -1 & 0 \\ 0 & 1 \end{smallmatrix}\right) = \left(\begin{smallmatrix} 0 & 1 \\ 1 & 0 \end{smallmatrix}\right)$, sending $\left(\begin{smallmatrix} v_1 \\ v_2 \end{smallmatrix}\right)$ to $\left(\begin{smallmatrix} v_2 \\ v_1 \end{smallmatrix}\right)$. Therefore, the outcome of Fig. 1 is a result of the fact that $\left(\begin{smallmatrix} 0 & -1 \\ -1 & 0 \end{smallmatrix}\right) \neq \left(\begin{smallmatrix} 0 & 1 \\ 1 & 0 \end{smallmatrix}\right)$.

One can cook up other, say higher dimensional, examples of the varying outcomes of composing linear transformations by exploiting the noncommutativity

C. Walton (✉)
The University of Illinois at Urbana-Champaign, Department of Mathematics, Urbana, IL, USA
e-mail: notlaw@illinois.edu

© The Author(s) and the Association for Women in Mathematics 2019
S. D'Agostino et al. (eds.), *A Celebration of the EDGE Program's Impact on the Mathematics Community and Beyond*, Association for Women in Mathematics Series 18, https://doi.org/10.1007/978-3-030-19486-4_23

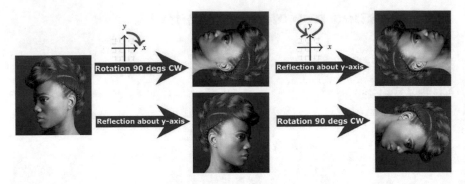

Fig. 1 The composition of rotation and reflection transformations is noncommutative

of matrix multiplication. This is all part of the general phenomenon that *functions* do not commute under composition typically. (Think of the myriad of outcomes of composing functions from everyday life—for instance, washing and drying clothes!)

Now let's turn our attention to special functions that we first encounter as children: **Symmetries**. To make this concept more concrete mathematically, consider the informal definition and notation below.

Definition 1 Take any object X. Then, a *symmetry* of X is an invertible, structure/property-preserving transformation from X to itself. The collection of such transformations is denoted by $\mathrm{Sym}(X)$.

Historically, the examination of symmetries in mathematics and physics served as one of the inspirations for defining a *group* as an abstract algebraic structure (see, e.g. [43, Section 1(c)]). Namely $\mathrm{Sym}(X)$ is a group with the identity element e being the "do nothing" transformation, with composition as the associative binary operation, and $\mathrm{Sym}(X)$ is equipped with inverse elements by definition.

Continuing the example above: Take $X = \mathbb{R}^2$ and $\mathrm{Sym}(\mathbb{R}^2)$ to be the collection of $\mathbb{R}$-linear transformations from $\mathbb{R}^2$ to $\mathbb{R}^2$ (so the origin is fixed). We get that $\mathrm{Sym}(\mathbb{R}^2)$ is the *general linear group* $\mathrm{GL}(\mathbb{R}^2)$, often written as $\mathrm{GL}_2(\mathbb{R})$ denoting the group of all invertible 2-by-2 matrices with real entries. Further, this group is *non-abelian*; thus, composition of $\mathbb{R}$-linear symmetries of $\mathbb{R}^2$ is noncommutative.

Another concept that is inherently noncommutative is that of a **representation**. We will see later in Sect. 3 that this is best motivated by elementary problem of finding *matrix solutions to equations* (which, in turn, can have physical implications). But for now let's think about the problem below.

Question 1 Which matrices $M \in \mathrm{Mat}_2(\mathbb{R})$ satisfy the equation $x^2 = 1$?

Now one could do the chore of writing down an arbitrary matrix $M = \left(\begin{smallmatrix} a & b \\ c & d \end{smallmatrix}\right)$ and solve for entries a, b, c, d that satisfy

$$\begin{pmatrix} a & b \\ c & d \end{pmatrix} \begin{pmatrix} a & b \\ c & d \end{pmatrix} = \begin{pmatrix} 1 & 0 \\ 0 & 1 \end{pmatrix}.$$

Not only is this boring, and it can be very tedious to find solutions to more general problems (e.g., taking instead $M \in \mathrm{Mat}_n(\mathbb{R})$ for $n > 2$). For a more elegant approach to Question 1, consider an abstract algebraic structure T defined by the equation $x^2 = 1$, and link T to $\mathrm{Sym}(\mathbb{R}^2)$ via a structure-preserving map ϕ. Then, a solution to Question 1 is produced in terms of an image of ϕ.

For example, we could take T to be the group $\mathbb{Z}_2$ as its presentation is given by $\langle x \mid x^2 = e \rangle$. An example of a structure-preserving map ϕ is given by

$$\phi : \mathbb{Z}_2 \to \mathrm{Sym}(\mathbb{R}^2), \quad e \mapsto \{\text{Do Nothing}\}, \quad x \mapsto \{\text{Reflection about } y-\text{axis}\}.$$

Indeed, $\phi(gg') = \phi(g) \circ \phi(g')$ for all $g, g' \in \mathbb{Z}_2$. For instance,

$$\phi(x) \circ \phi(x) = \{\text{Ref. about } y-\text{axis}\} \circ \{\text{Ref. about } y-\text{axis}\} = \{\text{Nothing}\} = \phi(e) = \phi(x^2).$$

Since $\phi(e)$ and $\phi(x)$ correspond, respectively, to $\left(\begin{smallmatrix} 1 & 0 \\ 0 & 1 \end{smallmatrix}\right)$ and $\left(\begin{smallmatrix} -1 & 0 \\ 0 & 1 \end{smallmatrix}\right)$, these matrices are solutions to Question 1. Further, other reflections of $\mathbb{R}^2$ produce additional solutions to Question 1 (Fig. 2). (Think about if *all* solutions to Question 1 can be constructed in this manner.)

Continuing this example, instead of using the abstract group $\mathbb{Z}_2$ we could have used the *group algebra* $T = \mathbb{R}\mathbb{Z}_2$, as it encodes the same information needed to address Question 1. We will discuss more about abstract algebraic structures in Sect. 2 (see Fig. 5); in any case, their representations are defined informally below.

Definition 2 Given an abstract algebraic structure T, we say that a *representation* of T is an object X equipped with a structure/property-preserving map $T \to \mathrm{Sym}(X)$.

An example of a representation of a group G is a vector space V equipped with a group homomorphism $G \to GL(V)$, where $GL(V)$ is the group of invertible linear transformations from V to itself (e.g., $GL(\mathbb{R}^2) = GL_2(\mathbb{R})$ as discussed above). Just as a representation of G is identified as a *G-module*, representations of rings and of algebras coincide with *modules* over such structures (see Fig. 12). See also [50, Chapters 1 and 3] for further reading and examples.

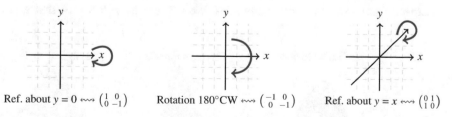

Fig. 2 Reflections of $\mathbb{R}^2$ and the corresponding solution to Question 1

Now *Representation Theory* is essentially a noncommutative area due to the following key fact. Take A to be a commutative algebra over a field $\Bbbk$ with a representation V of A, that is, a $\Bbbk$-vector space V equipped with algebra map $\phi : A \to GL(V)$. If (V, ϕ) is *irreducible* [Definition 14], then $\dim_{\Bbbk} V = 1$ [50, Section 1.3.2]. Therefore, representations of commutative algebras aren't so interesting.

Moreover, Representation Theory is a vital subject because the problem of finding matrix solutions to equations is quite natural. Since this boils down to studying representations of algebras that are generally noncommutative, the ubiquity of Noncommutative Algebra is conceivable. (Equations that correspond to representations of groups, like in Question 1, are special.)

To introduce the final notion in Noncommutative Algebra that we will highlight in this paper, observe symmetries and representations both occur under an *action* of a gadget T on an object X, but the difference is that symmetries form the gadget T (what is acting on an object), whereas representations are considered to be the object X (something being acted upon). What happens to these notions if we consider **deformations** of T and X? Consider the following informal terminology.

Definition 3 A *deformation of an object* X is an object X_{def} that has many of the same characteristics of X, possibly with the exception of a few key features. In particular, a *deformation of an algebraic structure* T is an algebraic structure T_{def} of the same type that shares a (less complex) underlying structure of T.

For example, a deformation of a ring R could be another ring R_{def} that equals R as abelian groups, possibly with a different multiplication than that of R (see Fig. 5).

Now if we deform an object X, is there a gadget T_{def} that acts on X_{def} naturally? On the other hand, if we deform the gadget T, is there a natural deformation X_{def} of X that comes equipped with an action of T_{def}? These are obvious questions, yet the answers are difficult to visualize. This is because, visually, symmetries of an object X are destroyed when X is altered, even slightly; see Fig. 3.

So we need to think beyond what can be visualized and consider a larger mathematical framework that handles symmetries under deformation. To do so, it is essential to think beyond group actions, because many classes of groups, including finite groups, do not admit deformations. However, *group algebras* or *function algebras on groups* do admit deformations, so we include these gadgets in the improved framework to study symmetries. We will see later in Sect. 4 that when

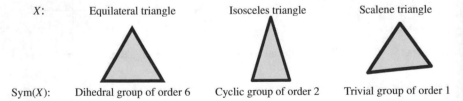

X:	Equilateral triangle	Isosceles triangle	Scalene triangle
Sym(X):	Dihedral group of order 6	Cyclic group of order 2	Trivial group of order 1

Fig. 3 Triangles and their respective symmetry groups

symmetries are recast in the setting where they could be preserved under deformation, other interesting and more general algebraic gadgets like *bialgebras* and *Hopf algebras* arise in the process. This is crucial in Noncommutative Algebra as some of the most important rings, especially those arising in physics, are noncommutative deformations of commutative rings; the symmetries of such deformations deserve attention.

> **Symmetries**, **Representations**, *and* **Deformations** *will play a key role*
> *throughout this article, just as they do in Noncommutative Algebra.*

The remainder of the paper is two-fold: first, we will review three historical snapshots of how Noncommutative Algebra played a prominent role in mathematics and physics. We will discuss William Rowan Hamilton's Quaternions in Sect. 2 and the birth of Quantum Mechanics in Sect. 3. We will also briefly discuss the emergence of Quantum Groups in Sect. 4, together with the concept of Quantum Symmetry. In Sect. 5 we present a couple of research avenues for further investigation. All of the material here is by no means exhaustive, and many references will be provided throughout.

2   Hamilton's Quaternions (1840s–1860s)

Can *numbers* be noncommutative? The best answer is, as always, "Sure, why not?" (Fig. 4).

In this section we will explore a number system that generalizes both the systems of real numbers $\mathbb{R}$ and complex numbers $\mathbb{C}$. The key feature of this new collection of numbers—the *quaternions* $\mathbb{H}$—is that they have a noncommutative multiplication! This feature caused a bit of ruckus for William Rowan Hamilton (1805–1865) after his discovery of the quaternions in the mid-nineteenth century.

Fig. 4 Numbers that we all know and love... but we should love more!

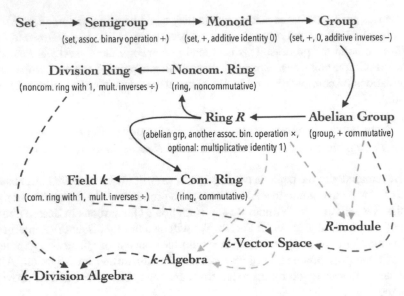

Fig. 5 Some algebraic structures. Straight arrows denote structures increasing in complexity. Dashed arrows denote structures merging compatibly to form another structure

> Quaternions came from Hamilton after really good work had been done; and, though beautifully ingenious, have been an unmixed evil to those who have touched them in any way...
> —Lord Kelvin, 1892

Now what do we mean by a *number system*? Loosely speaking, it is a *set of quantities* used to measure or count (a collection of) objects, which is equipped with an *algebraic structure* (Fig. 5).

Since we should be able to add, subtract, multiply, and divide numbers, we consider the following terminology.

Definition 4 Fix $n \in \mathbb{Z}_{\geq 1}$. An *n-dimensional division algebra D over $\mathbb{R}$* consists of the set of *n*-tuples of real numbers $\underline{a} := (a_1, a_2, \ldots, a_n)$, $a_i \in \mathbb{R}$, with $\mathbf{0} := (0, 0, \ldots, 0)$ and a unique element designated as $\mathbf{1}$ so that

- we can add and subtract two *n*-tuples $\underline{a}$ and $\underline{b}$ component-wise to form $\underline{a} + \underline{b}$ and $\underline{a} - \underline{b}$ in D, respectively;
- we can multiply $\underline{a}$ by a scalar $\lambda \in \mathbb{R}$ component-wise to form $\lambda * \underline{a}$ in D;
- there is a rule for multiplying $\underline{a}$ and $\underline{b}$ to form $\underline{a} \cdot \underline{b}$ in D (this is not necessarily done component-wise, nor does it need to be commutative); and
- there is a rule for dividing $\underline{a}$ by $\underline{b} \neq \mathbf{0}$ to form $\underline{a} \div \underline{b}$ in D;

in such a way that

(i) $(D, +, -, \mathbf{0})$ is an *abelian group*,
(ii) $(D, +, -, \mathbf{0}, *)$ is an $\mathbb{R}$-*vector space*,
(iii) $(D, \cdot, \mathbf{1})$ is an associative *unital ring*, and

(iv) $(D, +, -, *, \cdot, \mathbf{0}, \mathbf{1})$ is an associative $\mathbb{R}$- *algebra*

all in a compatible fashion (e.g., $\cdot$ distributes over $+$, etc.).

As one can imagine, there are not many of these gadgets floating around as they have a *lot* of structure. A 1-*dimensional division algebra D over $\mathbb{R}$* must be the field $\mathbb{R}$ itself. Moreover, a 2-*dimensional division algebra D over $\mathbb{R}$* is isomorphic to the field of complex numbers $\mathbb{C}$, where the pair (a_1, a_2) is identified with the element $a_1 + a_2 i$ for $i^2 = -1$. The algebraic structure for the pairs then follows accordingly, e.g., the multiplication of $\mathbb{C}$ is given by

$$(a_1, a_2) \cdot (b_1, b_2) = (a_1 b_1 - a_2 b_2, \ a_1 b_2 + a_2 b_1).$$

Note that the 1- and 2-dimensional real division algebras, $\mathbb{R}$ and $\mathbb{C}$, are commutative rings, and these can be viewed geometrically as in Fig. 6.

A natural question is then the following.

Question 2 What are the n-dimensional real division algebras for $n \geq 3$?

Hamilton obsessed over this question, especially the $n = 3$ case, for over a decade. Even his children would routinely ask him, "Papa, can you multiply triplets?" (Figs. 7 and 8).

His initial ideas were to use two imaginary axes i and j so that the 3-tuples (a_1, a_2, a_3) of a 3-dimensional number system correspond to $a_1 + a_2 i + a_3 j$.

Fig. 6 The real line, and the complex plane visualized as $\mathbb{R}^2$

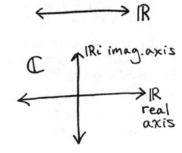

Fig. 7 A failed attempt at a 3-dimensional number system

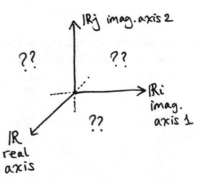

Fig. 8 Plaque on Brougham
Bridge in Dublin, recognizing
Hamilton's invention

However, he could not cook up rules that i and j should obey to make this collection of triples a valid division algebra [30, 55, 67]. According to some mathematicians, this obsession was quite "Mad" [4, 60].

Finally, on October 16th 1843, on a walk with his wife in Dublin, Hamilton had a moment of Eureka! In his words to his son Archibald,

> An *electric* circuit seemed to close; and a spark flashed forth, the herald, as I *foresaw immediately*, of many long years to come of definitely directed thought and work [...]
> Nor could I resist the impulse, unphilosophical as it may have been, to cut with a knife on the stone of Brougham Bridge, as we passed it, the fundamental formula with the symbols i, j, k; namely $i^2 = j^2 = k^2 = ijk = -1$, which contains the *solution* of the *problem...*
> —W. R. Hamilton, August 5th, 1865

Hamilton had discovered that day a number system generalizing both $\mathbb{R}$ and $\mathbb{C}$, consisting of *4-tuples* of real numbers, not constructed from triplets as he had imagined for so long [30].

Definition 5 The *quaternions* is a 4-dimensional real division algebra, denoted by $\mathbb{H}$, comprised of 4-tuples of real numbers $\underline{a} := (a_0, a_1, a_2, a_3)$, which are identified as elements of the form

$$a_0 + a_1 i + a_2 j + a_3 k, \quad \text{for } a_i \in \mathbb{R},$$

where addition, subtraction, and scalar multiplication are performed component-wise, and multiplication and division are governed by the rule

$$i^2 = j^2 = k^2 = ijk = -1.$$

Observe that $jk = i$, whereas $kj = -i$. Therefore, $\mathbb{H}$ is a noncommutative ring!

In any case, notice that the multiplication rule of $\mathbb{H}$ is a bit complicated:

$$(a_0, a_1, a_2, a_3) \cdot (b_0, b_1, b_2, b_3) = \begin{pmatrix} a_0b_0 - a_1b_1 - a_2b_2 - a_3b_3, \\ a_0b_1 + a_1b_0 + a_2b_3 - a_3b_2, \\ a_0b_2 - a_1b_3 + a_2b_0 + a_3b_1, \\ a_0b_3 + a_1b_2 - a_2b_1 + a_3b_0 \end{pmatrix} \tag{1}$$

... and let's not commit this rule to memory. To circumvent this issue Hamilton gave the quaternions a geometric realization that encodes their multiplication. Namely for $\underline{a} := a_0 + a_1 i + a_2 j + a_3 k \in \mathbb{H}$, let

> a_0 be the "scalar" component of $\underline{a}$, and
> $\vec{a} := a_1 i + a_2 j + a_3 k$ be the "vector" component of $\underline{a}$.

Then, the vector components are visualized as points/vectors in $\mathbb{R}^3$, whereas the scalar component is realized as an element of *time*. See, for instance, the footnote on page 60 and other parts of the preface of [29] for Hamilton's original thoughts on the connection between the quaternions and the laws of space and time. (Yes, yes, this was all very controversial back then!)

Hamilton then devised two vector operations, now known as the *dot product* ($\bullet$) and *cross product* ($\times$) to make the multiplication rule of $\mathbb{H}$ more compact:

$$\underline{a} \cdot \underline{b} = \left[a_0b_0 - \vec{a} \bullet \vec{b} \right] + \left[a_0\vec{b} + b_0\vec{a} + \vec{a} \times \vec{b} \right], \qquad \forall \underline{a}, \underline{b} \in \mathbb{H}. \tag{2}$$

Not only is formula (2) easier to retain than (1), the (commutative) dot product and (noncommutative) cross product have appeared in various parts of mathematics and physics throughout the years, including our multi-variable calculus courses (Fig. 9).

Geometrically, the operations in $\mathbb{H}$ capture symmetries of $\mathbb{R}^3$ [Definition 1]: addition/subtraction, scalar multiplication, and multiplication/division correspond, respectively, to translation, dilation, and rotation of vectors of $\mathbb{R}^3$; see, e.g. [29, page 272] and [45] for a discussion of rotation. To see rotations in action, first note that

Fig. 9 A successful attempt at a 4-dimensional number system

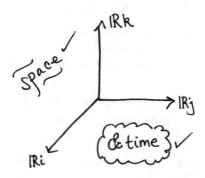

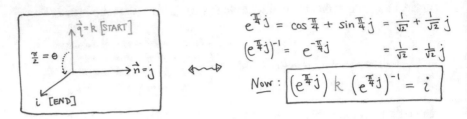

Fig. 10 Rotating vector k about axis j by $\frac{\pi}{2}$ radians $\leftrightsquigarrow$ Conjugating k by quaternion $e^{\frac{\pi}{4}j}$

the *length* of a quaternion $\underline{a}$ is given by

$$|\underline{a}| := \sqrt{a_0^2 + a_1^2 + a_2^2 + a_3^2}.$$

Next, fix an *axis of rotation* $\vec{n} := n_1 i + n_2 j + n_3 k$ with $|\vec{n}| = 1$, a quaternion of unit-length. Then, rotating a vector $\vec{q}$ about the axis $\vec{n}$ clockwise by θ radians (when viewed from the origin) corresponds to *conjugating* $\vec{q}$ by the quaternion $e^{\frac{\theta}{2}\vec{n}}$. It's helpful to use here an extension of Euler's formula, $e^{\frac{\theta}{2}\vec{n}} = \cos(\frac{\theta}{2}) + \sin(\frac{\theta}{2})\vec{n}$, to understand the quaternion $e^{\frac{\theta}{2}\vec{n}}$. An example is given in Fig. 10.

Moreover, rotations of $\mathbb{R}^3$ can be encoded as a representation [Definition 2] of the multiplicative subgroup $U(\mathbb{H})$ consisting of unit-length quaternions. Indeed, we have a group homomorphism

$$U(\mathbb{H}) \longrightarrow GL(\mathbb{R}^3) = GL_3(\mathbb{R}), \quad \text{given by}$$

$$
\begin{array}{c}
a_0 + a_1 i + a_2 j + a_3 k \\
\text{with} |\underline{a}| = 1
\end{array}
\mapsto
\begin{pmatrix}
a_0^2 + a_1^2 - a_2^2 - a_3^2 & 2a_1a_2 - 2a_0a_3 & 2a_1a_3 + 2a_0a_2 \\
2a_1a_2 + 2a_0a_3 & a_0^2 - a_1^2 + a_2^2 - a_3^2 & 2a_2a_3 - 2a_0a_1 \\
2a_1a_3 - 2a_0a_2 & 2a_2a_3 + 2a_0a_1 & a_0^2 - a_1^2 - a_2^2 + a_3^2
\end{pmatrix}.
$$

This geometric realization of $\mathbb{H}$ has many modern applications—we refer to the text [44] for a nice self-contained discussion of applications to computer-aided design, aerospace engineering, and other fields.

Returning to Question 2, its answer is now given below.

Theorem 1 ([26, 62], [46, Theorem 13.12], [8, 42, 71]) *The answer to Question 2 is Yes if and only if $n = 1, 2, 4, 8$. Such division algebras D are unique up to isomorphism in their dimension with isomorphism class represented by*

- the real numbers $\mathbb{R}$ for $n = 1$,
- quaternions $\mathbb{H}$ for $n = 4$,
- the complex numbers $\mathbb{C}$ for $n = 2$,
- octonions $\mathbb{O}$ for $n = 8$.

Here, D is commutative only when $n = 1, 2$, and is associative only when $n = 1, 2, 4$.

So, Hamilton discovered the last associative finite-dimensional real division algebra, but the price that he had to pay (at least mathematically) was the loss of commutativity. Perhaps this was not too high of a price—we are certainly willing to lose ordering when choosing to work with $\mathbb{C}$ instead of $\mathbb{R}$. If we are also willing to part with associativity, then the octonions $\mathbb{O}$ is a perfectly suitable number system; see [3] for more details. And, of course, there are further generalizations of number systems—see [19, 49, 70] to start, and go wild!

We return to the quaternions later in Sect. 5.1 for a discussion of potential research directions.

3 The Birth of Quantum Mechanics (1920s)

Another period that sparked an interest in Noncommutative Algebra was the birth of Quantum Mechanics in the 1920s. Three of the key figures during this time were Max Born (1882 1970), Werner Heisenberg (1901–1976), and Paul Dirac (1902–1984), who were all curious about the behavior of subatomic particles [7, 20, 32] (Fig. 11).

Along with their colleagues, Born, Heisenberg, and Dirac believed that important aspects of subatomic behavior are those that could be *observed* (or measured). However, the tools of classical mechanics available at the time (with *observables* corresponding to real-valued functions) were not suitable in capturing this behavior properly. A new type of mechanics was needed, leading to the development of quantum mechanics where observables are realized as linear operators. For a great account of how this transition took place (some of which we will recall briefly below), see Part II of the van der Waerden's text [68]. (For historical context of another figure, Pascual Jordan, who also played a role in these developments, see, e.g. [34].)

The two observables in which Born, Heisenberg, and Dirac were especially interested were the *position* and *momentum* of a subatomic particle, and they employed Niels Bohr's notion of *orbits* to keep track of these quantities. Mathematically,

Fig. 11 "More than anything, this photograph was really the result of a series of little accidents." —Billy Huynh, photographer . . . So is good mathematics!

this boils down to using matrices in order to book-keep data corresponding to the observables under investigation, thus initiating *matrix mechanics*. The surprising outcome of using this new matrix framework in studying subatomic particles was stated succinctly as follows [33]:

> The more precisely the position is determined, the less precisely the momentum is known, and vice versa.
> —Heisenberg's "Uncertainty Principle," 1927

More precisely, suppose that P and Q are square matrices of the same size representing the observables momentum and position, respectively. The fact that P and Q do not commute typically (as one expects in classic mechanics) led to the discovery of what Born dubbed as *The Fundamental Equation of Quantum Mechanics*:

$$PQ - QP = i\hbar * I, \tag{3}$$

Here i is the square root of -1, $\hbar$ is Planck's constant, and $i\hbar * I$ is the scalar multiple of the identity matrix I of the same size as P and Q. For physical reasons, it was known early in the theory of quantum mechanics that matrices P and Q that satisfy Eq. (3) should be of infinite size, and we will recall a well-known, mathematical proof of this fact later in Proposition 1.

As done in practice by many physicists and mathematicians, through rescaling let's consider a *normalized* version of The Fundamental Equation, as this still captures the spirit of Heisenberg's Uncertainty Principle:

$$PQ - QP = I, \tag{4}$$

Now with today's technology, one convenient way of studying matrix solutions P and Q to Eq. (4) (or to Eq. (3)) is to use the *theory of representations of (associative) algebras*. To see this connection, first let's fix a field $\Bbbk$ and for ease:

Standing Hypothesis. We assume in this section that $\Bbbk$ is a field of characteristic 0.

Then recall from Definition 4 (and Fig. 5) that a $\Bbbk$-*algebra* A is a $\Bbbk$-vector space equipped with the structure of a unital ring in a compatible fashion. In this case, $A = (A, +, -, *, \cdot, \mathbf{0}, \mathbf{1})$, where $(A, +, -, *, \mathbf{0})$ is the $\Bbbk$-vector space structure where $+$ is the abelian group operation and $*$ is scalar multiplication, and $(A, \cdot, \mathbf{1})$ is a unital ring with $\cdot$ denoting its multiplication. Next, we make our vague notion of representations in Definition 2 more precise in the context of $\Bbbk$-algebras.

Definition 6 Consider the following notions:

1. For a $\Bbbk$-vector space V, the *endomorphism algebra* $\mathrm{End}(V)$ on V is an $\Bbbk$-algebra consisting of endomorphisms of V with multiplication being composition $\circ$. (If V is an n-dimensional $\Bbbk$-vector space, then $\mathrm{End}(V)$ is isomorphic to the matrix algebra $\mathrm{Mat}_n(\Bbbk)$ with matrix multiplication. Here, n could be infinite.)

2. A *representation* of an associative $\Bbbk$-algebra A is a $\Bbbk$-vector space V equipped with a $\Bbbk$-algebra homomorphism $\phi : A \to \text{End}(V)$; say $\phi(a) =: \phi_a \in \text{End}(V)$ for $a \in A$. Namely for all $a, b \in A$, $\lambda \in \Bbbk$, and $v \in V$, we get that

$$\phi_{a+b}(v) = \phi_a(v) + \phi_b(v), \quad \phi_{\lambda * a}(v) = \lambda * \phi_a(v), \quad \phi_{ab}(v) = (\phi_a \circ \phi_b)(v).$$

3. The *dimension* of a representation (V, ϕ) of an associative $\Bbbk$-algebra A is the $\Bbbk$-vector space dimension of V, which could be infinite.

Representations of associative $\Bbbk$-algebras A go hand-in-hand with A-modules, as illustrated in Fig. 12.

Now for the purposes of finding matrix solutions of Eq. (4), consider the $\Bbbk$-algebra defined below.

Definition 7 The *(first) Weyl algebra* over a field $\Bbbk$ is the $\Bbbk$-algebra $A_1(\Bbbk)$ generated by noncommuting variables x and y, subject to relation $yx - xy = 1$. That is, $A_1(\Bbbk)$ has a $\Bbbk$-algebra presentation

$$A_1(\Bbbk) = \Bbbk\langle x, y \rangle / (yx - xy - 1),$$

given as the quotient algebra of the *free algebra* $\Bbbk\langle x, y \rangle$ (consisting of words in variables x and y) by the ideal $(yx - xy - 1)$ of $\Bbbk\langle x, y \rangle$. (This algebra is sometimes referred to as the *Heisenberg–Weyl algebra* due to its roots in physics.)

The Weyl algebra is also the first example of an *algebra of differential operators*—its generators x and y can be viewed as the differential operators on the polynomial algebra $\Bbbk[x]$ given by multiplication by x and $\frac{d}{dx}$, respectively. (Check that $\frac{d}{dx}x - x\frac{d}{dx}$ is indeed the identity operator on $\Bbbk[x]$.)

Returning to the problem of finding n-by-n matrix solutions to Normalized Fundamental Eq. (4)—this is equivalent to the task of constructing n-dimensional

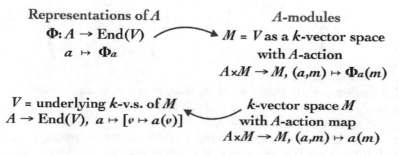

For an associative k-algebra A:

Representations of A   **A-modules**
$\Phi: A \to \text{End}(V)$ $\longrightarrow$ $M = V$ as a k-vector space
$a \mapsto \Phi_a$ with A-action
$A \times M \to M$, $(a, m) \mapsto \Phi_a(m)$

$V =$ underlying k-v.s. of M k-vector space M
$A \to \text{End}(V)$, $a \mapsto [v \mapsto a(v)]$ with A-action map
$A \times M \to M$, $(a, m) \mapsto a(m)$

Fig. 12 Connection between representations and modules of $\Bbbk$-algebras

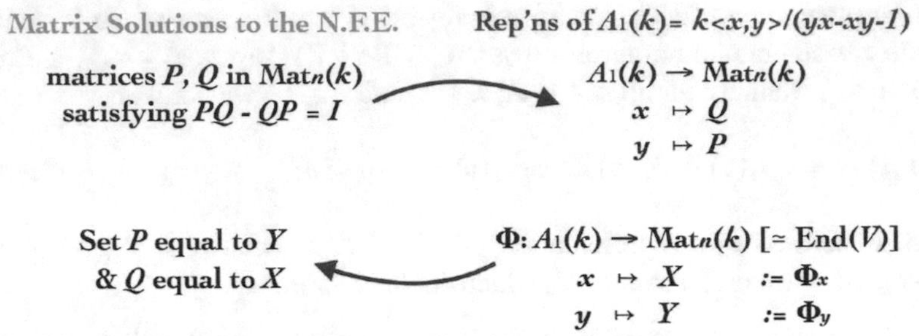

Fig. 13 Connection between matrix solutions to N.F.E. and representations of $A_1(\Bbbk)$

representations of $A_1(\Bbbk)$ as shown in Fig. 13. In fact, this is why $A_1(\Bbbk)$ is known as the *ring of quantum mechanics*.

Next, with the toolkit of matrices handy, we obtain a well-known result on the size of matrix solutions to (4). We need following facts about the *trace* of a square matrix X (which is the sum of the diagonal entries of X): $\operatorname{tr}(X \pm Y) = \operatorname{tr}(X) \pm \operatorname{tr}(Y)$ and $\operatorname{tr}(XY) = \operatorname{tr}(YX)$ for any $X, Y \in \operatorname{Mat}_n(\Bbbk)$.

Proposition 1 *The Normalized Fundamental Eq.* (4) *does not admit finite matrix solutions, i.e., representations of* $A_1(\Bbbk)$ *must be infinite-dimensional.*

Proof By way of contradiction, suppose that we have matrices $P, Q \in \operatorname{Mat}_n(\Bbbk)$ with $0 < n < \infty$ so that $PQ - QP = I$. Applying trace to both sides of this equation yields

$$0 = \operatorname{tr}(PQ) - \operatorname{tr}(PQ) = \operatorname{tr}(PQ) - \operatorname{tr}(QP) = \operatorname{tr}(PQ - QP) = \operatorname{tr}(I) = n,$$

a contradiction as desired.   □

On the other hand, the first Weyl algebra does have an infinite-dimensional representation. Take, for instance:

$$P = \begin{pmatrix} 0 & 1 & & & \\ & 0 & 2 & & \\ & & 0 & 3 & \\ & & & 0 & \ddots \\ & & & & \ddots \end{pmatrix} \quad \text{and} \quad Q = \begin{pmatrix} 0 & & & & \\ 1 & 0 & & & \\ & 1 & 0 & & \\ & & 1 & 0 & \\ & & & \ddots & \ddots \end{pmatrix}. \tag{5}$$

...And there are many, many more!

But finding *explicit* matrix solutions to equations is computationally difficult in general, especially when the most important representations of an algebra are infinite-dimensional. The power of representation theory, however, is centered on its tools to address more abstract algebraic problems that are (perhaps) related to computational goals. For instance, representation theory may address some of the following questions for a given $\Bbbk$-algebra A, which are all quite natural:

- Do representations of A *exist*? If so, what are their *dimensions*?

- When are two representations considered to be the *same* (or *isomorphic*)?
- Are (some of) the representations of A *parameterized* by a geometric object $\mathscr{X}$? Do isomorphism classes of representations correspond bijectively to points of $\mathscr{X}$?

We will explore a few of these questions and further notions in Sect. 5.2 towards a research direction in Representation Theory.

The representation theory of other algebras of differential operators has also been key in modeling subatomic behavior. This includes Dirac's *quantum algebra* that addresses the question of how several position observables $(Q_1, \ldots, Q_m)$ and momentum observables $(P_1, \ldots, P_m)$ commute, generalizing Heisenberg's Uncertainty Principle for $m = 1$ [20]. These days Dirac's algebra is now known as the *mth Weyl algebra* $A_m(\Bbbk)$, which has $\Bbbk$-algebra presentation:

$$
A_m(\Bbbk) = \frac{\Bbbk\langle x_1, \ldots, x_m, y_1, \ldots, y_m\rangle}{(x_i x_j - x_j x_i, \quad y_i y_j - y_j y_i, \quad y_i x_j - x_j y_i - \delta_{i,j})}. \tag{6}
$$

Here, $\delta_{i,j}$ is the Kronecker delta, and the generators x_i and y_i are viewed as elements of $\mathrm{End}(\Bbbk[x_1, \ldots, x_m])$ given resp. by multiplication by x_i and partial derivation $\frac{\partial}{\partial x_i}$.

Want more physics? We're in luck—the representation theory of numerous noncommutative $\Bbbk$-algebras plays a vital role in several fields of physics. *Some* of these algebras and a physical area in which they appear are listed below. Happy exploring!

Noncommutative $\Bbbk$-algebras	Appearance in physics	Reference (year)
$\mathscr{W}$-algebras	Conformal field theory	[9] (1993)
4-dimensional Sklyanin algebras	Statistical mechanics	[64] (1982)
3-dimensional Sklyanin algebras	String theory	[6] (2000)
Yang–Mills algebras	Gauge theory	[13] (2002)
Superpotential algebras	String theory	[25] (2006)
Various enveloping algebras of Lie algebras	* Everywhere *	Too many to list!

4 Quantum Groups (1980s–1990s) and Quantum Symmetries

Let's begin here with a question mentioned in the introduction on the ties between symmetries [Definition 1] and deformations [Definition 3].

Question 3 How do we best handle (i.e., axiomatize, or "make mathematical," the concept of) symmetries of deformations?

Several answers to this question lead us to use *Hopf algebras* [Definition 11]. But before we give the precise definition of this structure, we point out that Hopf algebras became prominent in mathematics in a few waves, including: its origins in Algebraic Topology [35], role in Combinatorics [38], and abstraction in Category Theory [40]. One tie to Noncommutative Algebra (in the context of Question 3) first appeared in the 1980s in statistical mechanics, especially in the *Quantum Inverse Scattering Method* for solving quantum integrable systems. The Hopf algebras that arose this way were coined *Quantum Groups* by Vladimir Drinfel'd [22], and have been a key structure in Noncommutative Algebra and physics ever since.

Instead of delving further into historical details, let's now discuss (quantum) symmetries of (deformed) algebras through concrete examples. Fix a field $\Bbbk$, and recall from Fig. 5 that an associative $\Bbbk$-algebra is a $\Bbbk$-vector space equipped with the structure of a (unital) ring; we consider their deformations below.

Definition 8 Fix a $\Bbbk$-algebra A. A $\Bbbk$-algebra A_{def} is a *deformation* of A if A_{def} and A are the same as $\Bbbk$-vector spaces, but their respective multiplication rules are not necessarily the same.

Example 1 Our running example of a $\Bbbk$-algebra throughout this section will be the *q-polynomial algebra*:

$$\Bbbk_q[x, y] = \Bbbk\langle x, y\rangle/(yx - qxy), \quad \text{for } q \in \Bbbk^\times,$$

which is the quotient algebra of the free algebra $\Bbbk\langle x, y\rangle$ by the ideal $(yx - qxy)$. Loosely speaking, $\Bbbk_q[x, y]$ is a q-deformation of $\Bbbk[x, y]$ as the former structure "approaches" the latter as $q \to 1$. More explicitly, note that $\Bbbk_q[x, y]$ and $\Bbbk[x, y]$ have the same $\Bbbk$-vector space basis $\{x^i y^j\}_{i,j\geq0}$, but their multiplication rules differ for $q \neq 1$.

Now let us examine symmetries of $\Bbbk_q[x, y]$ for $q \neq 1$ versus those of $\Bbbk[x, y]$. For this it is enough to consider *degree-preserving symmetries*, i.e., invertible transformations that send the generators x and y to a linear combination of themselves. Namely let $V = \Bbbk x \oplus \Bbbk y$ be the generating space of $\Bbbk_q[x, y]$ (or $\Bbbk[x, y]$ with $q = 1$). We want to pin down which invertible matrices in $GL(V) = GL_2(\Bbbk)$ also induce a symmetry of $\Bbbk_q[x, y]$, and to do so, we need to rewrite $\Bbbk_q[x, y]$ using the notion below. (From now on, we need an understanding of tensor products $\otimes$ and a nice discussion of this operation can be found in [15].)

Definition 9 Given a $\Bbbk$-vector space V, the *tensor algebra* $T(V)$ is the $\Bbbk$-vector space $\bigoplus_{i\geq0} V^{\otimes i}$, where $V^0 = \Bbbk$, and with multiplication given by concatenation, i.e., $(v_1 \otimes \cdots \otimes v_m)(v_{m+1} \otimes \cdots \otimes v_{m+n}) = v_1 \otimes \cdots \otimes v_{m+n}$.

Ideals I of tensor algebras $T(V)$ are defined as usual, and one can define a quotient $\Bbbk$-algebra given by $T(V)/I$.

Example 2 The free algebra $\Bbbk\langle x, y\rangle$ is identified with the tensor algebra $T(V)$ on the $\Bbbk$-vector space $V = \Bbbk x \oplus \Bbbk y$: for the forward direction insert $\otimes$ between variables, and conversely suppress $\otimes$ between variables. The q-polynomial algebra

$\Bbbk_q[x, y]$ is then identified as the quotient algebra of $T(\Bbbk x \oplus \Bbbk y)$ by the ideal $(y \otimes x - qx \otimes y)$.

Now take $g \in GL(V)$ for $V = \Bbbk x \oplus \Bbbk y$. We want to extend this symmetry on V to a symmetry of $\Bbbk_q[x, y]$ identified as $T(V)/(y \otimes x - qx \otimes y)$. Let's assume that, as in the case for group actions, g acts on $T(V)$ diagonally:

$$g(v \otimes v') := g(v) \otimes g(v'), \quad \forall v, v' \in V. \tag{7}$$

Now the question is: When is the ideal $(y \otimes x - qx \otimes y)$ preserved under this action? In fact it suffices to show that

$$g(y \otimes x - qx \otimes y) = \lambda(y \otimes x - qx \otimes y), \quad \text{for some } \lambda \in \Bbbk, \tag{8}$$

since the g-action is degree preserving. To be concrete, say $g \in GL(V)$ is given by

$$g(x) = \alpha x + \beta y \quad \text{and} \quad g(y) = \gamma x + \delta y, \quad \text{for some } \alpha, \beta, \gamma, \delta \in \Bbbk. \tag{9}$$

Then, $g(y \otimes x - qx \otimes y) = [g(y) \otimes g(x)] - q[g(x) \otimes g(y)]$, which is equal to

$$(1 - q)\alpha\gamma(x \otimes x) + (\beta\gamma - q\alpha\delta)(x \otimes y) + (\alpha\delta - q\beta\gamma)(y \otimes x) + (1 - q)\beta\delta(y \otimes y).$$

Therefore, the condition (8) is satisfied:

- Always,   if $q = 1$;
- Only when $\alpha = \delta = 0$ or $\beta = \gamma = 0$,   if $q = -1$;
- Only when $\beta = \gamma = 0$,   if $q \neq \pm 1$.

(Note that in the first case $\lambda = \alpha\delta - \beta\gamma$, the determinant of g when in matrix form.)

So, when we pass from the commutative polynomial algebra $\Bbbk[x, y]$ to its noncommutative deformation $\Bbbk_q[x, y]$ for $q \neq 1$, the amount of its degree-preserving symmetries shrinks abruptly. This is rather unsatisfying as passing "continuously" from $\Bbbk[x, y]$ to $\Bbbk_q[x, y]$ does not yield a "continuous passage" between their respective degree-preserving automorphism groups.

We need to think beyond group actions like those in (7). In general, we want to construct symmetries of a $\Bbbk$-algebra $T(V)/I$ by (i) considering symmetries of the generating space V, (ii) extending those to symmetries of $T(V)$, and then (iii) determining which symmetries in (ii) descend to $T(V)/I$. For step (i), take V to be a representation of an algebraic object H, e.g., H could be a group or a $\Bbbk$-algebra. (We often swap back and forth between using "representations" and "modules.") For (ii), one needs to tackle the issue of building a direct sum and tensor product of H-representations. The former is pretty straightforward—one can always construct the direct sum of H-representations to get another (the first guess is most likely the correct one!). But if we are given two vector spaces V_1 and V_2 that are H-modules,

Question 4 When is $V_1 \otimes V_2$ an H-module? [1]

If H were a group G, then one can give $V_1 \otimes V_2$ the structure of a (left) G-module via (7). We can extend this linearly to get that $V_1 \otimes V_2$ is a module over a group algebra on G. But if H were an arbitrary algebra, then the diagonal action on $V_1 \otimes V_2$ does not necessarily give it the structure of an H-module (as we will see in Remark 1). In fact, to have an action of H on $V_1 \otimes V_2$ we first need algebra maps

$$\Delta : H \to H \otimes H, \quad \Delta(h) \mapsto \sum h_1 \otimes h_2, \quad \text{and} \quad \epsilon : H \to \Bbbk.$$

Here, we use the *Sweedler notation* shorthand to denote elements of $\Delta(H)$. These maps should be compatible in a way that is dual to the manner that the multiplication map $m : H \otimes H \to H$ and unit map $\eta : \Bbbk \to H$ of an algebra are compatible (cf. $m(\eta \otimes \mathrm{id}_H) = \mathrm{id}_H = m(\mathrm{id}_H \otimes \eta)$). That is, after identifying $\Bbbk \otimes H = H = H \otimes \Bbbk$,

$$(\epsilon \otimes \mathrm{id}_H) \circ \Delta = \mathrm{id}_H = (\mathrm{id}_H \otimes \epsilon) \circ \Delta. \tag{10}$$

Definition 10 ([63, Chapter 5]) An associative $\Bbbk$-algebra $H = (H, m, \eta)$ is a $\Bbbk$-*bialgebra* if it is equipped with algebra maps Δ (*coproduct*) and ϵ (*counit*), so that (H, Δ, ϵ) is a *coassociative* $\Bbbk$-*coalgebra* with the structures (H, m, η) and (H, Δ, ϵ) being compatible.

To answer Question 4: If H is a bialgebra, the H-module structure on $V_1 \otimes V_2$ is

$$h(v_1 \otimes v_2) =: \sum h_1(v_1) \otimes h_2(v_2) \qquad \forall h \in H \text{ and } v_1, v_2 \in V.$$

We also get that $\Bbbk$ admits the structure of a *trivial* H-module via $h(1_\Bbbk) = \epsilon(h)1_\Bbbk$.

Remark 1 We cannot always use a diagonal action—sometimes a fancier coproduct is needed to address Question 4. To see this, take H to be the 2-dimensional associative $\Bbbk$-algebra $\Bbbk[h]/(h^2)$ (e.g., so that we are considering linear operators that are the zero map when composed with itself). If the coproduct of H is $\Delta(h) = h \otimes h$, then $\epsilon(h) = 1$ by (10). But this implies $0 = \epsilon(h^2) = \epsilon(h)^2 = 1$, a contradiction. To "fix" this, check that the coproduct $\Delta(h) = h \otimes 1 + 1 \otimes h$ with the counit $\epsilon(h) = 0$ gives $\Bbbk[h]/(h^2)$ the structure of a bialgebra over $\Bbbk$.

Moreover, one may be interested in (symmetries of) an algebra with generating space V^*, the *linear dual*; this will play a role later in Sect. 5.2. To get this, we want V^* to have the induced structure of an H-module, and we need an anti-algebra-automorphism $S : H \to H$ of H to proceed.

Definition 11 ([63, Chapters 6-7]) A $\Bbbk$-bialgebra $H = (H, m, \eta, \Delta, \epsilon)$ is a *Hopf algebra* over $\Bbbk$ if there exists anti-automorphism $S : H \to H$ (*antipode*) so that

---

[1] In categorical language, this is the question of whether the category of H-modules (or of representations of H) has a *monoidal* structure.

$$m \circ (S \otimes \mathrm{id}_H) \circ \Delta = \eta \circ \epsilon = m \circ (\mathrm{id}_H \otimes S) \circ \Delta.$$

If H is a Hopf algebra with H-module V, an action of H on V^* can be given by[2]

$$[h(f)](v) = f[S(h)(v)], \qquad \forall h \in H, \ f \in V^*, \ v \in V.$$

Examples of Hopf algebras are *group algebras* on finite groups $\Bbbk G$, *function algebras* on algebraic groups $\mathcal{O}(G)$, and *universal enveloping algebras* of Lie algebras $U(\mathfrak{g})$, which are all considered "classical" in the sense that they are *commutative* (as an algebra, $m \circ \tau = m$) or *cocommutative* (as a coalgebra, $\tau \circ \Delta = \Delta$), for $\tau(a \otimes b) = b \otimes a$. Indeed, these Hopf algebras capture the actions of a group on a $\Bbbk$-algebra by automorphism and actions of a Lie algebra on a $\Bbbk$-algebra by derivation. Moreover, deformations (or *quantized* versions) of these structures provide a setting to handle deformations of the aforementioned symmetries (cf. Question 3); refer to [1, 36, 51, 54] for examples of Hopf algebras arising in this fashion. We also recommend the excellent text on (actions of) Hopf algebras by Susan Montgomery [57].

Now we summarize a few frameworks for studying (quantum) symmetries of a $\Bbbk$-algebra A involving a group G or a Hopf algebra H. See [57] for more details.[3]

[G-ACT] *Group actions* on A: That is, A is a G-module with G-action map $G \times A \to A$ given by $(g, a) \mapsto g(a)$ satisfying $g(ab) = g(a)\, g(b)$ and $g(1_A) = 1_A$, for all $g \in G$ and $a, b \in A$.

[G-GRD] *Group gradings* on A: That is, A is G-graded if $A = \oplus_{g \in G} A_g$, for A_g a $\Bbbk$-vector space, with $A_g \cdot A_h \subset A_{gh}$. When G is finite, this is equivalent to A being acted upon by the *dual group algebra* $(\Bbbk G)^*$.

[H-ACT] *Hopf algebra* or *bialgebra actions* on A: That is, A is an H-module with H-action map $H \times A \to A$ given by $(h, a) \mapsto h(a)$ with $h(ab) = \sum h_1(a)\, h_2(b)$ and $h(1_A) = \epsilon(h)1_A$, for all $h \in H$ and $a, b \in A$, with $\Delta(h) = \sum h_1 \otimes h_2$.

Finally we end with an example of a Hopf algebra action on $\Bbbk_q[x, y]$, illustrating a scenario where Question 3 has a possible answer. For other (more general) examples in the literature, we refer to [41, Sections IV.7 and VII.3].

Example 3 (A Simplified Version of [41, Theorem VII.3.3]) For ease, we take $\Bbbk$ to be $\mathbb{C}$. Also, let q be a nonzero complex number that's not a root of unity. We aim to produce an action of a Hopf algebra H_q over $\mathbb{C}$ (whose structure depends on q) on the q-polynomial algebra $\mathbb{C}_q[x, y] = \mathbb{C}\langle x, y \rangle / (yx - qxy)$, so that

- the "limit" of H_q as $q \to 1$ is a "classical" Hopf algebra H (i.e., H is either commutative or cocommutative and H_q is a q-*deformation* of H), and
- the "limit" of the H_q-action on $\mathbb{C}_q[x, y]$ as $q \to 1$ is an action of H on $\mathbb{C}[x, y]$.

---

[2]In this case, the category of H-modules is a *rigid* monoidal category.

[3]For more settings of quantum symmetry, see, e.g. [63, Chapter 11] for a categorical framework.

We begin by defining a Hopf algebra H_q with algebra presentation,

$$H_q = \mathbb{C}\langle g, g^{-1}, h\rangle/(gg^{-1} - 1,\ g^{-1}g - 1,\ gh - q^2hg),$$

along with coproduct, counit, and antipode given by

$$\Delta(g) = g \otimes g, \quad \Delta(g^{-1}) = g^{-1} \otimes g^{-1}, \quad \Delta(h) = 1 \otimes h + h \otimes g,$$
$$\epsilon(g) = 1, \quad \epsilon(g^{-1}) = 1, \quad \epsilon(h) = 0,$$
$$S(g) = g^{-1}, \quad S(g^{-1}) = g, \quad S(h) = -hg^{-1}.$$

Next we define a *q-number* $[\ell]_q := \frac{q^\ell - q^{-\ell}}{q - q^{-1}}$ for any integer ℓ. Now for any element $p = \sum_{i,j \geq 0} \lambda_{ij} x^i y^j$ in $\mathbb{C}_q[x, y]$, the rule below gives us an action of H_q on $\mathbb{C}_q[x, y]$:

$$g(p) = \sum_{i,j \geq 0} \lambda_{ij} q^{i-j} x^i y^j,\ g^{-1}(p) = \sum_{i,j \geq 0} \lambda_{ij} q^{j-i} x^i y^j,\ h(p)$$
$$= \sum_{i,j \geq 0} \lambda_{ij} [j]_q\, x^{i+1} y^{j-1}.$$

To check this, it suffices to show that (i) the relations of H_q act on $\mathbb{C}_q[x, y]$ by zero, and that (ii) the relation space of $\mathbb{C}_q[x, y]$ is preserved under the rule above. We'll provide some details here and leave the rest as an exercise. We compute:

For(i), $(gh - q^2hg)(p) = g(\sum \lambda_{ij} [j]_q\, x^{i+1} y^{j-1}) - q^2 h(\sum \lambda_{ij} q^{i-j} x^i y^j)$
$$= \sum \lambda_{ij} [j]_q\, q^{i-j+2} x^{i+1} y^{j-1} - q^2 \sum \lambda_{ij} [j]_q\, q^{i-j} x^{i+1} y^{j-1} = 0;$$
For(ii), $h(yx - qxy) = [1(y)\, h(x) + h(y)\, g(x)] - qh(xy) = (x)(qx) - q(x^2) = 0.$

Now the "limit" of H_q as $q \to 1$ is $H = \mathbb{C}[x] \otimes \mathbb{C}\mathbb{Z}$, the *tensor product of Hopf algebras*, for $\mathbb{Z} = \langle g \rangle$ (see, e.g. [63, Exercise 2.1.19]); H is both commutative and cocommutative. Also, $H_q = H$ as $\mathbb{C}$-vector spaces. Moreover, as $q \to 1$, the generators g and g^{-1} (resp., h) of H act on $\mathbb{C}_q[x, y]$ as the identity (resp., by $\frac{\partial}{\partial y}$).

5 Research Directions in Noncommutative Algebra

We highlight a couple of directions for research in Noncommutative Algebra in this section, building on the discussions of Sects. 1–4. The material below could also serve as a topic for an undergraduate or Master's thesis project, or as a reading course topic. Finding a friendly faculty (or advanced graduate student) mentor to help with these pursuits is a good place to start...

5.1 On Symmetries

Continuing the discussions of Sects. 2 and 4, we propose the following avenue for research: Study of the symmetries of (algebraic structures that generalize) Hamilton's quaternions $\mathbb{H}$ [Problem 1]. One such generalization is given below.

Definition 12 ([12, Section 5.4]) Fix a field $\mathbb{k}$ with char $\mathbb{k} \neq 2$, with nonzero scalars $a, b \in \mathbb{k}$. Then a *quaternion algebra* $Q(a, b)_\mathbb{k}$ is a $\mathbb{k}$-algebra that has an underlying 4-dimensional $\mathbb{k}$-vector space with basis $\{1, i, j, k\}$, subject to multiplication rules

$$i^2 = a, \quad j^2 = b, \quad ij = -ji = k.$$

Note that $k^2 = ijk = -ab$, for instance.

Sometimes $Q(a, b)_\mathbb{k}$ is denoted by $(a, b)_\mathbb{k}$, by $(a, b; \mathbb{k})$, or even by (a, b) if $\mathbb{k}$ is understood. The structure above extends the construction of Hamilton's quaternions [Definition 5], namely $\mathbb{H} = Q(-1, -1)_\mathbb{R}$. Moreover, *split-quaternions*, $Q(-1, +1)_\mathbb{R}$, also appear frequently in the literature.

Fun fact: A quaternion algebra is either a 4-dimensional $\mathbb{k}$-division algebra [Definition 4], or is isomorphic to the matrix algebra $M_2(\mathbb{k})$! (The latter is called the *split* case.) Also, these cases are characterized by the *norm* of elements $Q(a, b)_\mathbb{k}$:

$$N(a_0 + a_1 i + a_2 j + a_3 k) := a_0^2 - a a_1^2 - b a_2^2 + a b a_3^2, \quad \text{for } a_0, a_1, a_2, a_3 \in \mathbb{k}.$$

Namely if $\mathbb{k}$ has characteristic not equal to 2, then $Q(a, b)_\mathbb{k}$ is a division algebra precisely when $N(a_0 + a_1 i + a_2 j + a_3 k) = 0$ only for $(a_0, a_1, a_2, a_3) = (0, 0, 0, 0)$ [18, Proposition 5.4.3]. For instance, $\mathbb{H} = Q(-1, -1)_\mathbb{R}$ is a $\mathbb{R}$-division algebra since

$$N(a_0 + a_1 i + a_2 j + a_3 k) = a_0^2 + a_1^2 + a_2^2 + a_3^2$$

for $a_0, a_1, a_2, a_3 \in \mathbb{R}$, and is 0 if and only if $(a_0, a_1, a_2, a_3) = (0, 0, 0, 0)$.

Quaternion algebras (in the generality of Definition 12) have appeared primarily in number theory [69] [56, Chapter 5] and in the study of quadratic forms [47, Chapter III]. They have also been used in hyperbolic geometry [52] [53, Chapter 2], and in various parts of physics and engineering; see, e.g. [5] and [61]. For more details about their applications and structure, see [14] and the references therein.

Recall from Sect. 4 that there are several frameworks for studying symmetries of a $\mathbb{k}$-algebra, including group actions [G-ACT], group gradings [G-GRD], and Hopf algebra actions [H-ACT]. Also, the latter symmetries are considered to be *quantum symmetries* if H is non(co)commutative, as discussed by Fig. 14.

Problem 1 Study the (quantum) symmetries of quaternion algebras. Namely pick a setting [G-ACT], [G-GRD], [H-ACT], a collection of structures (G or H) in this class, and classify all such symmetries of G or H on $Q(a, b)_\mathbb{k}$.

Fig. 14 Symmetries
deforming

Even if this problem is not addressed in full generality, a collection of examples would be quite useful for the literature. For instance, a group grading of $Q(-1, -1)_{\mathbb{R}} = \mathbb{H}$ was used in recent work of Cuadra and Etingof as a counterexample to show that their main result on *faithful* group gradings on division algebras fails when the ground field is not algebraically closed [18, Theorem 3.1, Example 3.4].

There are also other works that partially address Problem 1, such as on group gradings [17, 58, 59] and Hopf algebra (co)actions [21, 66]. These papers also contain work on (quantum) symmetries of some generalizations of quaternion algebras; **Problem 1 can also be posed for these generalizations of** $Q(a, b)_{\mathbb{k}}$ **as well.**

Moreover, **a second part of Problem 1** could include the study of two algebraic structures formed by the symmetries constructed above, namely the *subalgebra of (co)invariants*, and the *smash product algebra* (or, *skew group algebra* if [G-ACT] is used). See [57] for the definitions, examples, and a discussion of various uses of these algebraic structures. Overall, after one gets comfortable with the terminology, such problems are computational in nature ... and fun to do!

5.2   On Representations

In this section, $\mathbb{k}$ is a field of characteristic zero.

Towards a research direction in representation theory (continuing the discussion in Sect. 3) it is natural to think further about the representations of the first Weyl algebra $A_1(\mathbb{k})$. Since there are no finite-dimensional representations of $A_1(\mathbb{k})$ [Proposition 1], what are its infinite-dimensional representations? To get one, for example, identify $A_1(\mathbb{k})$ as a ring of differential operators on $\mathbb{k}[x]$ where the generators x and y act as multiplication by x and by $\frac{d}{dx}$, respectively. So, by fixing a basis $\{1, x, x^2, x^3, \dots\}$ of $\mathbb{k}[x]$, we get the (matrix form of) infinite-dimensional representation in (5). Producing explicit infinite-dimensional representations of $A_1(\mathbb{k})$ is tough in general. But there are many works on the *abstract* representation theory of $A_1(\mathbb{k})$ and of other rings of differential operators, and we recommend the student-friendly text of S.C. Coutinho on *algebraic D-modules* [16] for more information.

Now for a concrete research problem to pursue, we suggest working with deformations of Weyl algebras instead, particularly those that admit finite-dimensional representations (as this is more feasible computationally). One could:

Problem 2 Examine the (explicit) representation theory of *quantum Weyl algebras* (*at roots of unity*) [Definition 16].

Before we discuss quantum Weyl algebras, we introduce some terminology that will be of use later in order to make the problem above more precise. The text [23] (which, again, is student-friendly) is a nice reference for more details.

Definition 13 Take a $\Bbbk$-algebra A with a representation

$$\phi : A \to \mathrm{Mat}_n(\Bbbk) \ (\cong \mathrm{End}(V)) \quad \text{for } V = \Bbbk^{\oplus n}.$$

1. We say that ϕ is *decomposable* if we can decompose V as $W_1 \oplus W_2$ with $W_1, W_2 \neq 0$ so that $\phi|_{W_k} : A \to \mathrm{End}(W_k)$ are representations of A for $k = 1, 2$. Otherwise, we say that ϕ is *indecomposable*.
2. The representation ϕ is *reducible* if there exists a proper subspace W of V so that $\phi|_W : A \to \mathrm{End}(W)$ is a representation of A; here, $\phi|_W$ is called a *(proper) subrepresentation* of ϕ. If ϕ does not have any proper subrepresentations, then ϕ is *irreducible*; the corresponding A-module V is said to be *simple* (cf. Fig. 12).
3. Take another representation $\phi' : A \to \mathrm{End}(V')$ of A. We say that ϕ' is *equivalent* (or *isomorphic*) to ϕ if $\dim V = \dim V'$ and there exists an invertible $\Bbbk$-linear map $\rho : V \to V'$ so that $\rho(\phi_a(v)) = \phi'_a(\rho(v))$ for all $a \in A$ and $v \in V$.

Irreducible representations are indecomposable; the converse doesn't always hold.

To understand the notions above in terms of matrix solutions of equations (cf. Fig. 13), take a *finitely presented* $\Bbbk$-algebra A, that is, A has finitely many noncommuting variables x_i as generators, and finitely many words $f_j(\underline{x})$ in x_i as relations:

$$A = \frac{\Bbbk\langle x_1, \ldots, x_t \rangle}{\left(f_1(\underline{x}), \ldots, f_r(\underline{x})\right)}.$$

Let us also fix an n-dimensional representation of A, given by

$$\phi : A \to \mathrm{Mat}_n(\Bbbk), \quad x_i \mapsto X_i \ \text{for} \ i = 1, \ldots, t.$$

Definition 14 Retain the notation above. Suppose that we have a matrix solution $\underline{X} = (X_1, \ldots, X_t)$ to the system of equations $f_1(\underline{x}) = \cdots = f_r(\underline{x}) = 0$.

1. If each matrix X_i can be written as a direct sum of matrices $X_{i,1} \oplus X_{i,2}$, where

   • $X_{i,k} \in \mathrm{Mat}_{n_k}(\Bbbk)$ with $k = 1, 2$ for some positive integers n_1 and n_2, and

- $\underline{X_k} = (X_{1,k}, \ldots, X_{t,k})$ is a solution to $f_1(\underline{x}) = \cdots = f_r(\underline{x}) = 0$ for $k = 1, 2$,

then the matrix solution $\underline{X}$ is *decomposable*. Otherwise, $\underline{X}$ is *indecomposable*.

2. For $\mathrm{Mat}_n(\Bbbk)$ identified as $\mathrm{End}(V)$ with $V = \Bbbk^{\oplus n}$, suppose that there exists a proper subspace W of V that is stable under the action of each X_i. Then we say that $\underline{X}$ is *reducible*. Otherwise, $\underline{X}$ is *irreducible*.

3. We say that another matrix solution $\underline{X'} \in \mathrm{Mat}_{n'}(\Bbbk)^{\times t}$ to the system of equations $f_1(\underline{x}) = \cdots = f_r(\underline{x}) = 0$ is *equivalent* (or *isomorphic*) to $\underline{X}$ if $n = n'$ and there exists an invertible matrix $P \in \mathrm{GL}_n(\Bbbk)$ so that $P \ X_i \ P^{-1} = X_i$ for all i.

So two representations of A (or, two matrix solutions of $\{f_j(\underline{x}) = 0\}_{j=1}^r$) are equivalent precisely when they are the same up to change of basis of $V = \bigoplus_{i=1}^t x_i$. Therefore Problem 2 can be refined as follows.

Precise Version of Problem 2 Classify the explicit irreducible representations of the quantum Weyl algebras [Definition 16], up to equivalence.

Let's define the quantum Weyl algebras now. One way of getting these algebras is by deforming the mth Weyl algebras $A_m(\Bbbk)$ from (6) via the symmetry discussed below. (The reader may wish to skip to Definition 16 for the outcome of this discussion.)

Definition 15 Fix a $\Bbbk$-vector space V.

1. A $\Bbbk$-linear transformation $c : V \otimes V \to V \otimes V$ is a *braiding* if it satisfies the braid relation, $(c \otimes \mathrm{id}_V) \circ (\mathrm{id}_V \otimes c) \circ (c \otimes \mathrm{id}_V) = (\mathrm{id}_V \otimes c) \circ (c \otimes \mathrm{id}_V) \circ (\mathrm{id}_V \otimes c)$ as maps $V^{\otimes 3} \to V^{\otimes 3}$.
2. A braiding $\mathscr{H} : V \otimes V \to V \otimes V$ is a *Hecke symmetry* if it satisfies the Hecke condition, $(\mathscr{H} - q \ \mathrm{id}_{V \otimes V}) \circ (\mathscr{H} + q^{-1} \ \mathrm{id}_{V \otimes V}) = 0$ as maps $V \otimes V \to V \otimes V$, for some nonzero $q \in \Bbbk$.

Given a Hecke symmetry $\mathscr{H} \in \mathrm{End}(V \otimes V)$ one can form the $\mathscr{H}$-*symmetric algebra* $S_{\mathscr{H},q}(V) = T(V)/(\mathrm{Image}(\mathscr{H} - q \ \mathrm{id}_{V \otimes V}))$. For example, when $\mathscr{H} = $ flip (sending $x_i \otimes x_j$ to $x_j \otimes x_i$) and $q = 1$ we get that $S_{\mathrm{flip},1}(V)$ is the symmetric algebra $S(V)$ on V; this is isomorphic to the polynomial ring $\Bbbk[x_1, \ldots, x_m]$ for $V = \bigoplus_{i=1}^m \Bbbk x_i$.

Summarizing the discussion in [28], we now build a q-version of a Weyl algebra using a Hecke symmetry $\mathscr{H}$ as follows. Consider the dual vector space V^* and the induced $\Bbbk$-linear map $\mathscr{H}^* \in \mathrm{End}(V^* \otimes V^*)$. Then construct the algebra $A_{\mathscr{H},q}(V \oplus V^*)$ on $V \oplus V^*$, which is the tensor algebra $T(V \oplus V^*)$ subject to the relations: $\mathrm{Image}(\mathscr{H} - q \ \mathrm{id}_{V \otimes V})$, and $\mathrm{Image}(\mathscr{H}^* - q^{-1} \ \mathrm{id}_{V^* \otimes V^*})$, and certain relations intertwining generators from V with those from V^* by using $\mathscr{H}$. The resulting algebra $A_{\mathscr{H},q}(V \oplus V^*)$ is called the *quantum Weyl algebra associated with $\mathscr{H}$*.

For simplicity, we provide the presentation of $A_{\mathscr{H},q}(V \oplus V^*)$ for the standard 1-parameter Hecke symmetry given in [37, page 442] (provided in the form of an R-*matrix*). Here, $V = \bigoplus_{i=1}^m \Bbbk x_i$ and $V^* = \bigoplus_{i=1}^m \Bbbk y_i$ with $y_i := x_i^*$ (linear dual of x_i).

Definition 16 ([37, page 442] [28, Definition 1.4]) Take $m \geq 2$. The *1-parameter quantum Weyl algebra* is an associative $\Bbbk$-algebra $A_m^q(\Bbbk)$ with noncommuting generators $x_1, \ldots, x_m, y_1, \ldots, y_m$ subject to relations

$$x_i x_j = q x_j x_i, \quad y_i y_j = q^{-1} y_j y_i, \qquad \forall i < j$$
$$y_i x_j = q x_j y_i, \qquad\qquad\qquad\qquad \forall i \neq j$$
$$y_i x_i = 1 + q^2 x_i y_i + (q^2 - 1) \sum_{j>i} x_j y_j, \forall i.$$

By convention, we define $A_1^q(\Bbbk)$ to be $\Bbbk\langle x, y\rangle/(yx - qxy - 1)$. If q is a root of unity then we refer to these algebras as quantum Weyl algebras *at a root of unity*.

Notice that one gets the Weyl algebras $A_1(\Bbbk)$ [Definition 7] and $A_m(\Bbbk)$ [Eq. (6)] by taking the "limit" of $A_1^q(\Bbbk)$ and $A_m^q(\Bbbk)$ as $q \to 1$, respectively.

Fun fact: If q is a root of unity, say of order ℓ, then all irreducible representations of a quantum Weyl algebra $A_{\mathcal{H},q}(V \oplus V^*)$ are finite-dimensional! Moreover in this case, the dimension of an irreducible representation is $A_{\mathcal{H},q}(V \oplus V^*)$ is bounded above by some positive integer $N(\ell)$ depending on ℓ, and this bound is met most of the time. This is part of a general phenomenon for *quantum $\Bbbk$-algebras* with scalar parameters—they have infinite-dimensional irreducible representations in the generic case, and in the root of unity case all of their irreducible representations are finite-dimensional. Further, in the root of unity case, most irreducible representations of a quantum algebra A have dimension equal to the *polynomial identity (PI) degree* of A (see, for instance, the informative text of Brown–Goodearl [11]). For example, the PI degree of $A_1^q(\Bbbk)$ is equal to ℓ when q is a root of unity of order ℓ.

This leads us to discussion of **a partial answer to Problem 2**. Indeed, one was achieved for $A_1^q(\Bbbk)$, for q a root of unity of order ℓ, in two undergraduate research projects directed by Letzter [10] and by L. Wang [31]. The explicit irreducible matrix solutions (X, Y) to the equation $YX - qXY = 1$ were computed in these works (up to equivalence), the majority of which are ℓ-by-ℓ matrices.

Naturally, the **next case for Problem 2** is the representation theory of quantum Weyl algebras $A_{\mathcal{H},q}(V \oplus V^*)$, where $\dim_\Bbbk V = 2$ and q is a root of unity; this should build on the partial answer above. There are a few routes one could take, such as examining $A_m^q(\Bbbk)$ for $m \geq 2$, or more generally, addressing Problem 2 for *multi-parameter quantum Weyl algebras* as in [28, Example 2.1] [11, Definition 1.2.6].

Why care? One reason is that quantum Weyl algebras have appeared in numerous works in mathematics and physics, including Deformation Theory [27, 28, 37, 39], Knot Theory [24], Category Theory [48], Quantum mechanics and Hypergeometric Functions [65] to name a few. Therefore, any (partial) resolution to Problem 2 would be a welcomed addition to the literature. So let's have a go at this. :)

Photo and Figure Credits Figs. 2, 5–7, 9–10, 12–13: Author. (** = from unsplash.com)

Fig. 1: Tammie Allen, @tammeallen**. Fig. 8: Wikipedia, user: JP.
Fig. 3: GazzaPax (flickr.com). Fig. 11: Billy Huynh, @billy_huy**.
Fig. 4: Karolina Szczur, @thefoxis**. Fig 14. Dan Gold, @danielcgold**.

Acknowledgements C. Walton is partially supported by the US National Science Foundation with grants #DMS-1663775 and 1903192, and with a research fellowship from the Alfred P. Sloan foundation. The author thanks the anonymous referees and Gene Abrams for their valuable feedback.

References

1. Artin, M., Schelter, W., and Tate, J.: Quantum deformations of GL_n. Comm. Pure Appl. Math. **44**, no. 4–5, 879–895 (1991).
2. Axler, S.: Linear Algebra Done Right. Third edition. Undergraduate Texts in Mathematics, Springer International Publishing (2015).
3. Baez, J. C.: The octonions. Bull. Amer. Math. Soc. (N.S.) **39**, no. 2, 145–205 (2002).
4. Bayley, M.: Alice's adventures in algebra: Wonderland solved. New Scientist Magazine. December 16, 2009. https://www.newscientist.com/article/mg20427391-600-alices-adventures-in-algebra-wonderland-solved/. Cited July 26, 2018.
5. Baylis, W. E.: Clifford (Geometric) Algebras: with applications to physics, mathematics, and engineering. Birkhäuser Boston, Boston (1996).
6. Berenstein, D., Jejjala, V., and Leigh, R. G.: Marginal and relevant deformations of N=4 field theories and non-commutative moduli spaces of vacua. Nuclear Phys. B **589**, no. 1–2, 196–248 (2000).
7. Born, M. and Jordan, P.: Zur Quantenmechanik. Z. Phys. **34**, 858–888 (1925).
8. Bott, R. and Milnor, J.: On the parallelizability of the spheres. Bull. Amer. Math. Soc. **64**, 87–89 (1958).
9. Bouwknegt, P. and Schoutens, K.: $\mathcal{W}$ symmetry in conformal field theory. Phys. Rep. **223**, no. 4, 183–276 (1993).
10. Boyette, J., Leyk, M., Talley, J., Plunkett, T., and Sipe, K.: Explicit representation theory of the quantum Weyl algebra at roots of 1. Comm. Algebra **28**, no. 11, 5269–5274 (2000).
11. Brown, K. A. and Goodearl, K. R. Lectures on algebraic quantum groups. Advanced Courses in Mathematics. CRM Barcelona. Birkhäuser-Verlag, Basel (2002).
12. Cohn, P. M.: Further Algebra and Applications. SpringerLink: Bücher, Springer, London (2011).
13. Connes, A. and Dubois-Violette, M.: Yang-Mills algebra. Lett. Math. Phys. **61**, no. 2, 149–158 (2002).
14. Conrad, K.: Quaternion algebras. Available at http://www.math.uconn.edu/~kconrad/blurbs/ringtheory/quaternionalg.pdf. Cited July 28, 2018.
15. Conrad, K.: Tensor algebras. Available at http://www.math.uconn.edu/~kconrad/blurbs/linmultialg/tensorprod.pdf. Cited August 6, 2018.
16. Coutinho, S.C.: A primer of algebraic D-modules. Vol. 33. Cambridge University Press, Cambridge (1995).
17. Covolo, T. and Michel, J.-P.: Determinants over graded-commutative algebras, a categorical viewpoint. Enseign. Math. **62**, no. 3–4, 361–420 (2016).
18. Cuadra, J. and Etingof, P.: Finite dimensional Hopf actions on central division algebras. Int. Math. Res. Not. **2017**, no. 5, 1562–1577 (2016).
19. Dickson, L. E.: On quaternions and their generalization and the history of the eight square theorem. Ann. of Math. (2) **20**, no. 3, 155–171 (1919).
20. Dirac, P. A. M.: The fundamental equations of quantum mechanics. Proc. Roy. Soc. A **109**, no. 752, 642–653 (1925).
21. Doi, Y. and Takeuchi, M.: Quaternion algebras and Hopf crossed products. Comm. Algebra **23**, no. 9, 3291–3325 (1995).
22. Drinfel'd, V. G.: Quantum groups. Proceedings of the International Congress of Mathematicians **986**, Vol. 1, 798–820, AMS (1987).

23. Etingof, P., Golberg, O., Hensel, S., Liu, T., Schwendner, A., Vaintrob, D., and Yudovina, E.: Introduction to representation theory. With historical interludes by Slava Gerovitch. Student Mathematical Library, 59. American Mathematical Society, Providence (2011).
24. Fenn, R. and Turaev, V.: Weyl algebras and knots. J. Geom. Phys. **57**, no. 5, 1313–1324, (2007).
25. Franco, S., Hanany, A., Vegh, D., Wecht, B., and Kennaway, K.: Brane dimers and quiver gauge theories. J. High Energy Phys., no. 1, (2006).
26. Frobenius, G.: Über lineare Substitutionen und bilineare Formen. J. Reine Angew. Math. **84**, 1–63 (1878).
27. Gerstenhaber, M. and Giaquinto, A.: Deformations associated with rigid algebras. J. Homotopy Relat. Struct. **10**, no. 3, 437–458 (2015).
28. Giaquinto, A. and Zhang, J. J.: Quantum Weyl algebras. J. Algebra **176**, no. 3, 861–881 (1995).
29. Hamilton, W. R.. Lectures on quaternions. Hodges and Smith (1853).
30. Hamilton, W. R.: The mathematical papers of Sir William Rowan Hamilton. Vol. III: Algebra. Edited by H. Halberstam and R. E. Ingram. Cunningham Memoir No. XV. Cambridge University Press, London-New York (1967).
31. Heider, B. and Wang, L.: Irreducible representations of the quantum Weyl algebra at roots of unity given by matrices. Comm. Algebra **42**, no. 5, 2156–2162 (2014).
32. Heisenberg, W.: Über quantentheoretische Umdeutung kinematischer und mechanischer Beziehungen. Z. Phys. **33**, 879–893 (1925).
33. Heisenberg, W.: Über den anschaulichen Inhalt der quantentheoretischen Kinematik und Mechanik. Z. Phys. **43**, 172–198 (1927).
34. Hentschel, A. M. and Hentschel, K.: Physics and National Socialism: An Anthology of Primary Sources. Modern Birkhäuser Classics, Springer, (2011).
35. Hopf, H.: Ueber die Topologie der Gruppen-Mannifaltigkeiten und ihrcr Verallgemeinerungen, Ann. Math. **42**, 22–52 (1941).
36. Jimbo, M.: A q-difference analogue of $U(\mathfrak{g})$ and the Yang-Baxter equation. Lett. Math. Phys. **10**, no. 1, 63–69 (1985).
37. Jing, N. and Zhang, J.: Quantum Weyl algebras and deformations of $U(\mathfrak{g})$. Pacific J. Math. **171**, no. 2, 437–454 (1995).
38. Joni, S. and Rota, G.-C.: Coalgebras and bialgebras in combinatorics, Stud. Appl. Math. **61**, 93–139 (1979).
39. Jordan, D.: Quantized multiplicative quiver varieties. Adv. Math. **250**, 420–466 (2014).
40. Joyal, A. and Street, R.: Braided tensor catcgories. Adv. Math. **102**, no. 1, 20–78 (1993).
41. Kassel, C.: Quantum groups. Graduate Texts in Mathematics 155. Springer-Verlag, New York (1995).
42. Kervaire, M.: Non-parallelizability of the n sphere for $n > 7$, Proc. Nat. Acad. Sci. USA **44**, 280–283 (1958).
43. Kleiner, I.: The Evolution of Group Theory: A Brief Survey. Math. Mag. **59**, no. 4, 195–215 (1986).
44. Kuipers, J. B.: Quaternions and rotation sequences. A primer with applications to orbits, aerospace, and virtual reality. Princeton University Press, Princeton, NJ (1999).
45. Kuipers, J. B.: Quaternions and rotation sequences. Geometry, integrability and quantization (Varna, 1999), 127–143, Coral Press Sci. Publ., Sofia (2000).
46. Lam, T. Y.: A first course in noncommutative rings. Second edition. Graduate Texts in Mathematics, **131**. Springer-Verlag, New York (2001).
47. Lam, T. Y. Introduction to quadratic forms over fields. Graduate Studies in Mathematics **67**. American Mathematical Society, Providence (2005).
48. Laugwitz, R.: Comodule Algebras and 2-Cocycles over the (Braided) Drinfeld Double. Available at https://arxiv.org/pdf/1708.02641.pdf. Cited July 29, 2018.
49. Lewis, D.: Quaternion algebras and the algebraic legacy of Hamilton's quaternions. Irish Math. Soc. Bull. no. 57, 41–64 (2006).
50. Lorenz, M.: A Tour of Representation Theory (preliminary version). Available at https://www.math.temple.edu/~lorenz/ToR.pdf. Cited July 25, 2018.

51. Lusztig, G.: Introduction to quantum groups. Progress in Mathematics, 110. Birkhäuser Boston, Inc., Boston (1993).
52. Macfarlane, A.: Hyperbolic quaternions. Proc. Roy. Soc. Edinburgh **23**, 169–180 (1902).
53. Maclachlan, C. and Reid, A. W.: The arithmetic of hyperbolic 3-manifolds. Graduate Texts in Mathematics, 219. Springer-Verlag, New York (2003).
54. Manin, Y. I.: Quantum groups and noncommutative geometry. Université de Montréal Centre de Recherches Mathématiques, Montreal (1988).
55. May, K. O.: Classroom Notes: The Impossibility of a Division Algebra of Vectors in Three Dimensional Space. Amer. Math. Monthly **73**, no. 3, 289–291 (1966).
56. Miyake, T.: Modular forms. Springer Monographs in Mathematics. Springer-Verlag, Berlin (2006).
57. Montgomery, S.: Hopf algebras and their actions on rings. No. 82. American Mathematical Soc. (1993).
58. Morier-Genoud, S. and Ovsienko V.: Well, papa, can you multiply triplets? Math. Intelligencer **31**, no. 4, 1–2 (2009).
59. Morier-Genoud, S. and Ovsienko, V.: Simple graded commutative algebras. J. Algebra **323**, no. 6, 1649–1664 (2010).
60. National Public Radio: Weekend Edition Saturday. The Mad Hatter's Secret Ingredient: Math. March 13, 2010. https://www.npr.org/templates/story/story.php?storyId=124632317. Cited July 26, 2018.
61. Onsager, L.: Crystal statistics. I. A two-dimensional model with an order-disorder transition. Phys. Rev. **65**, no. 2, 117–149 (1944).
62. Palais, R. S.: The classification of real division algebras. Amer. Math. Monthly **75**, 366–368 (1968).
63. Radford, D. E.: Hopf algebras. Series on Knots and Everything 49. World Scientific Publishing Co. Pte. Ltd., Hackensack (2012).
64. Sklyanin, E. K.: Some algebraic structures connected with the Yang-Baxter equation. Funkt-sional. Anal. i Prilozhen. 16, no. 4, 27–34 (1982).
65. Spiridonov, V.: Coherent states of the q-Weyl algebra. Lett. Math. Phys. **35**, no. 2, 179–185 (1995).
66. Van Oystaeyen, F. and Zhang, Y.: The Brauer group of Sweedler's Hopf algebra H_4. Proc. Amer. Math. Soc. **129**, no. 2, 371–380 (2001).
67. van der Waerden, B. L.: Hamilton's discovery of quaternions. Translated from the German by F. V. Pohle. Math. Mag. **49**, no. 5, 227–234 (1976).
68. van der Waerden, B. L.: Sources of Quantum Mechanics. Dover Books on Physics. Dover (2007).
69. Vignéras, M.-F.: Arithmétique des algèbres de quaternions. Lecture Notes in Mathematics, **800**. Springer, Berlin (1980).
70. Wikipedia contributors: Cayley-Dickson construction. Wikipedia, The Free Encyclopedia. Wikipedia, The Free Encyclopedia, Cited July 26, 2018.
71. Zorn, M.: Theorie der alternativen Ringe, Abh. Math. Sem. Univ. Hamburg **8**, 123–147 (1930).

Part V
Mathematical Lives

Taking a Leap Off the Ivory Tower: Normalizing Unconventional Careers

Karoline P. Pershell

Abstract I explain my own winding journey to shed light on how paths in mathematics can unfold. I share this personal story to remind us that you can start at any point and get to any other point, to share my own litmus test for when to move forward bravely and boldly, and to encourage each of us to keep searching for the space where we are productive.

1 Introduction

I love my job.

As Executive Director of the Association for Women in Mathematics, I get to support and celebrate women across the career spectrum *individually* with programming, prizes, grant opportunities, and *collectively* with activities that promote institutional change. I empower volunteers to direct their efforts toward projects that personally or professionally advance their careers and passions. I develop strategic initiatives to address the social problems surrounding the advancement of women in mathematics. And I constantly meet inspirational people at every stage of their careers.

In mathematics, the gold standard for a successful career has been a tenure-track job at a research institution with a light teaching load. We are raised and trained by our PhD or master's advisors who are exactly in such positions and often groom us to reach similar positions. This inherited value system of what matters in math had unjustifiably narrowed my perceived career paths and distorted my own measures of success. The result was that I consistently undervalued my worth and abilities outside of research mathematics early on in my career.

This perspective on careers (and on the mythical "right" career) is something I have internally fought as I traipsed back and forth across the country and around

K. P. Pershell (✉)
Association for Women in Mathematics, Washington, DC, USA
e-mail: karoline@awm-math.org

© The Author(s) and the Association for Women in Mathematics 2019
S. D'Agostino et al. (eds.), *A Celebration of the EDGE Program's Impact on the Mathematics Community and Beyond*, Association for Women in Mathematics Series 18, https://doi.org/10.1007/978-3-030-19486-4_24

the world. Along the way, I built a unique skill set that is authentic to me. Most importantly, I was aware of my growth, and that consciousness allowed me to seek out jobs that a younger me would have assumed were out of my scope or unattainable to me. My own unconventional career emphasizes that careers don't have to be trajectories, that there is no single "right" starting place, and that one can be a mathematician across academia, government, business, industry, and the nonprofit sector. So how did I start my own unconventional career? By saying yes to the most unconventional opportunity: bull riding.

2   The Not-so-Typical Start to a Math Career

I came to the University of Tennessee at Martin (UTM) for an undergraduate degree in mathematics on a pretty non-traditional path: by joining the men's rodeo team. Sometimes, when I say it out loud, it sounds silly, but—at that time and based on my current 19-year-old experiences—I felt it was the right decision for me.

I grew up training and showing horses to make money in the summers. I actually started college at Saint Mary's (the all-women's sister-school of Notre Dame in South Bend, Indiana), just as Notre Dame was starting a rodeo team. I didn't know anything about rodeo, but I joined thinking I would ride saddle broncs. Through a series of unexpected events, I got unceremoniously ushered into bull riding.

I pursued bull riding on my own for another year, competing in a handful of rodeos, including one at UTM. It was then that I decided I needed to get some professional help for my bull riding addiction. Since UTM had and has a national championship rodeo program, I decided to transfer there in my sophomore year. I don't exactly remember, but I think the conversation with my parents went something like:

> Mom, Dad, I want to leave my full-ride academic scholarship at a private women's university to attend a school you have never heard of, in a state you have never visited, to pursue something I am not very good at.[1]

I may be creatively paraphrasing, but what I heard my parents say was: "Karoline, that is the smartest decision you have ever made." And it turns out they were right.

I was the first woman to compete in Bull Riding in the National Intercollegiate Rodeo Association (NIRA) Eastern region, the first woman nationally to complete a full year and to compete for subsequent seasons in Bull Riding in NIRA. I also joined the Professional Women's Rodeo Association and won Rookie of the Year for Bull Riding and Bareback Broncs in 2001. This meant driving cross-country on long weekends and school holidays to compete in rodeos, while completing my major in math and a minor in physics with a perfect GPA and working in the UTM

---

[1] Excerpts from a speech given by Karoline Pershell to the University of Tennessee at Martin Development Committee (Spring 2012) and reported on here http://www.utm.edu/facesofutm/facesbio.php?personNum=47&pageNum=5.

Math Tutoring Lab. (I didn't tell my rodeo friends I did math, and I didn't talk to my math friends about rodeo. They just seemed like two separate crowds.)

During my college rodeo career, my riding drastically improved and the small victories I had when competing—when I simultaneously knew what I was supposed to do *and* was able to execute the move—were some of the most rewarding moments. I competed for "personal bests" but was never great and there was no clear path in rodeo for me past college.

Unconventional paths can be lonely: you may not have peers or mentors and there is never an obvious plan for next steps. Rather, it feels like you are walking at night on a dark path with a weak flashlight. You only can see a few steps in front of you. You know there is risk, but you just have to keep walking to see what is next. It doesn't matter that you can't see the whole path. Just keep walking.

3 Next Steps

3.1 Grad School

I went to get my master's and PhD in low-dimensional topology from Rice University under the brilliant and patient direction of Dr. John Hempel. I lived many of the canonical grad school horror stories, with feelings of failure exacerbated, because I believed I was supposed to eliminate all other facets of my life until I was *just* math. I now realize this was a terrible plan: things will go wrong with your math research and math career. When I was multifaceted, I could rebound from a failure in one area of my life because I was more than just that one thing. But, when my entire identity was wrapped up in math, failure in math translated to my complete failure as a person.

Near the end of my time in graduate school, I was "gifted" an opportunity to expand my identity. Due to poorly managed finances and a poorly chosen ex, at the beginning of my fifth year of grad school, I had a quarter million dollars in debt to my name, a deconstructed house that was worth about a fifth of what I owed, and a dissertation problem that had just been neatly solved and published . . . by someone else. My advisor helped me refocus on a different problem, and on the side, I also became my own general contractor. I was pulling permits and hiring subcontractors when necessary, becoming quite skilled in tile work and drywall and competent at residential-level electrical and plumbing upgrades. I owe many thanks to helpful neighbors, my parents who came out to lend some elbow grease, and to all of the collective expertise that could be found on the internet. I was able to sell at a profit, clear my debts, have funding toward my next home purchase, and finish my dissertation. Ten years later, I now have five home renovations under my belt, each aiding in creating some financial stability that has allowed me to focus on personal growth.

3.2   The Tenure-Track Position

As I proceeded down the next step in my mathematical journey, I jumped at the chance to return to my undergrad institution (the University of Tennessee at Martin) as an Assistant Professor. This was a friendly place to land after graduate school and figure out "what next?" in a supportive environment. I taught my classes, pursued research, and kept reaching up and out to the opportunities that interested me:

Student Clubs I initiated a student club for researching green technologies, which petitioned the student body to get a vote to raise their own tuition and use the funds toward energy efficiency improvements on campus. I also served as a faculty advisor for the Alpha Kappa Alpha sorority and worked to revitalize a local chapter of the National Society for Black Engineers.

NSF EPSCoRE I applied for and received an NSF grant to build a cross-disciplinary team to conduct research on reversible hydrogen fuel cells. I worked with a brilliant and eccentric retired British physicist, who had started his own winery in west Tennessee, as well as UTM faculty and students.

Qingdao Summer Teaching I accepted an email offer to teach college-level math courses for a summer in China. I had very little information before I got there and realized only later that this international exchange was probably not following the intent of the law with regard to work visas, as it was actually a for-profit company that was running courses under the umbrella of an academic institution. With the other visiting faculty, we quickly had to organize and require administrative structures around courses, attendance, final exams, and grades.

Fulbright Scholar I was awarded a Fulbright and jetted off to India to teach proof-writing and logic to master's-level students, while also collecting data on cultural perceptions of Indian women in mathematics. Though feeling completely out of my element, I presented my preliminary findings at a sociology conference in Delhi that focused on gender harassment and disparities in academia.

STEM Study Abroad Program With a chemistry colleague, I co-led a 10-day study abroad program for UTM students to examine the STEM industrial applications in Germany.

Despite a welcoming department, I still did not yet feel like I belonged in math. I did enjoy working with students and clearly capitalized on the flexibility that academia offered, pursuing a variety of other life experiences as I tried to find the right balance that would make this tenure-track job everything I thought I was supposed to want in an academic career.

3.3   AAAS Science & Technology Policy Fellowship

While in India, I was accepted as a Science & Technology Policy Fellow for a 2-year appointment in Washington, DC. So after 4 years at UTM, I left my tenure-track job, sold my renovated dream house on 9 acres, and headed to DC with a moving truck and 1.75 dogs. (I have two dogs and one has 3 legs. It would also be mathematically correct to guess that I had one, seven-legged dog. No one has yet done so.)

On the drive, I got a call that my intended position had been cut due to the government sequester.

Let's revisit the part that says, I just quit my tenure-track job, sold my dream home, and am sitting alone in a truck with 1.75 stinky dogs so that I could pursue an adventure that had now evaporated. The next several hours were spent in tears and self-pity. Despite my inner-dialogue that could only sarcastically congratulate myself on ruining the rest of my life, I *kept driving* north to DC. I had no other options, so I still showed up, made calls, and positively pitched myself as capable of any opportunities they had left. I worked with AAAS to review what appointments were still available that would take a mathematician. (It turns out, many of them!)

I left academia so that I could pursue a position that would have placed me in governmental labs, helping me transition to more applied lab sciences and engineering. Well, that was my intention, anyway. Instead, I landed at the Department of State's Foreign Service Institute. I became their Evaluation Coordinator, working under the Director of the Institute (a former Ambassador) and her Deputy, both brilliant and strategic individuals who taught me how to play the long-game for creating institutional change. This role provided me a variety of experiences, such as:

The Day Job   As the Evaluation Coordinator, I was responsible for understanding the large swath of training and determining the effectiveness of training toward execution of national priorities. I was responsible for building a cooperative community of evaluators and leading them to develop Department of State-adopted policies and standards around training evaluation.

Assess, then Develop and Implement Policy   I traveled to five embassies in the Middle East to work on issues relating to how the US teaches Arabic dialects and help develop policies for training such dialects. I also traveled to the United Nations and the US's European training headquarters to assess how the US was training in multilateral negotiations. In both instances, I learned how to assess costs to time, personnel, and morale when implementing change and developing implementation strategies that met the needs of those affected.

Some Math!   I did a tour at the State Department's United Nations office to look at UN voting trends using qualitative clustering techniques. More than half of my time at this post was devoted to assessing how the appropriate personnel perceived, reviewed, and applied the information I presented them. In many ways, this pulled on my previous life in the classroom: I had to convey what the data meant and why they should care.

Side Projects on Behalf of the Ambassador I was tasked with bringing some great (but nascent) ideas to fruition, like the establishment of an educational arm to review the implementation of diplomatic initiatives, and leading a cross-discipline committee on developing guidelines for evaluation that would reach across all Department of State training programs.

I learned a lot about how to get things done in a large bureaucracy, how to create coalitions, lead projects, and then get out of the way when others have better ideas. I worked with brilliant people from across the social sciences and yet just like in the math world, I was painfully aware that I was not from their world. This was not my "normal" and this was not where I wanted my career to stay for the long haul.

3.4 From Visiting Embassies in Suits and Heels to Working from Home in Yoga Pants

In the last 4 months of my Fellowship, I got a message from a recruiter who offered me half-time contract work with a Silicon Valley start-up. The offer was to work remotely on algorithm optimization for social media text mining and with a definite pay cut. But I would get my *time* back. Working in government, I realized I sorely missed autonomy over my own time, and now prioritized that above all else in this next career transition.

I tested the waters with this start-up, continuing with the Department of State while working nights and weekends because I didn't believe this company actually needed me. But I quickly identified and solved organizational-level problems regarding accuracy of the algorithms, management of the remote technical team, and evaluation of customer needs. I was building off the skills I had just gotten in my last job! I moved on to manage a team that developed sentiment-recognition classifiers, and I worked on corporate expansion strategy. I know I can learn, and I know I like to be challenged to solve new problems, but this was my first experience where I didn't feel like I was playing catch-up . . . and it was amazing! I had a skill set that was needed in this world that overlapped math and leadership.

It was awesome. I mattered. I was needed. There was a space for someone like me. This is what I needed out of a job, and jobs like this existed!

3.5 Capitalizing on Our Own Ideas

My husband and I jumped all-in when we won an NSF small business grant to bring our robotics and Internet of Things software integration framework to market. I left the Silicon Valley start-up and he left his work as a space roboticist at the US Naval Research Lab to build our own start-up, Service Robotics & Technologies (SRT). At SRT, I do everything from recruit and manage staff to oversee optimizing Kahlman

Filters that reduce signal processing noise. We made the early years of SRT work by renting our DC homes, moving into a home we owned by the Johnson Space Center (Houston), renovating the home, running it as a start-up space with four different computer programmers crashing with us for nearly a year, and then selling the home as we moved back to DC. (Those home renovation skills just keep getting used!) At SRT, I work with people I like to develop cool technology. It is awesome! So why did I still not feel like it was enough?

3.6 AWM

As I learned about the Association for Women in Mathematics' search for a new Executive Director, I realized what I was still looking for in a career: a clear path for giving back. My work as Executive Director brings together research mathematics, teaching, project management, monitoring and evaluation, program design, organizational development, implementation of institutional change, network building, grant writing, and fundraising in such a way, that . . . well . . . my career *almost* looks connected. Or even planned!

In my current role, I use skills I learned from leading negotiations at State, from fundraising and grant writing for the robotics start-up, serving as a technical liaison between customers and programmers at the Silicon Valley start-up, and from understanding the math world so I can communicate with members, and tap into my earliest loves and fears of math so that I am driven to listen to other's journey. Working as Executive Director is the culmination of my past experiences, and yet I know there is still room to grow.

4 Devising My Career Litmus Test

So how did my cobbled-together "career" happen? How did I decide to make these transitions? Because at each juncture, I was hit with the fact that I wanted to do *MORE.*

- But I didn't know what *MORE* was.
- Or where you find *MORE.*
- Or how you actually do *MORE* when you find it.
- I just wanted to do *MORE.*

That's not a lot to go on, but luckily, I have never felt the need to make decisions based on the clarity of outcomes. I balance the risk and decide if I am going to jump.
So I took chances.
When an opportunity came up, I learned from a fabulous mentor to ask two things of myself: does this use my skills and does the work look interesting? I did *not* ask, "Where do I see myself in 10 years?" or "What's the right path to get there?"

Instead, I had a nagging inner voice that said, "You can do *MORE*" and a level of judgment that merely said, "Yup. This sure looks like *MORE*. You should go do that."

Let's be clear that *"MORE"* for me could have been anything: a different academic position, a new institution, motherhood, astronaut, or rodeo cowboy. (Wait, not that last one. I already did that one.) It just so happened that the doors that I knocked on, that opened, and that I chose to walk through took me through a path that led away from the Ivory Tower.

5   Conclusion

I am a mathematician.

I was in academia, government, industry, and now am in the nonprofit sector. I have followed a very unconventional career trajectory that may not work for everyone, but it has worked for me: at each stage, I added meaningful (often unplanned!) experiences that were crucial in the next stage of my career. Each position showed me what I wanted (or didn't want!) in my next job. Most importantly, an understanding of how I add value and how I want to grow as a professional means I was aware of the potential for opportunities as they came along.

A career is only labeled unconventional if we don't see other people doing it. As such, these careers seem improbable and even intimidating. By shedding light on the different ways that professional life can actually unfold, I hope to help normalize such careers.

My unconventional career path included a climb up the Ivory Tower, and a lot of uncertainty when I decided I wanted to do something outside academia. In my current role with the Association for Women in Mathematics, I have an amazing opportunity to empower women to pursue whatever their dream job might be, however, conventional or unconventional it may seem.

A Mathematician's Journey to Public Service

Carla D. Cotwright-Williams

Abstract The author discusses her path from a childhood growing up in a family that valued helping others, to earning a PhD in Mathematics, to working as an assistant professor of mathematics, to serving as a American Mathematical Society/American Association for the Advancement of Science Congressional Fellow, to public service work as a federal employee keeping data safe at the US Social Security Administration and conducting research on autonomous vehicles at the NASA Ames Research Center. During the author's work for the federal government, she has held the titles of "computer scientist," "senior research analyst," and "data scientist." Nonetheless, she has discovered that a mathematician by another name is still a mathematician. Further, she argues that mathematicians who are interested in using their knowledge and skills to help improve the lives of others and to help find solutions to the government's toughest problems should consider public service careers.

1 Helping Others

Growing up in L.A., I saw people in need all around. I observed my family helping the community and those less fortunate in their personal and professional lives. My dad was a police officer, helping protect the citizens of our city. My mom volunteered with the children's ministry at church. I was exposed to many instances of helping others. I recall once, in a routine trip to the grocery store, when my mom brought groceries for two brothers who were asking for money outside the store to buy food. Rather than simply give them the money, she brought them everyday staples like milk, cereal, bread, and lunch meat—so they could eat for more than a day or so. My family was not affluent nor well off. I learned helping others doesn't require a lot of money, but it calls for humility.

C. D. Cotwright-Williams (✉)
U.S. Department of Defense, Joint Artificial Intelligence Center (JAIC), Pentagon, Arlington, VA, USA

© The Author(s) and the Association for Women in Mathematics 2019
S. D'Agostino et al. (eds.), *A Celebration of the EDGE Program's Impact on the Mathematics Community and Beyond*, Association for Women in Mathematics Series 18, https://doi.org/10.1007/978-3-030-19486-4_25

377

From an early age, I developed a heart to help others. I volunteered with the Skid Row Ministry at my church to help sort through donated food and clothing for those in need. I was a youth leader and active in a local youth fellowship, using my computer skills to prepare handouts and programs for participants. In high school, I was a member of the Student Government. For two summers while in high school, I attended a minority engineering summer institute at a local university. Despite the program being hosted by the School of Engineering, it was designed to provide STEM enrichment for students from surrounding public schools. Exposure to those helping underprivileged and underrepresented students cultivated my appreciation for helping others through organized programs.

Throughout undergrad, I looked for opportunities to improve the lives of others and support my community. I continued as a youth leader. I was a Civil Engineering major learning to build safe buildings and roads. I ultimately changed to mathematics and took a few courses for future teachers. It was important to me to not only study something I enjoyed, but also I wanted to earn a degree where I might help others.

During my PhD program, I was not a traditional grad student. My days weren't solely filled with research and teaching. I mentored middle school girls in a STEM enrichment program in north Mississippi with a local public service organization. I was an officer in the University of Mississippi's Graduate Student Council. As Secretary and ultimately Vice President, I represented my fellow graduate students around campus in various capacities. I had the opportunity to travel to Washington, D.C. to lobby on behalf of UM graduate students regarding the taxes required for graduate student stipends. I was amazed of the immediate access I, or anyone, could have to our elected officials. I saw first-hand how I could have my voice be heard, especially as I represented my graduate student community. This helped to further peak my interest in public service.

2 Public Service in a Formal Role

Notwithstanding my unique public service exposure in graduate school, I continued on the road to an academic career. I sought a tenure-track position. After graduate school, I accepted a visiting assistant professor position at Wake Forest University and worked to improve my teaching and research. While attending the SACNAS (Society for the Advancement of Chicano and Native Americans in Science) conference that fall, I learned of the American Association for the Advancement of Sciences (AAAS). The AAAS is a long-standing international non-profit organization dedicated to advancing science for the benefit of all people. The AAAS sponsors a Science and Technology Policy Fellowship to provide the nation's outstanding scientists and engineers a hands-on opportunity to learn and engage in policymaking as well as to provide them knowledge and analytical skills in the federal government.

For over 40 years, the AAAS S&T Fellowship program has hosted fellows in the Legislative, Executive, and Judicial Branches of the US Government. Close to 900 applications are received once a year from PhD scientists for approximately 250 placements. (Engineers can apply with a masters.) Depending on a fellow's placement, they will spend 12 months working among federal employees at one of the approximately 430 federal departments, agencies, and sub-agencies in the DC-metro. Fellow's experiences range from working with the United States Agency for International Development (USAID) to provide technology assistance after a national disaster in a third-world country to informing the strategic planning of international multifaceted partnerships with the Department of Energy (DOE). Exposure to this unique learning experience inspired me to begin to explore my options outside of academia.

I took steps to prepare myself to apply for this fellowship. While in my tenure-track position at Norfolk State University, I took public policy courses at a local university. I completed requirements for a graduate certificate in public policy analysis. I preferred to take classes in person because it gave me the opportunity to interact and engage with peers outside of my math network. I expanded my research portfolio. For two summers, I conducted research with NASA and the US Navy, respectively. I wanted to learn and do math in areas which had an impact on the public in some form. These research experiences exposed me to real-world applications of math, expanded my professional network, and gave an initial view of working for the federal government. I became even more active in my local community. I continued working to provide STEM enrichment to underrepresented girls. I also volunteered on several local, state, and national political campaigns. I believed it was important to branch out from the STEM academic community.

The American Mathematical Society (AMS), in conjunction with the AAAS, award one PhD mathematician to serve as AMS/AAAS Congressional Fellow. In 2012, I was awarded the AMS Congressional Fellowship. I took leave from my tenure-track position at Norfolk State University to work on Capitol Hill—the extraordinary opportunity I'd been working toward.

My work with Congress was a whirlwind. The work of Congress can be fast paced despite the intentionally slow legislative process. I served as a staffer supporting a policy portfolio which included cybersecurity, science, and homeland security. I worked in both a personal House office and on the majority staff of the US Senate Homeland Security and Government Affairs Committee (HSGAC) prepping the chair for two congressional hearings.

There is a great deal of analytical thinking in the Halls of Congress by members and their staff. Critical thinking and problem solving are the required activities no matter the party affiliation. Math, as most of us know it, doesn't really appear every day in policymaking. Not because math isn't necessary to make important decisions, but because at that level of policymaking, only the final outcomes matter. Legislators must make tough decisions on behalf of their constituents who care about the bottom line—how the policy will impact their lives. My time on "The Hill" opened my eyes to the impact of science and mathematics on policy and decision-making. This rare experience helped solidify my desire for a math career outside of academia.

3 Goodbye Academia

During my year as AMS congressional fellow, I resigned from my position in academia. I looked for work so I could continue to pursue and fulfill my desire to use math in a way to impact my community and the world outside the classroom. I faced a number of personal challenges in doing so. But with a newfound focus and supportive network, I have been successful working in the public sector. My path has encouraged others to consider doing the same.

3.1 You've Got the Right Stuff, Baby

My first hurdle in transitioning from academia was me. I didn't always *feel* like I was good enough in my new line of work. It took me time to feel comfortable and confident in my abilities. Finding work outside of academia wasn't always determined by whether I had a PhD or not. Many people I have worked with didn't have a PhD. Being successful outside of academia is about being a problem solver. As a PhD mathematician, I'd learned and become an expert problem solver. I could synthesize mass amounts of information quickly and use the knowledge to expand my problem-solving capabilities. I learned how to abstract key aspects from a problem necessary to develop a meaningful solution.

To this end, I found it critical to re-brand myself. It was vitally important to highlight these KSAs (knowledge, skills and abilities). During my congressional fellowship, I interacted with other scientists who had completed the Science and Technology Policy Fellowship and chose not to return to their academic positions. As fellows, we were offered workshops on effective communication as a scientist and remarketing our diverse skill set acquired during the PhD process. I re-marketed my experiences to connect with government hiring managers and fit job descriptions. Academic jobs have a standard job description—effective teaching, produce research/publications and grantsmanship. Government positions have standard job qualifications—effective written and oral communication, strategic problem solving, ability to plan, analyze, coordinate, and evaluate priorities. While expressed differently, a critical analysis yields the abilities sought after in academia are similarly sought in the government.

I took advantage of opportunities to engage leadership and other stakeholders on projects I worked on. While working on a data analytics project for the US Citizenship and Immigration Services (USCIS), I talked to a seasoned contractor working on our project. I asked about their career progression and sought career advice for this new environment. The response had become a familiar one. "You're smarter than me, you easily could do this." I was readily impressed by their skills and capabilities as we completed project tasks. It was a confidence builder knowing they were impressed with what I brought to the table. I found that despite being in a new environment, under new circumstances, I had the right skills and abilities to be successful.

3.2 Finding Solace

Another hurdle I dealt with was feeling like I had I betrayed the math community by not remaining in the classroom and continuing research in an academic position. It felt like if I didn't remain to do research, teach, and apply for and earn tenure I was not a good mathematician. Primarily, I was encouraged to follow the path of academia because of my pure math training. It was a natural progression. In addition, given the growing diversity of students, I've grown to believe in the significance of having proportional numbers of women and underrepresented groups in academic positions.

My graduate research appeared not to have real-world applications. In pure math, initially, I just didn't study or look for applications of my research. There always seems to be two paths for a mathematician: pure and applied. Working in academia, I believed I was expected to seek to expand math only through research. It took me several years to become openly comfortable with exploring real-world applications of my research. I realize that not all math needs to have everyday applications to be meaningful. I greatly respect those who do math for math's sake. It is important and necessary to expand the mathematical world. I could not however continue to deny my own desire to do math for the public's sake.

I found solace in a small but growing part of my math community who supported me in my effort to establish a different kind of math career, one I was defining. As I became more comfortable in expressing my new math focus, I received support from close math friends and mentors. One mentor gave me a book on financial math after I enthusiastically shared a recent experience and considered pursing work in the field. I've found in recent years growing efforts within the math community to support those seeking careers in the mathematical sciences outside of academia. One organization is the BIG Math network (Business, Industry, and Government) who has partnered with the AMS, MAA, and SIAM, among others, to bring together the math community to expand awareness of BIG careers for mathematical scientists.

4 Real-World Math and Policy

As I continue on this career path, I seek to remain aware and expand my scope of the use math in the public sector. It is important to me to be aware of the applications of math in the real world. In addition to increasing my real-world math awareness, I wanted to know and understand potential policy outcomes of these applications. Here are two instances of how math impacts the world around us. These two examples have led to discussions, development, and implementation of policies, which protect and support the lives of citizens and residents across the USA and the world.

4.1 Keeping Data Safe

The federal government collects and uses administrative data every day. When we apply for passports or financial aid, we provide our personal information in order to complete our desired transactions. Data are collected and stored for processing. Our data must be kept safe and secure. My time at the US Social Security Administration included working in the Office of Information Security, where I examined the methods of securing the data of the 65 million people who receive social security benefits and the 165 million people who pay into social security each year.

The National Institute of Standards and Technology (NIST) establishes the standards and frameworks for federal government information security capabilities. One method to secure electronic communication, developed by the scientists and mathematicians at NIST, are hash functions. Hash functions are used for computer security applications like message authentication and password verification. A critical characteristic of hash functions is that they are not easily invertible. It is extremely difficult to determine the original inputs for advanced hash functions.

In the mid-2000s, through the use of mathematics and brute computer force, hackers were able to crack early hash function encryption. Soon after, NIST moved to use and recommend the use of a stronger hash functions in protecting the data stored in federal information systems. Legislation such as the Federal Information Security Modernization Act of 2014 (FISMA 2014), was enacted to ensure appropriate techniques and technology are implemented to secure federal information systems.

4.2 Autonomous Vehicles

A few years ago, I conducted research at the NASA Ames Research Center. I worked with a team of engineers and computer scientists to develop and improve Avionics— the electrical systems of airplanes and space vehicles. The team examined avionic systems health in an effort to create methods and technology to diagnose the systems health of avionic and autonomous vehicles.

These electrical power systems can be modeled with graphs or networks. Bayesian networks are frequently used to construct diagnostic and prognostic systems. Given my background in graph theory, I examined the connections between random graphs and Bayesian networks to help identify graph analytic properties which could be used to minimize the propagation time of these analytic systems.

Autonomous vehicles, using this technology, have gained notoriety in recent years with the potential of the availability to the public. Under laboratory conditions and in computer models, predictive models work well in determining how these vehicles will function. However, predicting human behavior in the real world, a critical variable in the use of self-driven vehicles, can be extremely difficult resulting in performance uncertainty and safety concerns. Prior to the public release of this technology, entities such as the US Department of Transportation (USDOT) are developing policies and guidance for the development and testing of automated vehicle technology to ensure safe deployment of self-driven vehicles for public use.

5 A Mathematician by Any Other Name Is Still a Mathematician

Since changing career paths, I've held the titles "computer scientist," "senior research analyst," and "data scientist" but I have not been formally called "mathematician." I was the technical lead on the data quality project with US Citizenship and Immigration Services (USCIS). I have worked with the Social Security Administration, (SSA), on a number of the agency's high-profile information technology projects, such as the launch of SSA's cloud infrastructure and developing anti-fraud data analytics. My work involved various aspects of my math ability, but even more so, my problem solving and communication skills.

I now appreciate that I am just as much a contributor to the math community as I would be in the classroom or on a university campus. I continue to research, teach, and mentor but the goals and final outcomes are measured differently. It's not always for a grade or evaluation, but I do it to help others be better at their jobs so that we all do a better job of providing services and safety to improve the daily lives of the public we serve.

I have discovered I am a better mathematician than I give myself credit for being. After many years of personal challenges and self-doubt, I am confident in what I bring to my organization as a public servant. I knew I wanted to have an impact on the world in a different way—outside of academia. I continued to seek out opportunities to use mathematics, critical thinking, and analytical skills to make a broader impact on society.

Mathematicians may not always find themselves solving actual math problems. They often find themselves facing large sums of information, an unclear problem, and the urgent need for a solution to the problem. As a PhD mathematician and former member of academia, I found that the skills I gained as a graduate teaching assistant and assistant professor prepared me to take on the most abstract and vague problems to find possible solutions in the real world. Attacking a problem systematically and synthesizing vast amounts of data to identify viable solutions are not skills the average person possesses, but mathematicians do!

The world is a big place. There is a substantial amount of important work to do to make the world a better place. I want to do my part in making it better. I have unique exposure, experiences and a perspective of public service, which I find deeply satisfying and rewarding. Local, state, and federal governments need a variety of skilled workers to do the work of the people. Mathematicians who are interested in using their knowledge and skills to help improve the lives of others should consider a career in public service.

Mathematicians are equipped with skills to address and help find solutions to the toughest of problems faced by the government.

Reflections on the Challenges of Mentoring

Carol Wood

Abstract Mentoring mathematics students continues to be a challenge, for me and surely for others. I taught in the EDGE Program at Spelman in the summer of 2004. To help students make the transition from undergraduate to graduate school, I decided to offer a sample of a first year graduate algebra course. While this was not a terrible approach, it missed the mark on some of the key goals of the program. In this article I describe the challenges of mentoring young women at this stage, with more questions than answers. Along the way I mention some of my own experiences, and my observations of the power of the EDGE Program.

Before preparing this article, I learned from the EDGE website that nine of the fourteen students in EDGE 2004 had completed PhDs in the mathematical sciences, seven of these within 6 years. This strikes me as a great result, and reassures me that I did not do substantial harm that summer. Of course these were—and are—exceptional women, of whom one expects success. But I too often meet a young woman and think "this one cannot fail" and then she disappears off the radar.

In 2004 I had been on the faculty at Wesleyan University for over 30 years, and had taught graduate algebra often. My research area is model theory, a branch of logic, in my case heavily skewed toward applications in other parts of mathematics, of which algebra was my first love. I was invited to teach in EDGE on the recommendation of Robert Bozeman; he and I had served together writing the GRE Math subject test, and he had noticed my fondness for algebra questions. I was happy to accept, having admired the work that Sylvia Bozeman and Rhonda Hughes did with the EDGE Program, plus I had a good friend Teresa Edwards who was at Spelman at the time. In addition, I am an advocate for women, including having served as a president of the Association for Women in Mathematics. One memorable

C. Wood (✉)
Edward Burr Van Vleck Professor of Mathematics Emerita, Wesleyan University, Middletown, CT, USA
e-mail: cwood@wesleyan.edu

© The Author(s) and the Association for Women in Mathematics 2019
S. D'Agostino et al. (eds.), *A Celebration of the EDGE Program's Impact on the Mathematics Community and Beyond*, Association for Women in Mathematics Series 18, https://doi.org/10.1007/978-3-030-19486-4_26

if trivial moment came during my AWM term when I tilted with the CEO of Mattel, a woman, when talking Barbie said "math class is hard." I'm not sure the CEO got it, but the general outcry made Barbie's talking career mercifully brief.

Wesleyan has a small doctoral program, with a number of US based students coming from small colleges. Like many EDGE participants, they are good at mathematics but have limited experience with material beyond the undergraduate curriculum. The traditional first year graduate courses in algebra and analysis can pose difficulties. EDGE made a smart strategic choice in providing courses in these two areas as part of its summer program.

I studied at a liberal arts college for women, where I had the great luck to find there two exceptional women in the mathematics department, Evelyn Wiggins Casner and M. Gweneth Humphreys, both with doctorates from the University of Chicago. In light of my ambition, they advised me to learn more mathematics than the curriculum in the major offered, and even suggested I transfer to a university after 2 years. But I dug in my heels, staying for four happy years, taking courses in many subjects, and adding mathematics via reading courses and summer programs at other institutions. I entered graduate school a bit deficient in analysis but well prepared in algebra and topology, and with an interest in logic also. My first year experience was the proverbial drinking from a fire hose. I aver that all graduate students should learn as much mathematics as possible, most especially during the first year, and without asking why. It's professional preparation, that's why!

I decided to cast my teaching in the EDGE Program as professional training, with fast-moving formal lectures in group theory: theorem, proof, examples—in essence, algebra boot camp. Their understanding of the lectures was tested with problem sets. It soon became apparent that this was very different from what the students had been led to expect by the first half of the program. Nonetheless, I stubbornly chose for them to be blown out of the water by me, instead of by their professors in the coming year. US graduate programs are international, with foreign students who arrive with strong backgrounds in mathematics. I had seen many small college students doubt their own talents, when the only problem was that they were not as well prepared as their international peers.

Even though I came from a small college myself, I was super confident, for which I thank and blame my mother. As my sister once put it, our mother told us we could do anything; she was the smartest and most competent person we knew and so we took her word as gospel. In mathematics, confidence can be crucial to success, all the more if one differs from the typical mathematician in some way such as gender or color. My goal at EDGE was to build that confidence in a friendly environment by helping the students see they could handle tough work.

The first clue that I was overdoing my approach came when the graduate mentors asked me how to do the homework. I was grateful that they were willing to admit what they did not know, a good sign for any mathematician. We scheduled faculty-mentor sessions before they met with the students. I mentored the mentors, who in turn mentored the students. These sessions were fun for me, and the mentors told me both then and later that it was a good learning experience for them. It puzzled me that most of the students did not approach me directly. Too scary I suppose. I should add that a couple of the participants were clearly on top of the material and not at all reluctant to talk with me, but others struggled. With fourteen students it should have been possible for me to adapt, but it was hard to figure out how in real time.

Fortunately the EDGE students were resilient. They did what I demanded, mostly tolerant of my taking a big chunk of their time. They made it clear they had not expected to work this hard, and I began to realize that the bonding time and supportive environment of the program were at least as important as their comprehension of group theory. But I was a woman on a mission! I had experience in having to stand my ground in the world of mathematics. I had learned to play by what I call "boys' rules," and wanted to give them a sense of how to navigate that world. I had also served on visiting committees at small colleges where I was shocked at how women students were coddled. I was appalled when faculty explained that the women were not capable of facing serious challenges. This did them no favor; indeed, some women realized they were hothouse plants and told me that they had developed strategies for hardening off, such as taking courses at nearby graduate programs to sample the real world.

Despite my social ineptitude, I was able to share some of my own stories, including my first crisis of confidence. Qualifying exams in my graduate program were oral. There was no special syllabus and no information about the committee other than it would be one person from each of three general areas: algebra, analysis, and topology. During the 2 weeks leading up to my exam I began to panic, and concluded that I knew almost nothing. I recalled a strategy from memorizing piano music, namely to overlearn the material. There was no way to overlearn everything, so I picked one thing: Galois theory. I learned the theorems and their proofs down cold. This was calming for me: at least I knew something. In a stroke of serendipity, the first question at my exam was "can you tell me something about Galois theory?" The examining committee had a hard time getting me to stop talking Galois theory in order to move on to other topics. By then, I was on my way!

There was another down side to my approach to teaching in EDGE. Rhonda, Ami, and Sylvia were lively and fun, with social skills that I observed with admiration but with no hope of emulating. My tough love turned out to be isolating, plus I'm a lousy dancer. This is not to say that I did no socializing, but I had established myself as someone other than a pal.

Luckily I underestimated the power of the EDGE Program and the perseverance of the women involved. The students were tolerant and kind, and experienced in navigating demands placed on them throughout their lives. Over the years I have felt their warmth toward me when we encounter each other at the annual math meetings. I continue to be in awe of these remarkable women.

What is my takeaway from this one time experience at EDGE? There is a fine line to be walked in encouraging students in mathematics and at the same time in preparing them to succeed in mathematics. Make it too hard and they shut down. Make it too easy and they get clobbered later. Finding a balance is not a natural skill for most mathematicians, certainly not for me. I have tried and succeeded at times, and have failed at other times. Thus it was at EDGE.

Another balancing act involves providing helpful signposts and warnings. I wanted to guide the EDGE women as they figured out how to learn mathematics, whom to trust, and whom to ignore. In such matters there are no easy answers, at least none that I have found. No one questioned my right or ability to do mathematics until I entered graduate school, and then only a couple of peers, far too late to dissuade me. Still, unpleasant experiences or even poor advice can be a huge distraction, another reason it is important to be optimistic and not bitter. One benefit of the passage of time is that I now feel much more free to tell my own horror stories, since they are so far in the past that they are met with disbelief—and also because I can laugh about them myself. Perhaps the best I can do now is to show up. I am always astonished—but pleased—when a former student or colleague tells me it mattered to have me as a role model (warts and all).

Thinking long and hard about the challenges of mentoring has brought me no clear solutions although the EDGE experience heightened my awareness of the problems. Some successful mathematical mentors barely recognize that the students are a different gender or color; I have seen students relax when they sense this detachment. Luckily, much wisdom went into the design of the EDGE Program, and the effect produced by the program is profound. The EDGE Program takes a proactive approach, in providing a network and a safety net for a student who—like almost all of us—is bound to hit stumbling blocks. Having both peer group and mentors matters a lot. We all need to be able to share bizarre encounters, if only to learn that we are not among the crazy persons in the story. EDGE provides a network of colleagues who expect a young woman to get up and dust herself off, and who will cheer her on. It would take much more than my boot camp insanity to derail that!

I took a colleague along to an EDGE reunion at the Joint Mathematics Meetings. Over dessert we heard reports of successes such as "passed my prelims" and "defended my thesis." The spirit in that room was inspiring. He, as an outsider and experienced teacher, was dazzled. Me too.

Author Index

A
Alvarado, Alejandra, 55, 79

B
Bryant, Lance, 89
Bryant, Sarah, 9, 19, 67, 89
Buchmann, Amy, 19, 67, 99

C
Cotwright-Williams, Carla D., 377
Craig, Erin, 289
Curry, Jamye, 197
Czaja, Wojciech, 213

D
D'Agostino, Susan, 1, 19, 67
Dang, Xin, 197
Danielli, Donatella, 55
Davis, Rachel, 55
DeCoste, Rachelle C., 47, 119
Duong, Yen, 99

E
Evans, Zenephia, 55

F
Fefferman, Nina H., 317

G
Goins, Edray Herber, 55
Graham, Erica J., 237
Guinn, Michelle Craddock, 19, 67, 149

H
Harris, Leona A., 19, 31, 67

J
Jordan, Jill E., 167

K
Karaali, Gizem, 129

L
Lewis, Torina, 277

O
Otto, Carolyn, 175

P
Pershell, Karoline P., 369
Poimenidou, Eirini, 289
Price, Candice R., 79, 317

© The Author(s) and the Association for Women in Mathematics 2019
S. D'Agostino et al. (eds.), *A Celebration of the EDGE Program's Impact on the Mathematics Community and Beyond*, Association for Women in Mathematics Series 18, https://doi.org/10.1007/978-3-030-19486-4

389

R
Radunskaya, Ami, 99, 237

S
Sang, Hailin, 197
Schleben, Bradford, 149
Spayd, Kimberly, 327
Spott, Jessica, 9
Swanson, Ellen R., 327

W
Walton, Chelsea, 339
Ward, Farrah J., 31
White, Diana, 89
Wood, Carol, 385

Y
Yacoubou Djima, Karamatou, 213

Printed in the United States
By Bookmasters